Impact Studies

Series Editor: Christian Koeberl

Charles Cockell
Christian Koeberl
Iain Gilmour
(Editors)

Biological Processes Associated with Impact Events

With 102 Figures, 5 in colour

ESF IMPACT

Springer

EDITORS

PROFESSOR CHARLES COCKELL
DR. IAIN GILMOUR
PLANETARY AND SPACE SCIENCES
RESEARCH INSTITUTE
OPEN UNIVERSITY
MILTON KEYNES
UNITED KINGDOM

Email: c.s.cockell@open.ac.uk
i.gilmour@open.ac.uk

PROFESSOR CHRISTIAN KOEBERL
Department of Geological Sciences
University of Vienna
Althanstrasse 14
1090 Vienna
Austria

Email: christian.koeberl@univie.ac.at

ISSN 1612-8338
ISBN 10 3-540-25735-7 **Springer Berlin Heidelberg New York**
ISBN 13 978-3-540-25735-6 **Springer Berlin Heidelberg New York**

Library of Congress Control Number: 2006920520

Springer is a partof Springer Science+Business Media
springeronline.com

Printed in Germany

Cover design: E. Kirchner, Heidelberg
Production: A. Oelschläger
Typesetting: Camera-ready by the Editors
Printing: Stürtz, Germany
Binding: Stürtz, Germany

Printed on acid-free paper 32/2132/AO 5 4 3 2 1 0

Preface

The biological effects of asteroid and comet impacts have been widely viewed as primarily destructive. The role of an impactor in the K/T boundary extinctions has had a particularly important influence on thinking concerning the role of impacts in ecological and biological changes.

During the 10th and final workshop of the ESF IMPACT program during March 2003, we sought to investigate the wider aspects of the involvement of impact events in biological processes, including the beneficial role of these events from the prebiotic through to the ecosystem level.

The ESF IMPACT programme (1998-2003) was an interdisciplinary effort that is aimed at understanding impact processes and their effects on the Earth environment, including environmental, geological and biological changes. The IMPACT programme has 15 member states and the activities of the programme range from workshops to short courses on topics such as impact stratigraphy, shock metamorphism, etc. The program has also awarded mobility grants and been involved in the development of teaching aids and numerous publications, including this one.

The 10th workshop was held at King's College, Cambridge, from March 29 to April 1, 2003. The theme of the workshop 'Biological Processes associated with Impact Events'. Some of the questions that were addressed during this workshop included: What beneficial effects can impact events have for life? What are the environmental changes on the local level as well as the global level? How does microbial life take advantage of impact craters? What role do impact events have in the origins and evolution of life? From prebiotic molecules to complex metazoans, the workshop took a snapshot of the diversity of fields of biology and ecology that intersect with impact studies. Within this milieu of discussions the workshop also considered astrobiological aspects of cratering and their relevance for life elsewhere, if it exists.

The diversity of papers presented in this volume attest to the fact that impact cratering is very much a biologic process. It is not often that biologists, ecologists, and astrobiologists rub shoulders with the impact and planetary science communities, but the ESF IMPACT programme achieved an important contribution to this discussion at this meeting.

The papers within this volume cover some important aspects of biological processes. The potential for survival of organics at impact sites is investigated. The survival of microbes during impact and their subsequent colonization of impact rocks is examined. The biotic potential of hydrothermal systems is addressed. Extinctions caused by impacts are discussed at two extinction boundaries, while another paper compares impact processes to the processes of ecological disturbance caused by volcanoes. The effects of marine impacts on ecosystems are discussed and in another chapter, the biotic consequences of post-impact wildfires are described. A good balance is provided of local environmental effects of impacts compared to global effects. In totality this volume provides a useful view, but by no means a complete one, of impact cratering as a biologic process. We hope that it will encourage further debate, discussion and collaboration.

Charles Cockell
Open University
Milton Keynes, UK

Iain Gilmour
Open University
Milton Keynes, UK

Christian Koeberl
University of Vienna
Vienna, Austria

Acknowledgements

The editors are grateful to the ESF-IMPACT programme for making the King's College, Cambridge workshop and the preparation of this volume possible. The workshop was partly funded by the NASA Astrobiology Institute and we would like to express our gratitude for NASA's generous support.

We thank the following referees for their assistance: Philip Bland, Madelaine Böhme, Eric Buffetaut, Phillipe Claeys, Rodolfo Coccioni, Fabio Donadini, Henning Dypvik, Alan Fitzsimmons, Darlene Lim, Gerta Keller, Simon Kelley, Wolfgang Kiessling, David King, Kalle Kirsimae, Frank Kyte, Tõnu Meidla, Jared Morrow, Jens Ormo, John Parnell, Victoria Pearson, Uwe Reimold, Jan Smit, Morten Smelror, Frances Westall.

Contents

List of Contributors

Doreen E. Ames
Natural Resources Canada
Geological Survey of Canada
601 Booth Street
Ottawa, ON, K1A 0E8
Canada
(dames@nrcan.gc.ca)

Martin Baron
Department of Geology and Petroleum Geology
University of Aberdeen
Aberdeen, AB24 3UE
U.K.

Clare M. Belcher
Department of Geology
Royal Holloway University of London
Egham
Surrey, TW20 0EX
U.K.
(c.belcher@gl.rhul.ac.uk)

Daniel Benest
Observatoire de la Côte d'Azur
Observatoire de Nice
B.P. 4229
F-06304 Nice Cedex 4
France
(benest@obs-nice.fr)

Eric Buffetaut
CNRS
16 cour du Liégat
75013 Paris
France
(eric.buffetaut@wanadoo.fr)

Rodolfo Coccioni
Instituto di Geologia e Centro di Geobiologia dell'Università "Carlo Bo"
Campus Scientifico
Località Crocicchia
61029 Urbino
Italy
(cron@info-net.it)

Charles Cockell
Open University
Milton Keynes, MK7 6AA
U.K.
(c.s.cockell@open.ac.uk)

Eveline de Vos
School of Earth, Ocean & Planetary Sciences
Cardiff University
P.O. Box 914
Cardiff, CF10 3YE
U.K.

Fabio Donadini
Division of Geophysics
University of Helsinki
PO Box 64
FIN-00014 Helsinki
Finland
(fabio.donadini@helsinki.fi)

Natalia I. Dykan
Institute of Geological Sciences,
National Academy of Sciences of Ukraine,
55–b Oles Gonchar Street,
Kiev 01054
Ukraine

Henning Dypvik
Department of Geology
University of Oslo
PO Box 1047
Blindern
N-0316 Oslo
Norway
(henning.dypvik@geo.uio.no)

David A. Fike
Department of Earth, Atmospheric and Planetary Sciences
Massachusetts Institute of Technology
77 Massachusetts Ave. 54-812
Cambridge, MA 01239
USA

Andrew S. Gale
Department of Earth & Environmental Sciences
Medway School of Science
University of Greenwich
Chatham Maritime
Kent, ME4 4TB
U.K.

Harold L. Gibson
Department of Earth Sciences
Laurentian University
Ramsey Lake Rd
Sudbury, P3E 2C6
Canada

Eugene P. Gurov
Institute of Geological Sciences,
National Academy of Sciences of Ukraine,
55–b Oles Gonchar Street,
Kiev 01054
Ukraine

Gerda Horneck
German Aerospace Center DLR
Institute of Aerospace Medicine
D-51170 Koeln
Germany
(gerda.horneck@dlr.de)

Gordon J. Irvine
School of Earth, Ocean & Planetary Sciences
Cardiff University
P.O. Box 914
Cardiff, CF10 3YE
U.K.

Ian R. Jonasson
Natural Resources Canada
Geological Survey of Canada
601 Booth Street
Ottawa, ON K1A 0E8
Canada

Simon P. Kelley
Department of Earth Sciences,
Open University,
Milton Keynes MK 6AA,
U.K.
(s.p.kelley@open.ac.uk)

David T. King
Department of Geology
Auburn University
Auburn, AL 36849-5305
USA
(kingdat@auburn.edu)

Christian Koeberl
Department of Geological Sciences
University of Vienna
Althanstrasse 14
A-1090 Vienna
Austria
(christian.koeberl@univie.ac.at)

Pascal Lee
M/S 245
NASA Ames Research Center
Moffett Field, CA 94035-1000
USA

Martti Lehtinen
Division of Geophysics
University of Helsinki
PO Box 64
FIN-00014 Helsinki
Finland

Valeria Luciani
Dipartimento di Scienze della Terra dell'Università
Corso Erecole I d'Este, 32
44100 Ferrara
Italy
(lcv@unife.it)

Iain McDonald
School of Earth, Ocean & Planetary Sciences
Cardiff University
P.O. Box 914
Cardiff, CF10 3YE
U.K.
(iain@earth.cf.ac.uk)

Thornton L. Neathery
Neathery and Associates
1212-H Veterans Parkway
Tuscaloosa, AL 35404
USA
(tlneathery@prodigy.net)

Adriana Ocampo
Jet Propulsion Laboratory
4800 Oak Grove Dr. ms 183-301
Pasadena CA 91109
USA
(adriana.c.ocampo@jpl.nasa.gov)

Gordon Osinski
Planetary and Space Sciences Center
Department of Geology
University of New Brunswick
Bailey Drive
Fredricton, New Brunswick E3B 5A3
Canada

John Parnell
Department of Geology and Petroleum Geology
University of Aberdeen
Aberdeen, AB24 3UE
Scotland
(j.parnell@abdn.ac.uk)

Lauri Pesonen
Division of Geophysics
University of Helsinki
PO Box 64
FIN-00014 Helsinki
Finland

Lucille Petruny
Astra-Terra Research
Auburn, AL 36831-3323
USA

and

Department of Curriculum and Teaching
Auburn University
Auburn, AL 36849
USA
(lpetruny@att.net)

Jüri Plado
Institute of Geological Sciences
University of Tartu
Tartu
Estonia

Kevin O. Pope
Geo Eco Arc Research
16305 St Mary's Church Road
Aquasco, MD 20608
USA

Michael R. Rampino
Earth and Environmental Science Program
New York University
100 Washington Square East
New York, NY 10012
USA
(mmr1@nyu.edu)

Wolf Uwe Reimold
Impact Cratering Research Group
School of Geosciences
University of the Witwatersrand
P.O. Wits 2050
South Africa
(remoldw@geosciences.wits.ac.za)

Johanna Salminen
Division of Geophysics
University of Helsinki
PO Box 64
FIN-00014 Helsinki
Finland

Morten Smelror
Geological Survey of Norway
N-7491 Trondheim
Norway
(morten.smelror@ngu.no)

Sten Suuroja
Geological Survey of Estonia
Kadaka tee 82
Tallinn 12168
Estonia
(s.suuroja@egk.ee)

and

Department of Mining
Tallinn Technical University
Kopli 82
Tallinn
Estonia

Kalle Suuroja,
Geological Survey of Estonia
Kadaka tee 82
Tallinn 12168
Estonia

Vivi Vajda
GeoBiosphere Science Centre
University of Lund
Sölvegatan 12
223 62 Lund
Sweden
(vivi.vajda@geol.lu.se)

Stephanie Werner
Freie Universität of Berlin
Institute for Geosciences
Malteserstrasse 74-100
12249 Berlin
Germany

Helen Wycherley
Department of Geology and Petroleum Geology
University of Aberdeen
Aberdeen, AB24 3UE
UK
(h.wycherley@abdn.ac.uk)

The Potential for Survival of Organic Matter in Fluid Inclusions at Impact Sites

John Parnell, Martin Baron, Helen Wycherley

Department of Geology and Petroleum Geology, University of Aberdeen, Aberdeen, AB24 3UE, U.K. (j.parnell@abdn.ac.uk)

Abstract. Fluid inclusions within rocks have potential for the preservation of organic molecules. Trace quantities of biomolecules could be entrapped inside micrometer-scale inclusions during the growth of surface precipitates, reflecting any ambient life in the surrounding waters. Developing technologies for the high-resolution detection of biomolecules offer encouragement for the future detection of these trace biomolecules. The terrestrial geological record shows that organic molecules can survive relatively high temperatures within inclusions, including the temperatures of hydrothermal systems in impact craters.

1 Introduction

A potential consequence of the high temperatures involved in impact events is that organic matter in the target is destroyed. Where the temperatures are so high that rocks are volatilized, destruction of the organic matter is inevitable, although conceivably some methane generated by the process could become entrapped in adjacent rocks. Where rocks are simply heated and deformed by the impact, some organic matter may survive, but heating above 200 to 300 ºC is likely to leave this matter as a refractory solid from which little biomolecular data can be extracted. However a small fraction of organic material may be entrapped within fluid inclusions in rocks, where it has a greater chance of survival. In this account, we review the potential for survival of organic matter in fluid inclusions, with special reference to impact sites.

2
Fluid inclusions

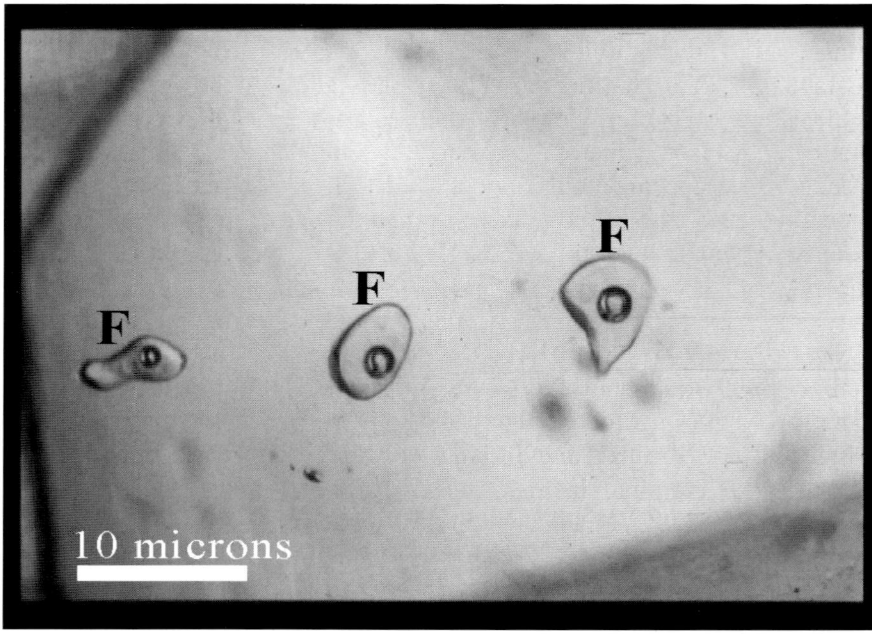

Fig. 1. Fluid inclusions (F, showing liquid with vapour bubble) in hydrothermal quartz, viewed in transmitted light. Most inclusions are <10 micrometers size.

Fluid inclusions are micrometer-scale volumes of fluid entrapped during mineral growth (Fig. 1), which in sedimentary rocks may be during cementation or during healing of microfractures. They are therefore representative of the ambient fluid present during these processes. They can be regarded as sealed vessels of constant volume, whose contents adjust to the changing pressure and temperature conditions experienced. They can be studied in detail by the preparation of doubly polished wafers (unsupported slices of rock, 100 to 150 micrometers thick) that are examined using optical microscopy in transmitted light. A basic distinction is made between primary and secondary inclusions. Primary inclusions, which formed at the time of mineral precipitation, are identified by their relationship to growth zones in the host mineral. After mineral precipitation, deformation can cause the development of microfractures, which entrap fluid during healing. These secondary inclusions cut across growth zones, and so are distinguishable from primary inclusions. Most inclusions are entrapped during burial, at higher temperature and pressure than at the surface. On cooling at the surface, a vapour bubble commonly nucleates, so that the fluid becomes two-phase (vapour, liquid). Important information from inclusions is recorded by the practice of microthermometry, which involves the observation of their behaviour during heating and cooling in the laboratory.

Established microthermometric techniques allow reconstruction of the initial temperature conditions of fluid entrapment. Upon heating, the vapour bubble is eliminated, i.e. the fluid homogenizes back to a single phase. This homogenization temperature is taken as a minimum trapping temperature for the fluid during mineral precipitation. Monophase inclusions, which have no vapour bubble, suggest entrapment at low temperatures, probably at less than 50 °C. Upon cooling, the freezing behaviour of the inclusions reflects their chemistry. Aqueous fluids freeze at lower temperatures with increasing concentrations of dissolved salts. The behaviour of aqueous fluids is well understood, but fluids dominated by methane and/or carbon dioxide can also be recognised by their response during laboratory heating and cooling. Further details of the methodology of non-destructive fluid inclusion studies are given by Roedder (1984) Shepherd et al. (1985), and Goldstein and Reynolds (1994).

Additionally, the fluids within inclusions can be extracted in bulk for direct geochemical analysis. Extraction could be undertaken by crushing, heating or dissolution of the host mineral in acid. Extraction from inclusions by crushing is favoured to retain unaltered organic molecules, and can be achieved by hammer, ball-milling or screw action (Kazahaya and Matsuo 1985, Stuart et al. 1994, Dennis et al. 2001). Organic geochemistry on extracted inclusion fluids is a proven technique, mostly on oil-bearing samples where organic molecules are highly concentrated (e.g., Etminan and Hoffmann 1989, Pang et al. 1998, Parnell et al. 2001b), but also on organic components in aqueous inclusions (Ruble et al. 1998).

3 Preservation of organic matter within fluid inclusions

Fluid inclusions contain several forms of organic matter (Roedder 1984), including hydrocarbon liquids, hydrocarbon gases, bituminous solids and even recognisable remains of organisms. The preservation of evidence for life inside fluid inclusions could be in the form of trapped microbes or, more likely, biochemical signatures in the water. Microbes are found in terrestrial fluid inclusions, particularly in rock salt (halite) where the inclusions can be relatively large, up to millimetre-scale (Reiser and Tasch

1960, Dombrowski 1963, Norton and Grant 1988, Norton et al. 1993, Denner et al 1994). In some cases, RNA extracted from salt deposits in the geological record is interpreted to represent fossil organic material, which is presumed to be located particularly in fluid inclusions (Grant et al. 1998, Radax et al. 2001, Fish et al. 2002), although microbial remains can also be preserved in solid mineral matrix. Detailed genetic data indicate that archael strains in salt are ancient, rather than modern contamination (Stan-Lotter et al. 1999). Laboratory experiments show that bacteria can be entrapped in artificially-grown halite crystals (Fig. 2, Wilkins et al. 2002).

The water in inclusions represents a potential source of organic molecules that may be a more direct signature of former life (Parnell et al. 2002). The metabolism and degradation of bacteria and algae contribute to all of the major groups of organic compounds that are dissolved in natural waters, including hydrocarbons, carbohydrates, carboxylic acids, amino acids and humic acids (Thurman 1985). A few of these products have long-term stability. Natural waters, therefore, contain traces of compounds that reflect the occurrence of microbial life in the environment. Analytical sensitivities are high enough to detect these compounds in samples of terrestrial waters. There is wide expertise in their detection, from contamination studies, environmental studies, diagnostic microbiology, and analysis of trace organics in oilfield waters (Ruble et al. 1998). There is, therefore, a substantial value in trying to develop the capability to extract such compounds from inclusion fluids, both in terrestrial settings and in the surface/near-surface deposits of Mars (Parnell 2002). A range of surface environments could include mineral precipitates (Ellery et al. 2002) that contain fluid inclusions representing the ambient water and any accompanying traces of life.

Fluid inclusions have a potential for survival over billions of years. Terrestrial studies of fluid inclusions of Archean age show that they have preserved organic compounds for over 2 billion years, including abiotic organic compounds based upon mantle methane (Bray et al. 1991) and oils derived from Archean microbial matter (Dutkiewicz et al. 1998). Meteorite data similarly show that inclusion waters can be preserved over this time scale (Zolensky et al. 1999).

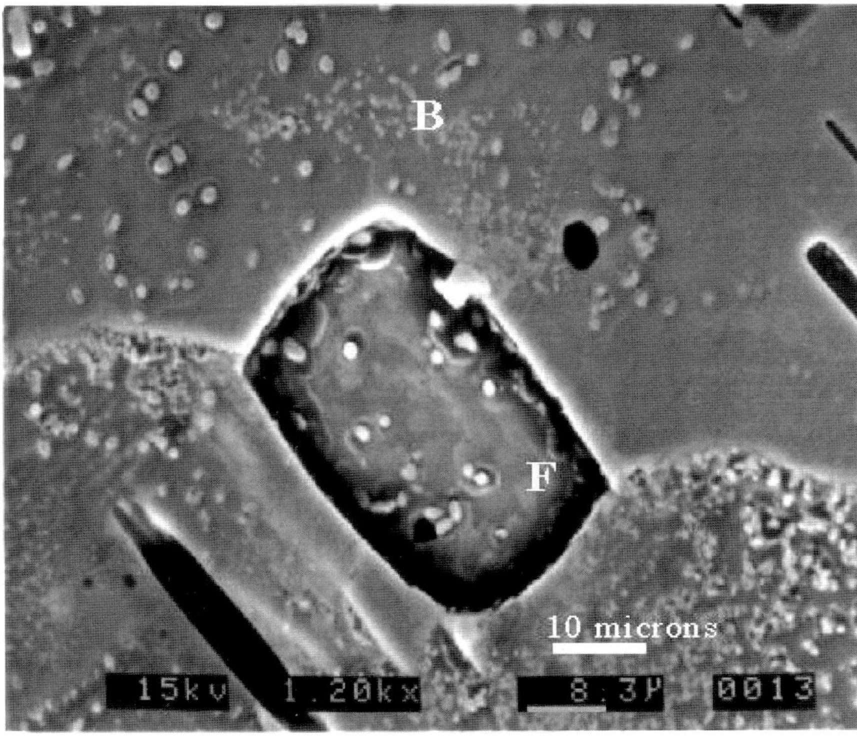

Fig. 2. Bacteria (B) within exhumed fluid inclusion (F) and below fracture surface of laboratory-grown halite (from Wilkins et al. 2002).

4 Survival of organic matter in inclusions during heating

There are important parallels with the preservation of fluid inclusions, including hydrocarbon inclusions, in terrestrial settings where petroleum systems have encountered temperatures and pressures higher than normally encountered in sedimentary basins. Inclusions containing high molecular weight hydrocarbons are preserved in sites where microthermometry indicates temperatures of up to 200 °C or higher (e.g., Pearcy and Burruss 1993, Kontak and Sangster 1998, Parnell et al. 2001a). Hydrocarbon inclusions are regularly recorded in ore deposits which have been formed by anomalous hydrothermal activity (e.g., Thomson et al. 1992, Rasmussen and Buick 2000). In one of the most extreme examples documented, hydrocarbons up to C_{33} have been extracted from fluid inclusions in an ore deposit that experienced temperatures of 500-600 °C (Hoffmann et al. 1988). Thermal cracking of the large molecules is inhibited in the 'sealed vessel' environment of the inclusions.

The stability of hydrocarbons at these high temperatures is commonly regarded as anomalous, which it would be at the surface, but under elevated pressures in the sub-surface, stability fields are extended to high temperatures (Price 1993, Price and De Wit 2001). Thus, during impact events the high pressures which are developed (Robertson and Grieve 1977) actually help to preserve hydrocarbons from thermal degradation.

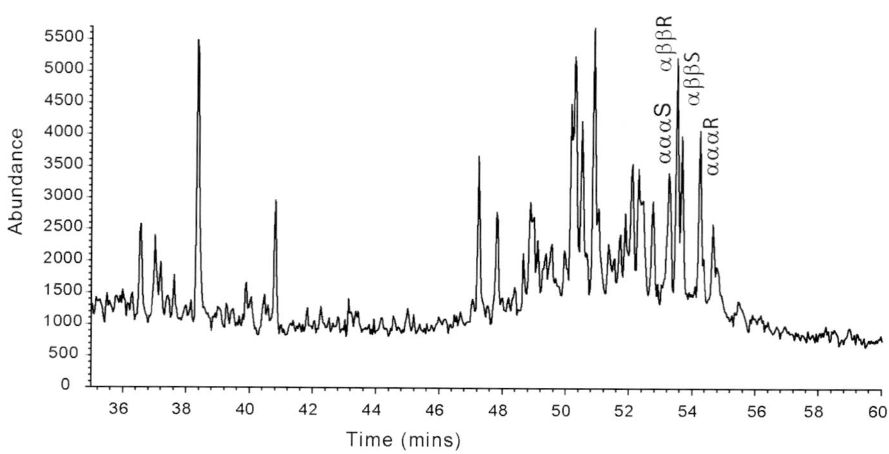

Fig. 3. Mass fragmentogram (m/z 217) for solid bitumen sample from fractured granite, Lockne impact crater. $C_{29}\alpha\alpha\alpha$ and $\alpha\beta\beta$ 20R and 20S peaks highlighted. $\alpha\alpha\alpha$ S/S+R ratio is 0.43. Obtained by gas chromatography-mass spectrometry using a Hewlett Packard HP 5970 MSD attached to a HP5890 gas chromatograph. Abundance units are relative, dependent on operating conditions.

There is potential to extract information on the degree of heating through the use of quantitative maturation parameters, either based on hydro-carbons in inclusions, or on hydrocarbon residues in the host rocks. For example, gas chromatography-mass spectrometry of solid bitumen (migrated hydrocarbons) in fractured granite in the vicinity of the Lockne impact crater, Sweden (see Sturkell et al. 1998), yields a sterane profile (Fig. 3), from which a ratio of biomarkers can be measured and compared against a standard maturity scale. The $C_{29}\alpha\alpha\alpha$ 20R and 20S peaks highlighted in Figure 3 represent compounds of biological and geological origin respectively, and the ratio of S/S+R consequently increases with thermal alteration. This parameter is kinetically dependent (e.g., Grantham 1986), i.e., it changes in response to heat to a degree controlled by the duration of heating. Thus a given ratio could represent a range of combinations of temperature and time from high temperature/brief time to low temperature/prolonged time.

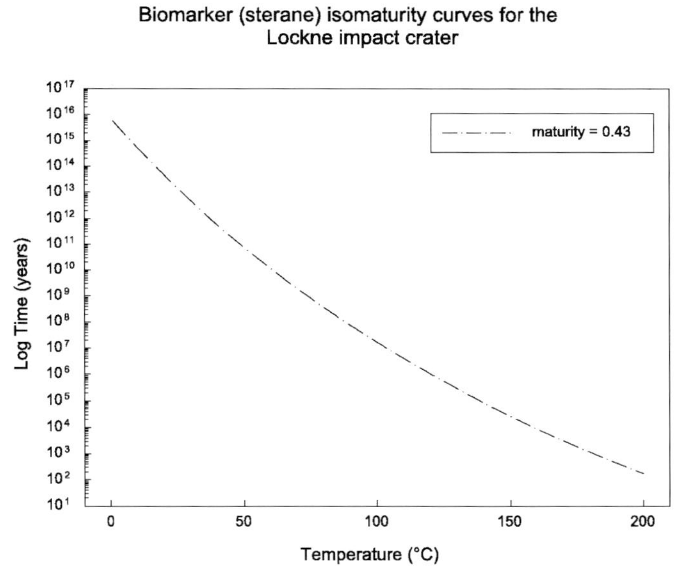

Fig. 4. Isomaturity curve, showing conceivable combinations of time and temperature that could yield a sterane ratio of 0.43, as measured at Lockne impact crater. Obtained using reaction constants for sterane isomerization of Marzi and Rullkötter (1992). In reality, heating at Lockne was a multi-stage process and the contribution of impact heat to the measured maturity requires more sophisticated modelling.

The isomaturity curve in Figure 4 shows this range of combinations for the $C_{29}\alpha\alpha\alpha$ 20S/20S + 20R ratio of 0.43 measured from the Lockne crater sterane profile in Figure 3, using reaction constants determined for sterane isomerization by Marzi and Rullkötter (1992). If we assumed that all the thermal maturity represented by this ratio was achieved during a single episode of heating, it could be the result of heating at 160 °C for 10,000 years or 75 °C for a billion years. However, the thermal history will always be more complex than a single-stage process. Rocks achieve a degree of thermal maturity due to burial and heating by the ambient geothermal gradient, and may be further subjected to thermal pulses by magmatic activity or tectonic (structural) events. In the case of Lockne, the rocks have been heated by burial, the impact event and deformation related to the Caledonian Orogeny (Sturkell et al. 1998). Nevertheless, where a large data set is available, it may be possible to determine the increment of thermal maturity due to impact and constraints on the duration of heating.

5 Survival of organic matter in inclusions during impact events

Several aspects of meteorite studies contribute to our confidence that at least some fluid inclusions and included biomolecules are not destroyed by impacts. The very fact that pre-impact fluid inclusions occur in meteorites shows that they survive the impact ejection process and subsequent interplanetary travel. Mineralogical evidence from meteorites shows that beneath the millimetre-scale fusion crust, the interiors contain pre-impact low-temperature mineral assemblages which have not been extensively heated, i.e. there are very high thermal gradients and the interiors have experienced temperatures no greater than 100 °C (Weiss et al. 2000, Bridges and Grady 2000a). Such temperatures allow preservation of complex organic molecules There is also direct evidence from fluid inclusion microthermometry that temperature histories were not extreme: observations in both carbonaceous chondrites (Bodnar and Zolensky 2000, Saylor et al. 2001) and Martian meteorites (Zolensky et al. 1999, Bridges and Grady 2000b) suggest that fluids were entrapped in inclusions at less than 100 °C. Hydrothermal deposits in terrestrial impact craters also yield fluid inclusion temperatures that are not very high: Data from the Charlevoix, Manson, Siljan, Roter Kamm, Lockne, Kärdla and Sudbury craters all yield temperatures below 350 °C (Fig. 5, based on data in Pagel and Poty 1975, Puura and Suuroja 1992, Boer et al. 1996, Sturkell et al. 1998, Komor et al. 1988, Koeberl et al. 1989, and Marshall et al. 1999). Clearly at the time of the impact itself, the much higher temperatures generated at the crater centre (1000 °C+) may be enough to destroy pre-existing inclusions, but inclusions outside the craters survive. For example, samples from the Siljan crater, Sweden, contain a mixture of inclusions related to an earlier granite-driven hydrothermal system, and a syn-impact hydrothermal system (Komor et al. 1988). The impact process may also eject blocks recording other diverse fluid-related events, including earlier episodes of hydrothermal activity or low-temperature precipitation of minerals by the evaporation of water. When pre-existing rocks are fractured due to impact, healing of the fractures would entrap a new generation of fluid. Thus ejecta could provide a variety of included waters for sampling.

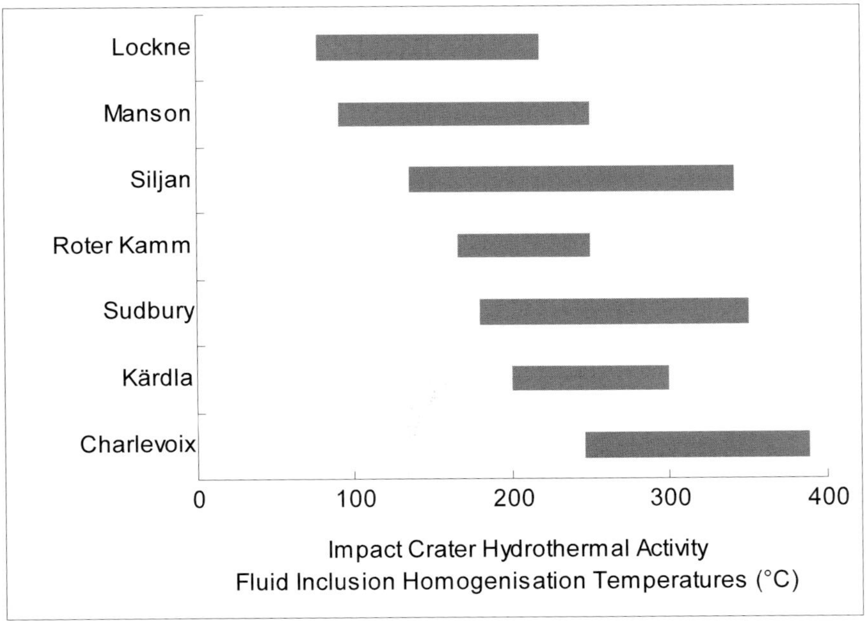

Fig. 5. Ranges of fluid inclusion homogenization temperature (minimum fluid trapping temperatures) determined from hydrothermal mineralization in impact craters (data from Pagel and Poty 1975, Komor et al. 1988, Koeberl et al. 1989, Puura and Suuroja 1992, Boer et al. 1996, Sturkell et al. 1998, Marshall et al. 1999).

Some craters preserve hydrocarbon fluid inclusions, representing either hydrocarbons that were present before the impact event and have survived destruction, or new hydrocarbons generated by the heat of the impact event and subsequent hydrothermal activity. An example of pre-impact hydrocarbons is recorded in the Tertiary-age Haughton Crater, Devon

Island, Canada, where the dolomite target rocks contain oil inclusions in crystal fabrics that pre-date the impact event (Parnell et al. 2003; Fig. 6). Hydrocarbon gases (predominantly methane) generated from organic matter by the heat of impact events may be preserved in inclusions, in the crater rocks and the surrounding region, as documented in the Gardnos, Norway (Andersen and Burke 1996), Lockne, Sweden (Sturkell et al. 1998), and Chicxulub, Mexico (Luders et al. 2003), craters.

Fig. 6. Hydrocarbon inclusions (largest at F) fluorescing under U-V light in thin section of calcite from hydrothermal mineral vein, Haughton Crater. Each inclusion 5-10 micrometers size.

An impressive example of the survival of organic matter in a large impact structure over a long time period is the Vredefort structure, South Africa, where organic matter of Archean origin (depositional age about 2700 to 3000 Ma; Robb et al. 1997) has been preserved as both a refractory residue in the country rocks and as hydrocarbon fluids within inclusions (Dutkiewicz et al. 1998, Drennan et al. 1999, Parnell 2001), despite the large-scale impact which occurred at about 2025 Ma (Robb et al. 1997).

6 Detection of organic molecules in inclusion fluids

Where organic molecules are abundant in an inclusion fluid, as in an inclusion oil sample on Earth, their detection is relatively easy (Munz 1991). However, where traces of biomolecules are sought in aqueous fluids, very high resolution analyses are required. Unfortunately DNA and RNA degrade over time, and except in special circumstances (e.g., at very low temperatures in ice; Hansen et al. 2002, or preserved in amber, Bada et al. 1994) will not survive on time scales greater than a few million years (Lindahl 1993; Bada et al. 1999). Other biomarkers common in recent sediments, including amino acids, sugars and alcohols, also degrade relatively quickly. Experience in petroleum geochemistry (Hunt 1996) show that hopanes and steranes have much greater stability. These compounds are derived from prokaryotic and eukaryotic membranes respectively, and are routinely sought in samples of age hundreds of millions of years or even billions of years (Brocks et al. 1999, 2003; Summons et al. 1999). Linear and simple branched alkanes also have high long-term survival potential (Simoneit et al. 1998). Possible approaches to the high-resolution detection of these biomolecules include:

(1) *High-resolution gas chromatography-mass spectrometry (GC-MS)*. GC-MS is the traditional main tool of organic geochemistry, whereby solvent extracts of organic matter are separated into distinct compounds by chromatography and ionized mass fragments can be analyzed for detailed distribution of biomarkers. This approach has been developed to very high resolution for work on traces of organic matter in meteorites (e.g., Sephton et al. 2000), and is the basis of organic detection experiments in some Mars mission proposals (Cabane et al. 2001).

(2) *Electrospray ionization (ESI)*. ESI involves the spraying of liquid samples (solvent containing organic compounds extracted from rock or sediment, or diluted inclusion fluids) into an ion source where organic molecules become charged. The mobility of the ions is measured in an ion mobility spectrometer, where time of flight is dependent upon ion size and shape as they collide with other molecules in an electric field. Some ions can be further trapped for mass measurement. This three-stage process is being integrated for a Mars mission proposal (Kanik et al. 2003) for which ppb-level detection is promised. The effectiveness of the ion

mobility spectrometer in resolving organic molecules is described by Beegle et al. (2001).

(3) *Time of Flight-Secondary Ion Mass Spectrometry (ToF-SIMS)*. ToF-SIMS allows the detection of biomolecules from material that is being imaged, and hence morphologically characterized, so that chemistry can be mapped out over complex structures (Toporski et al. 2002). This technique is applicable to both organic and inorganic species. It has been applied to the detection of hopanes (Steele et al. 2001b) and is proposed for the BepiColombo mission to Mercury (Whitby et al. 2003). The application of ToF-SIMS to organic compounds in fluid inclusions has been demonstrated by Li and Parnell (2003).

(4) *Molecular probes*. A variety of probes have been designed to detect particular target molecules, including nucleic acids and antibodies. Nucleic acid probes involve engineered molecules that give an identifiable response (e.g., fluorescence) upon combination with a target molecule. Probes can detect DNA at the pM level (Svanvik et al. 2000), and theoretically could detect single molecules (Hansen et al. 2002). Fluorescence responses from target capture can be made evident by incorporating the probes within fibre-optic systems. Potentially up to millions of different targets for specific organic molecules can be arranged in microarrays on a glass slide, with positive reactions causing a colour change or fluorescence (e.g., Gibson 2002). Antibodies exist for a variety of molecular targets, and detection based on immunological reactions to them is widely proposed for astrobiology (Steele et al. 2001a; Briones et al. 2002, Warmflash et al. 2002). Specific antibodies have been developed for hopanes (Maule et al. 2003), making this a realistic approach for ancient samples. Artificial molecular receptors have also been proposed, and have the advantage of being highly robust (Henry et al. 2002). In some cases where biomolecules from the geological record have been detected by immunological techniques (De Jong et al. 1974, Lowenstein 1981), these are suggested to be atypical as they are protected inside the mineralized tissues of macrofossils and are therefore of limited value as bacterial biomarkers (Toporski and Steele 2002). However, fluid inclusions similarly offer protection from alteration processes, and could be a more widespread source of biomolecular information in the future if contamination can be prevented.

(5) *Raman spectroscopy*. Raman spectroscopy, based on the inelastic scattering of incident light by molecules, is applicable to

both liquids and solids, and measures responses from both organic and inorganic compounds. The analytical resolution is low compared to some other techniques, but it is a useful approach for specific objectives, such as analyzing endolithic communities (Edwards et al. 1999), and it has the important advantage that it can be undertaken on unprepared rock *in situ*. It is listed here because of the good spatial resolution, which approaches the micron-level. Accordingly it is being miniaturized for mission proposals (Dickensheets et al. 2000). It can even be applied beneath the surface of translucent minerals, so can obtain data from fluid inclusions without opening them (Pasteris et al. 1988). Surface-enhanced resonance Raman scattering (SERRS) *does* have the potential to detect fmol-level concentrations of organic compounds, using on-chip technology (Keir et al. 2002)

These techniques might be applied to samples from the terrestrial geological record, and those returned from future planetary exploration. If they are to be used *in situ* during extraterrestrial missions, they require miniaturization. The development of microfluidic handling techniques (Freemantle 1999, Mitchell 2001) will be an important tool in manipulating and analysing samples at the necessary scale.

7 Conclusions

Organic matter may be preserved within fluid inclusions in rocks within and around impact craters, and has a better chance of survival compared to organic matter that is not sealed within inclusions. The evidence for this high survival potential includes:

(i) Fluid inclusions in general are known to contain biomolecules and even microbial life.

(ii) Organic matter sealed within fluid inclusions can survive high temperatures without loss of volatile components.

(iii) Fluid inclusions in meteorites survive the ejecta and travel processes.

(iv) Hydrothermal systems in impact craters are generally at temperatures where inclusions survive intact.

(v) Pre-impact hydrocarbons survive in some terrestrial craters, especially within inclusions.

(vi) Hydrocarbons generated by the heat of impact in terrestrial craters are preserved within inclusions.

These observations not only indicate that organic matter can survive at impact sites within fluid inclusions, but also that inclusions in hydrothermal systems in craters may trap samples of fluid in which we might seek evidence for primitive, thermophile life. Development of a range of high-resolution techniques will help to detect this evidence.

References

Andersen T, Burke EAJ (1996) Methane inclusions in shocked quartz from the Gardnos impact breccia, South Norway. European Journal of Mineralogy 8: 927-936

Bada JL, Wang XS, Poinar HN, Paabo S, Poinar GO (1994) Amino acid racemization in amber-entombed insects: implications for DNA preservation. Geochimica et Cosmochimica Acta 58: 3131-3135

Bada JL, Wang XS, Hamilton H (1999) Preservation of key biomolecules in the fossil record: current knowledge and future challenges. Philosophical Transactions of the Royal Society of London, Series B 354: 77-87

Beegle LW, Kanik I, Matz L, Hill HH (2001) Electrospray ionization high-resolution ion mobility spectrometry for the detection of organic compounds, 1. Amino acids. Analytical Chemistry 73: 3028-3031

Bodnar RJ, Zolensky ME (2000) Fluid inclusions in meteorites: Are they useful and why are they so hard to find? Meteoritics and Planetary Science 35: A29

Boer RH, Reimold WU, Koeberl C, Kesler SE (1996) Fluid inclusion studies on drill core samples from the Manson impact crater: Evidence for post-impact hydrothermal activity. In: Koeberl C, Anderson RR (eds) The Manson Impact Structure, Iowa: Anatomy of an Impact Crater. Geological Society of America Special Paper 302: 377-382

Bray CJ, Spooner ETC, Thomas AV (1991) Fluid inclusion volatile analysis by heated crushing, on-line gas chromatography; applications to Archean fluids. Journal of Geochemical Exploration 42: 167-193

Bridges JC, Grady MM (2000a) Evaporite mineral assemblages in the nakhlite (martian) meteorites. Earth and Planetary Science Letters 176: 267-279

Bridges JC, Grady MM (2000b) Petrography and fluid inclusion studies of Zag. [abs.] Meteoritics and Planetary Sciences 35: A33

Briones C, Parro V, Perez-Mercader J (2002) Development of microarray assays for biomarker identification in astrobiology. International Journal of Astrobiology 1: 115-116

Brocks JJ, Logan GA, Buick R, Summons RE (1999) Archean molecular fossils and the early rise of eukaryotes. Science 285: 1033-103

Brocks JJ, Buick R, Summons RE, Logan GA (2003) A reconstruction of Archean biological diversity based on molecular fossils from the 2.78 to 2.45 billion-year-old Mount Bruce Supergroup, Hamersley Basin, Western Australia. Geochimica et Cosmochimica Acta 67: 4321-4355

Cabane M, Coll P, Rodier C, Israel G, Raulin F, Sterberg R, Niemann H, Mahaffy P, Jambon A, Rannou P (2001) In situ organic and inorganic analysis (Pyr/CD-GC/MS) of the Martian soil, on the 2005 mission. Planetary and Space Science 49: 523-531

De Jong EW, Westbroek P, Westbroek JF, Bruning JW (1974) Preservation of antigenic properties in macromolecules over 70 million years old. Nature 252: 63-64

Denner E, McGenity T, Busse H-J, Wanner G, Stan-Lotter H (1994) *Halococcus salifedinae* sp. nov., an archael isolate from an Austrian salt mine. International Journal of Systematic Bacteriology 44: 774-780

Dennis PF, Rowe PJ, Atkinson TC (2001) The recovery and isotopic measurement of water from fluid inclusions in speleothems. Geochimica et Cosmochimica Acta 65: 871-884

Dickensheets DL, Wynn-Williams DD, Edwards HGM, Schoen C, Crowder C, Newton EM (2000) A novel miniature Confocal microscope/Raman spectrometer system for biomolecular analysis on future Mars missions after Antarctic trials. Journal of Raman Spectroscopy 31: 633-635

Dombrowski HJ (1963) Bacteria from Palaeozoic salt deposits. Annals of the New York Academy of Sciences 108: 453-460

Drennan GR, Boiron M-C, Cathelineau M, Rob LJ (1999) Characteristics of post-depositional fluids in the Witwatersrand Basin. Mineralogy and Petrology 66: 83-109

Dutkiewicz A, Rasmussen B, Buick R (1998) Oil preserved in fluid inclusions in Archaean sandstones. Nature 395: 885-887

Edwards HGM, Farwell DW, Grady MM, Wynn-Williams DD, Wright IP (1999) Comparative Raman microscopy of a Martian meteorite and Antarctic lithic analogues. Planetary and Space Science 47: 353-362

Ellery A, Kolb C, Lammer H, Parnell J, Edwards H, Richter L, Patel M, Romstedt J, Dickensheets D, Steele A, Cockell C (2002) Astrobiological instrumentation for Mars – the only way is down. International Journal of Astrobiology 1: 365-380

Etminan H, Hoffmann CF (1989) Biomarkers in fluid inclusions: A new tool in constraining source regimes and its implications for the genesis of Mississippi Valley-type deposits. Geology 17: 19-22

Fish SA, Shepherd TJ, McGenity TJ, Grant WD (2002) Recovery of 16S ribosomal RNA gene fragments from ancient halite. Nature 417: 432-436

Freemantle M (1999) Downsizing chemistry: Chemical analysis and synthesis on microchips promise a variety of potential benefits. Chemical and Engineering News 77 (8): 27-36

Gibson G (2002) Microarrays in ecology and evolution: a preview. Molecular Ecology 11: 17-24

Goldstein RH, Reynolds TJ (1994) Systematics of Fluid Inclusions in Diagenetic Minerals. Society of Economic Paleontologists and Mineralogists Short Course 31: 199pp

Grant WD, Gemmell RT, McGenity TJ (1998) Halobacteria: the evidence for longevity. Extremophiles 2: 279-287

Grantham PJ (1986) Sterane isomerization and moretane/hopane ratios in crude oils derived from Tertiary source rocks. Organic Geochemistry 9: 293-304

Hansen AJ, Willersley E, Mørk S, Hedegaard MM, Rønn R, Jeffares DC (2002) JAWS: Just add water system – A device for detection of nucleic acids in Martian ice caps. Proceedings of the Second European Workshop on Exo/Astrobiology. European Space Agency Special Publication 518: 309-312

Henry OYF, Piletsky S, Grant WD, Sims WD, Cullen DC (2002) Robust molecular imprinted polymer thin-films for an astrobiology biomimetic sensor array. Proceedings of the Second European Workshop on Exo/Astrobiology. European Space Agency Special Publication 518: 513-514

Hoffmann CF, Henley RW, Higgins NC, Solomon M, Summons RE (1988) Biogenic hydrocarbons in fluid inclusions from the Aberfoyle tin-tungsten deposit, Tasmania, Australia. Chemical Geology 70: 287-299

Hunt JM (1996) Petroleum Geochemistry and Geology (2nd edition). W.H. Freeman and Company, New York, 743pp

Kanik I, Johnson PV, Beegle LW, Cooks RG, Laughlin BC, Hill HH (2003) Electrospray ionization/ion mobility spectrometer/cylindrical ion trap mass spectrometer system for in-situ detection of organic compounds. Lunar and Planetary Science XXXIV, abstract 1292 (CD-ROM)

Kazahaya K, Matsuo S (1985) A new ball-milling method for extraction of fluid inclusions from minerals. Geochemical Journal 19: 45-54

Keir R, Igata E, Arundell M, Smith WE, Graham D, McHugh C, Cooper JM (2002) SERRS. In situ substrate formation and improved detection using microfluidics. Analytical Chemistry 74: 1503-1508

Koeberl C, Fredriksson K, Götzinger M, Reimold WU (1989) Anomalous quartz from the Roter Kamm impact crater, Namibia: Evidence for post-impact hydrothermal activity? Geochimica et Cosmochimica Acta 53: 2113-2118

Komor SC, Valley JW, Brown PE (1988) Fluid-inclusion evidence for impact heating at the Siljan Ring, Sweden. Geology 16: 711-715

Kontak DJ, Sangster DF (1998) Aqueous and liquid petroleum inclusions in barite from the Walton Deposit, Nova Scotia, Canada: A Carboniferous, carbonate-hosted Ba-Pb-Zn-Cu-Ag deposit. Economic Geology 93: 845-868

Li R, Parnell J (2003) In situ microanalysis of petroleum fluid inclusions by Time of Flight-Secondary Ion Mass Spectrometry as an indicator of evolving oil chemistry: a pilot study in the Bohai Basin, China. Journal of Geochemical Exploration 78-79: 377-384

Lindahl T (1993) Instability and decay of the primary structure of DNA. Nature 362: 709-715

Lowenstein JM (1981) Immunological reactions from fossil materials. Philosophical Transactions of the Royal Society of London 292: 143-149

Luders V, Horsfield B, Kenkmann T, Mingram B, Wittmann A (2003) Hydrocarbons and aqueous fluids in Cretaceous sediments of the ICDP-Chicxulub drill core YAX-1. Lunar and Planetary Science XXXIV, abstract 1378 (CD-ROM)

Marshall D, Watkinson D, Farrow C, Molnar F, Fouillac A-M (1999) Multiple fluid generations in the Sudbury igneous complex: fluid inclusion, Ar, O, H, Rb and Sr evidence. Chemical Geology 154: 1-19

Marzi R, Rullkötter J (1992) Qualitative and quantitative evolution and kinetics of biological marker transformations – Laboratory experiments and application to the Michigan Basin. In: Moldowan JM, Albrecht P, Philip RP (eds) Biological Markers in Sediments and Petroleum, Prentice Hall, Englewood Cliffs, pp 18-41

Maule J, Steele A, Toporski J, McKay DS (2003) A new antibody for category 1 biomarker detection. Lunar and Planetary Science XXXIV, abstract 2131 (CD-ROM)

Mitchell P (2001) Microfluidics: Downsizing large-scale biology. Nature Biotechnology 19: 717-721

Munz IA (2001) Petroleum inclusions in sedimentary basins: systematics, analytical methods and applications. Lithos 55: 195-212

Norton CF, Grant WD (1988) Survival of halobacteria within fluid inclusions in salt crystals. Journal of General Bacteriology 134: 1365-1373

Norton CF, McGenity TJ, Grant WD (1993) Archaeal halophiles (halobacteria) from two British salt mines. Journal of General Microbiology 139: 1077-1081

Pagel M, Poty B (1975) Fluid inclusions in the rocks of the Charlevoix structure (Quebec, Canada). Fortschritte der Mineralogie 52: 479-489

Pang LSK, George SC, Quezada RA (1998) A study of the gross compositions of oil-bearing fluid inclusions using high performance liquid chromatography. Organic Geochemistry 29: 1149-1161

Parnell J (2001) Paragenesis of mineralization within fractured pebbles in Witwatersrand conglomerates. Mineralium Deposita 36: 689-699

Parnell J (2002) Sampling of palaeo-water and biomolecules from surface deposits on Mars. Proceedings of the Second European Workshop on Exo/Astrobiology. European Space Agency Special Publication 518: 395-398

Parnell J, Chen H, Klubov B (2001a) Hot oil in the Russian Arctic: Precipitation of vanadiferous bitumens, Novaya Zemlya. In: Piestrzynski A (ed) Mineral Deposits at the Beginning of the 21st Century, Balkema, Lisse, pp 71-74

Parnell J, Middleton D, Chen H, Hall D (2001b) The use of integrated fluid inclusion studies in constraining oil charge history and reservoir compartmentation: examples from the Jeanne d'Arc Basin, offshore Newfoundland. Marine and Petroleum Geology 18: 535-549

Parnell J, Mazzini A, Chen H (2002) Fluid inclusion studies of chemosynthetic carbonates: strategy for seeking life on Mars. Astrobiology 2: 43-57

Parnell J, Osinski GR, Lee P, Baron M, Pearson MJ, Feely M (2003) Hydrocarbons in the Haughton impact structure, Devon Island, Nunavut, Canada. Lunar and Planetary Science XXXIV, abstract 1118 (CD-ROM)

Pasteris JD, Wopenka B, Seitz JC (1988) Practical aspects of quantitative laser Raman microprobe spectroscopy for the study of fluid inclusions. Geochimica et Cosmochimica Acta 52: 979-988

Pearcy EC, Burruss RC (1993) Hydrocarbons and gold mineralization in the hot-spring deposit at Cherry Hill, California. In: Parnell J, Kucha H, Landais P (eds) Bitumens in Ore Deposits, Springer, Heidelberg, pp 117-137

Price LC (1993) Thermal stability of hydrocarbons in nature: Limits, evidence, characteristics, and possible controls. Geochimica et Cosmochimica Acta 57: 3261-3280

Price LC, De Wit E (2001) Evidence and characteristics of hydrolytic disproportionation of organic matter during metasomatic processes. Geochimica et Cosmochimica Acta 65: 3791-3826

Puura V, Suuroja K (1992) Ordovician impact crater at Kärdla, Hiiumaa Island, Estonia. Tectonophysics 216: 143-156

Radax C, Gruber C, Stan-Lotter H (2001) Novel haloarchaeal 16S rRNA gene sequences from Alpine Permo-Triassic rock salt. Extremophiles 5: 221-228

Rasmussen B, Buick R (2000) Oily old ores: Evidence for hydrothermal petroleum generation in an Archean volcanogenic massive sulphide deposit. Geology 28: 731-734

Reiser R, Tasch P (1960) Investigation of the viability of osmophile bacteria of great geological age. Transactions of the Kansas Academy of Sciences 63: 31-34

Robb LJ, Charlesworth EG, Drennan GR, Gibson RL, Tongu EL (1997) Tectono-metamorphic setting and paragenetic sequence of Au-U mineralization in the Archean Witwatersrand Basin. Australian Journal of Earth Sciences 44: 353-371

Robertson PB, Grieve RAF (1977) Shock attenuation at terrestrial impact structures. In: Roddy DJ, Pepin RO, Merrill RB (eds) Impact and Explosion Cratering, Pergamon Press, New York, pp 687-702

Roedder E (1984) Fluid Inclusions. Mineralogical Society of America, Reviews in Mineralogy 12, 644 pp

Ruble TE, George SC, Lisk M, Quezada RA (1998) Organic compounds trapped in aqueous fluid inclusions. Organic Geochemistry 29: 195-205

Saylor J, Zolensky M, Bodnar R, Le L, Schwandt C (2001) Fluid inclusions in carbonaceous chondrites. Lunar and Planetary Science XXXII, abstract 1875 (CD-ROM)

Sephton MA, Pillinger CT, Gilmour I (2000) Aromatic moieties in meteoritic macromolecular materials: Analysis by hydrous pyrolysis and $\delta^{13}C$ of individual compounds. Geochimica et Cosmochimica Acta 64: 321-328

Shepherd TJ, Rankin AH, Alderton DH (1985) A Practical Guide to Fluid Inclusion Studies. Blackie, Glasgow, 239pp

Simoneit BRT, Summons RE, Jahnke LL (1998) Biomarkers as tracers for life on early Earth and Mars. Origins of Life and Evolution of the Biosphere 28: 475-483

Stan-Lotter H, McGenity TJ, Legat A, Denner EBM, Glaser K, Stetter KO, Wanner G (1999) Very similar strains of *Halococcus salifodinae* are found in geographically separated Permo-Triassic salt deposits. Microbiology 145: 3565-3574

Steele A, McKay D, Schweitzer M (2001a) Biotechnology approaches to life detection. In General Meeting of the NASA Astrobiology Institute (pp. 206-208), Carnegie Institute of Washington, Washington DC

Steele A, Toporski JKW, Avci R, Guidry SA, McKay DS (2001b) Time of Flight-Secondary Ion Mass Spectrometry (ToF-SIMS) of a number of bacterial hopanoids. Organic Geochemistry 32: 905-911

Stuart F, Turner G, Taylor R (1994) He-Ar isotope systematics of fluid inclusions: Resolving mantle and crustal contributions to hydrothermal fluids. In: Matsuda J (ed) Noble Gas Geochemistry and Cosmochemistry. Terra Scientific Publishing Company, Tokyo, pp 261-277

Sturkell EFF, Broman C, Forsberg P, Torssander P (1998) Impact-related hydrothermal activity in the Lockne impact structure, Jämtland, Sweden. European Journal of Mineralogy 10: 589-606

Summons RE, Jahnke LL, Hope JM, Logan GA (1999) 2-Methylhopanoids as biomarkers for cyanobacterial oxygenic photosynthesis. Nature 400: 554-557

Svanvik N, Westman G, Wang D, Kubista M (2000) Light-up probes: Thiazole orange-conjugated peptide nucleic acid for detection of target nucleic acid in homogeneous solution. Analytical Biochemistry 281: 26-35

Thomson ML, Mastalerz M, Sinclair AJ, Bustin RM (1992) Fluid source and thermal history of an epithermal vein deposit, Owen Lake, central British Columbia: evidence from bitumen and fluid inclusions. Mineralium Deposita 27: 219-225

Thurman EM (1985) Organic Geochemistry of Natural Waters. Nijhoff, Dordrecht, 497pp

Toporski J, Steele A (2002) The relevance of bacterial biomarkers in astrobiological research. Proceedings of the Second European Workshop on Exo/Astrobiology. European Space Agency Special Publication 518: 239-242

Toporski J, Steele A, Westall F, Avci R, Martill DM, McKay DS (2002) Morphologic and spectral investigation of exceptionally well-preserved bacterial biofilms from the Oligocene Enspel Formation, Germany. Geochimica et Cosmochimica Acta 66: 1773-1791

Warmflash D, Larios-Sanz M, Fox GE, McKay DS (2002) Progress in the use of rapid molecular techniques to detect life forms in soil: implications for interplanetary astrobiology missions. Lunar and Planetary Science XXXIII, abstract 1963 (CD-ROM)

Weiss BP, Kirschvink JL, Baudenbacher FJ, Vali H, Peters NT, MacDonald FA, Wikswo JP (2000) Reconciliation of magnetic and petrographic constraints in ALH84001? Panspermia lives on! Lunar and Planetary Science XXXI, abstract 2078 (CD-ROM)

Whitby JA, Rohner U, Benz W, Wurz P (2003) Laser-ablation mass spectrometer for the surface of Mercury. Lunar and Planetary Science XXXIV, abstract 1624 (CD-ROM)

Wilkins AD, Wright A, Parnell J, Artz R (2002) Astrobiological use of model crystals containing biomolecules and microbes: Testing analytical techniques and space exposure experiments. Abstracts of the Second European Workshop on Exo/Astrobiology, Graz, Austria, European Space Agency Special Publication 518: 567-568

Zolensky ME, Bodnar RJ, Gibson EK, Nyquist LE, Reese Y, Shih C-Y, Wiesmann H (1999) Asteroidal water within fluid inclusion-bearing halite in an H5 chondrite, Monahans. Science 285: 1377-1379

Geomicrobiology of Impact-Altered Rocks

Charles S. Cockell[1], David A. Fike[2], Gordon R. Osinski[3], Pascal Lee[4]

[1]Open University, Milton Keynes, MK7 6AA, U.K. (c.s.cockell@open.ac.uk)
[2]Department of Earth, Atmospheric and Planetary Sciences, Massachusetts Institute of Technology, 77 Massachusetts Avenue, 54-812, Cambridge, MA 01239, USA.
[3]Planetary and Space Sciences Center, Department of Geology, University of New Brunswick, Bailey Drive, Fredricton, New Brunswick E3B 5A3. Canada.
[4]NASA Ames Research Center, Moffett Field, CA 94035-1000, USA.

Abstract. Rocks shocked by asteroid or comet impact events can be made more porous by the shock volatilization of minerals, and they can be fractured by the intense heat and pressures of impact. New spaces within the rock provide access points and surfaces for the growth of microbial communities, illustrating an example of how shock metamorphism can generate new habitats for microbial colonization. We review data on the colonization of shocked gneiss from the Haughton impact structure by phototrophs and heterotrophs. Shocked rocks can preferentially trap water and protect against wind-induced desiccation. The interior of shocked rocks is often warmer than the air temperature, and protects against ultraviolet radiation. Because impact events are a ubiquitous process on solid planetary surfaces, the shocked-rock habitat may be important on other planets, and it may have been important on the early Earth when primitive microorganisms lived under a much higher impact flux than today.

1
Introduction

The large-scale geologic rearrangements associated with asteroid and comet impacts change the hydrology of the local region and, thus, the availability of liquid water, an essential requirement for life. Macroscopically obvious manifestations of these changes are impact lakes, which form in the depression resulting from impact excavation. Although there is no fresh crater on Earth today, many structures do still retain lakes within the excavation cavity. Examples include Bosumtwi (Ghana), Tswaing (S.

Africa), New Quebec (Canada), and the Clearwater craters (Canada). The biota of some impact lakes has been characterized previously (Schoeman and Ashton 1982, Bouchard 1989, Gronlund et al. 1990, Cremer and Wagner 2003).

A macroscopically less obvious change in the hydrology occurs at the scales of centimeters down to microns, where impact-induced fractures and pore formation can localize liquid water, potentially for the benefit of microbial life. Water may be trapped by capillary action, or by direct filtering of the water into the cracks caused by impact, particularly in biomes that experience precipitation in the form or snow or rain.

Impact-induced fractures and pores are habitats that can potentially last on the time scale of billions of years. Other habitats that provide liquid water in the post-impact environment are generally more short-lived. For example, hydrothermal systems, which would provide a habitat for heat-loving organisms, last for thousands to a few millions of years depending on the scale of the impact (Newsom et al. 1986, McCarville and Crossey 1996, Osinski et al. 2001). Impacts with an energy of 10^6 Mt, for instance, are predicted to generate hydrothermal ecosystems for several thousand years, whereas impacts of 10^8 Mt are believed to induce hydrothermal systems that can last up to a million years (Zurcher and Kring 2003). Eventually, as the structure cools, these systems cease to be active and the microbial habitat associated with them is no longer available.

Impact crater lakes can last for longer than their hydrothermal systems, but their lifetime is set by the infilling of the bowl, the breach of the rim or the total erosional removal of the surface expression of the impact structure. Their maximum life times are potentially tens to hundreds of millions of years, because so long as the excavation cavity remains, the crater can pond water. A crater will not necessarily harbour a lake continuously if the climatic regimen (particularly precipitation patterns) is not always favourable for a lake to exist.

Impact fractures and pore spaces, however, will last for as long as some of the original affected target material still survives. Impact-induced fracturing is still evident, for example, at the 2-billion year old Vredefort structure in South Africa (see Gibson and Reimold 2001). As a habitat for microorganisms, this longevity makes impact-altered rocks of great interest in the context of early Earth and astrobiology, as after an impact event these habitats could potentially be available for a significant fraction of the planetary life time.

The Haughton impact structure, centered at 75°22'N and 89°41'W on Devon Island in the Canadian High Arctic, has abundant well-preserved breccia outcrops (Redeker and Stöffler 1988, Osinski and Spray 2001) and therefore provides a potentially important insight into post-impact microbial colonization of shocked rocks. The clasts within the breccia have been examined previously (Metzler et al. 1988) and thus a geological context exists in which to place biological studies. Because of the structure's high latitude location, vegetation is sparse (Cockell et al. 2001) (less than 1% cover) and the region is dominated by microbial communities (Walker and Peters 1977, Svoboda and Freedman 1981), thus making the identification and study of microbial patterns of crater colonization much easier than would be the case with craters covered in soils and heavy vegetation. The crater has a diameter of 23-24 km and was formed during the Miocene ~23 million years ago (Jessberger 1988). It is located within the Ordovician-Silurian Allen Bay formation, consisting of dolomites and limestones that overlay Precambrian basement gneisses (Frisch and Thorsteinsson 1978).

In this paper we review data on the microbial colonization of impact-altered rocks, specifically shock-metamorphized gneiss. We describe the advantages that may be obtained by microorganisms that live inside impact-shocked rocks (as opposed to living on the surface of rocks – a habitat available on all rock surfaces, whether they are shocked or not).

2 Characteristics of the shocked gneiss

Metzler et al. (1988) examined the characteristics of shocked crystalline and sedimentary rocks at Haughton. They recognized 13 different classes of crystalline rocks comprised of 12 categories of gneiss and one category of basalt. The gneiss clasts that they studied, and which we used for our studies on colonization, are exposed on the surface of the breccia outcrops within the crater, which cover an area of ~ 60 km^2. These clasts are derived from the Precambrian basement, from an approximate depth of 1,700 m and were excavated and shocked during the impact.

The chemical composition of the rocks varies widely. From a sample of 38 rocks, Metzler et al. (1988) found that quartz content varied from 40-85 wt%, Al_2O_3 varied from 7-20 wt% and P_2O_5 from <0.01 to 5 wt%. The shock levels of the rocks also vary, with a maximum shock pressure of around 60 GPa. The crystalline rocks conspicuously lack whole melting and this is attributed to the fact that they are excavated from the deeper basement where the 60 GPa isobar only barely penetrated. An additional factor might have been the formation of the breccia at temperatures insufficient to cause substantial rock melting. The consequences are that the rocks are heavily fractured and pitted, but not melted, and this has subsequent effects on the availability of subsurface spaces for microbial attachment and biofilm formation, discussed below.

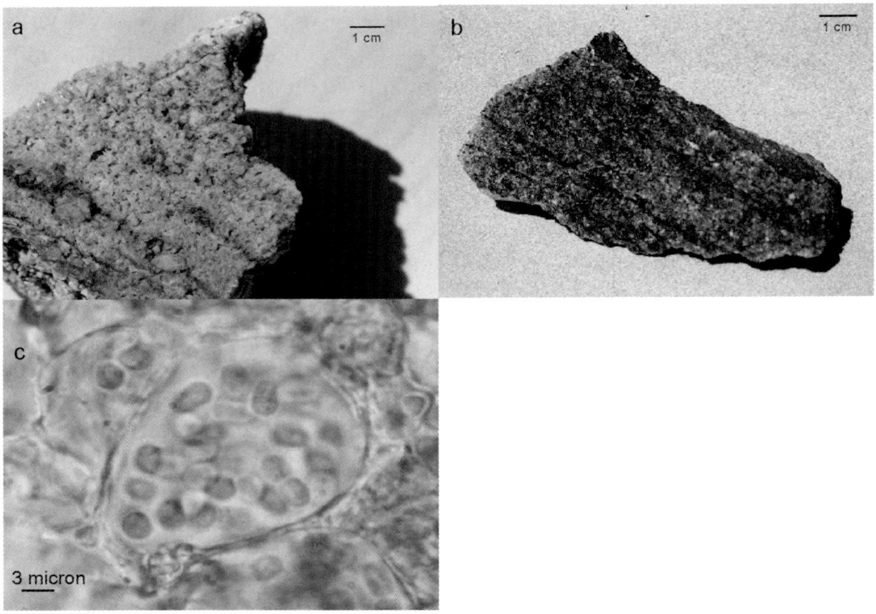

Fig. 1. (a) Shocked gneiss. Material shocked to 25-45 GPa. **(b)** Unshocked gneiss. Difference in texture compared to shocked material is evident. **(c)** Colony of coccoid cyanobacteria in a colonial form typical of those found growing as endoliths and chasmoliths within shocked gneiss. Material obtained from the Haughton crater at location 75°24.53'N and 89°49.76'W.

The shock pressures experienced by the crystalline rock vary from shock stage 0 (unshocked) to III/IV (55-60 GPa), but approximately half of the crystalline rocks studied by Metzler et al. (1988) were within the shock range 25-45 GPa. The highest shock levels are found near Anomaly Hill, the gravitational anomaly presumed to be close to the point of impact. At this location rocks with shock levels up to 60 GPa are found.

For the geomicrobiological study presented here we collected rocks of shock pressures 25-45 GPa distributed across the melt hills of the crater (Cockell et al. 2002). This material is conspicuous as a white pumice-like material (Figure 1a) Material referred to as 'low shocked' underwent shock pressures less than 10 GPa. This material was conspicuous on account of its dark colour and higher density than the shocked material (Figure 1b) and was used for comparisons with the shocked material of pore space and light penetration.

The mean pore surface area for pores of >1 mm diameter (the size range that is relevant for microorganisms) within shocked material was 0.10 m^2g^{-1} and it was 25 times less than this for low-shocked material (Cockell et al. 2002), illustrating the substantial increase in colonizable surface area caused by shock processing. The mean porosity of the shocked rocks was 22 vol%.

In this study we were concerned with microbial colonization of the fractures and pore spaces created by shock processing and so we did not distinguish between different chemical compositions of gneiss described by Metzler et al. (1988). Thus, the study we report here focused on the influence of the altered *physical* environment on microbial colonization.

We cannot rule out effects of impact-shock on the chemical composition of the rocks and thus colonization. As pointed out previously, shock volatilization of many macronutrients important for life may have occurred (Cockell et al. 2003b).

3
The environment within impact-altered rocks

3.1
Temperature and water

The micro-environmental conditions within the shocked rocks can be very different from the macro-climatic conditions outside of the rock, or even on the rock surface. Temperatures within a gneissic rock located at 75°24.53'N and 89°49.76'W of size 6 x 6 x 7 cm. Figure 2 shows the temperature profile during nine days during July 2002, measured using a Campbell CR-10X datalogger (Cockell et al. 2003a). The mean rock surface temperature was 5.49 °C and the mean air temperature (1 m height) was 4.51 °C, demonstrating that living even on the surface of the rock can provide thermal advantages to being exposed to air temperatures. This is partly caused by the absorption of solar heat by the rock and also some protection afforded from wind by being close to the ground in the boundary layer. In the endolithic habitat (the habitat beneath the rock surface -

see Golubic et al. 1981 for a more detailed terminological discussion) at 2 mm depth in the rock, the mean temperature was 5.93 °C, 0.44 °C higher than even the rock surface. This temperature difference may not appear to be substantial. However, during the 'night' (the sun is still above the horizon at this latitude during July at midnight, but the solar insolation is approximately one order of magnitude lower than during midday), the temperature of the rock surface and interior equilibrates with the air temperature, reducing the mean difference. Greatest temperature differences were recorded after midday. The endolithic temperature can also exceed the air temperature and rock surface temperature substantially. At 13:17 on July 20, for instance, the endolithic temperature was 21.8 °C, the air temperature was 10.6 °C and the rock surface temperature was 16.1° C.

We did not measure rapid freeze-thaw cycles on the surface of the rocks in the arctic. Freeze-thaw cycles were observed on the surface of rocks in the Antarctic Dry Valleys (McKay and Friedmann 1985). The lack of regular freeze-thaw cycles may be one factor accounting for the presence of surface (epilithic) colonization on many of the rocks on Devon Island.

The effects of the high midday temperatures in the endolithic zone on microbial populations are not clear. For psychrophilic organisms that need to grow between 0 and 15 °C, the higher temperatures recorded in the endolithic habitat on some days compared to the surface of the rock could inhibit growth. However, on many days the rapid warming of the endolithic habitat from nearly freezing temperatures early in the morning is likely to be beneficial for growth compared to the surface, whatever the metabolic temperature optima of the organisms.

We also recorded the temperature in the endolithic habitat from 14 August 2001 to 10 July 2002 (Cockell et al. 2003a). The mean temperature during the year was -20.3 °C. The lowest temperature recorded was -40.3 °C during late February 2002 (Figure 3) showing that for approximately nine months of the year the rocks and their enclosed biota are frozen and during the winter they are in complete darkness. During this period, the microbiota is not likely to be metabolically active. The period of highest activity will be between mid-May and August when temperatures rise above freezing and liquid water from snow melt and, later in the season, when rain is available.

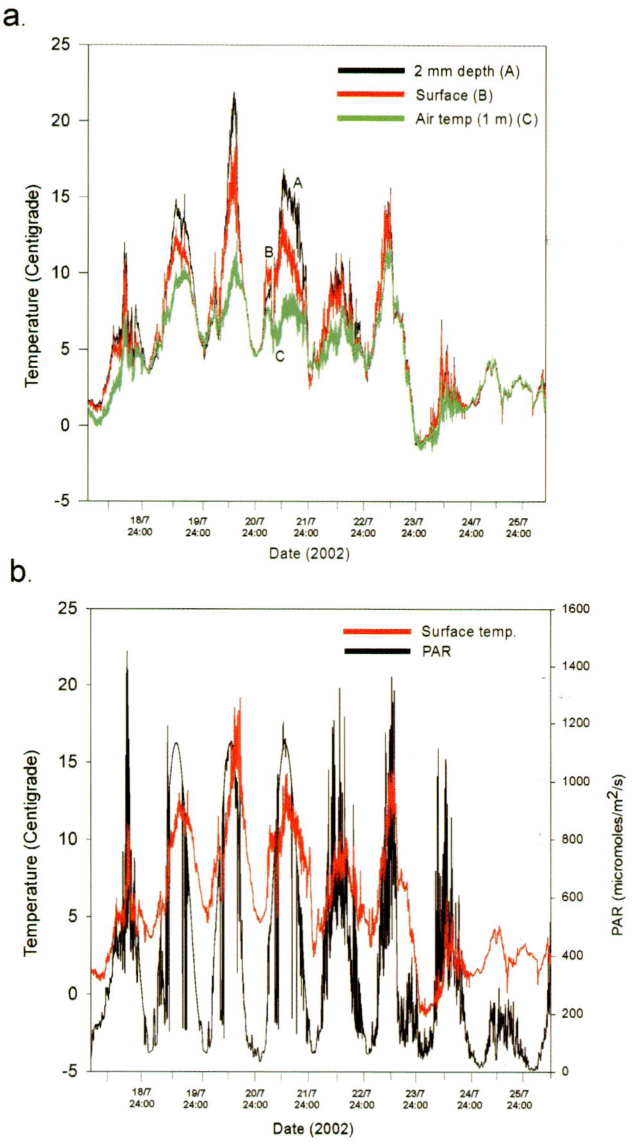

Fig. 2. (a) Temperatures recorded in an impact-shocked rock from 18 July to 26 July 2002. Upper curve shows temperature recorded within the endolithic habitat at 2 mm depth. Middle curve shows rock surface temperature and lower curve shows air temperature. **(b)** Levels of photosynthetically active radiation (PAR) incident on the rock over the period of measurements. Rock surface temperature is also plotted to show relationship to insolation.

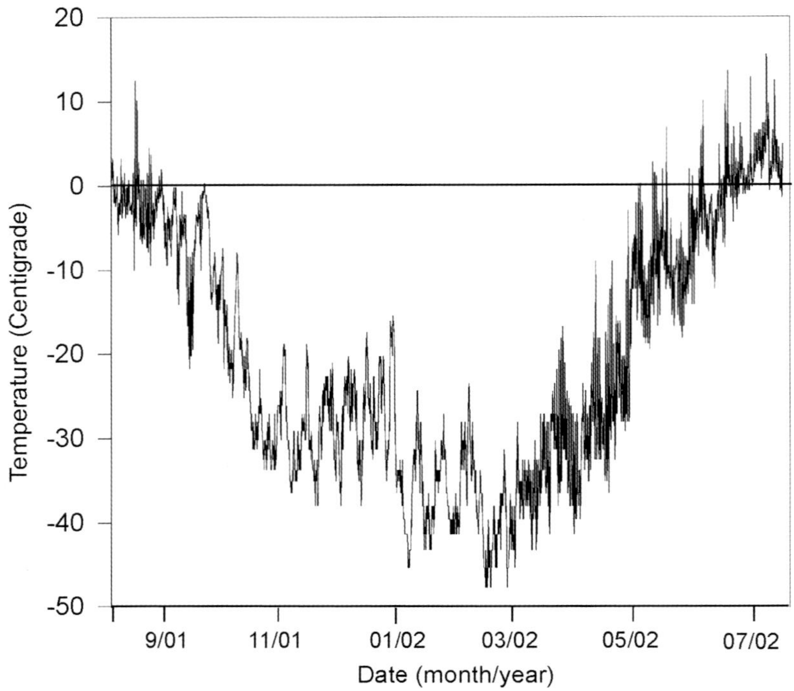

Fig. 3. Rock surface temperature from 14 August 2001 to 10 July 2002 (same sample as that used to obtain data in Figure 2).

The pore spaces and fractures of the shocked rocks retain moisture for many days after rain events. We measured the mass of two selected rocks over a period of nine days (Figure 4). The two pieces of shocked gneiss with different mass (a, 207.2 g, b, 68.9 g, see Figure 4) were selected from the melt rock outcrop at 75°24.53'N and 89°49.76'W. The two rocks were placed next to the rock used for temperature measurements. Following a 1 hour rain event that delivered 1 mm of rain, the larger (207.2 g) rock had increased the percentage of its dry mass in water from 1.3 to 4.8 wt% and the smaller from 0.3 to 5.2 wt%. This water filters into the pore spaces and fractures and soaks into the interior of the rock. The water can be retained for some days within the rock until the next rain event again provides moisture (Figure 4, between first and second rain event). The distribution of this water will vary. The near-surface environment will dry out more quickly than deeper regions in the rock between rain events. However, the data show that the structure of the rock can trap meltwater from snow and ice and rain water for the enclosed microbiota. Thus, the shocked rock habitat provides advantages for water availability as well as thermal advantages.

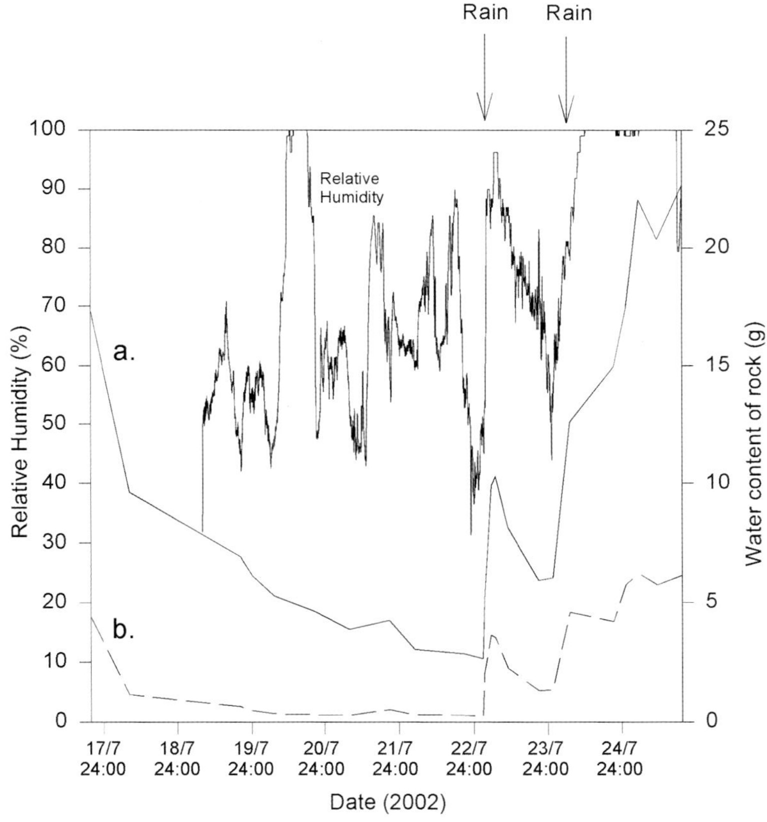

Fig. 4. Relative humidity (left axis) from 17 July to 25 July 2002. The right axis shows water mass gained by two impact-shocked gneiss samples (sample a, 207.2 g, sample b, 68.9 g). In the first rain event 1 mm of rain was deposited in ~1 hour on 23 July and in the second rain event 4 mm of rain was deposited from 24 July to 25 July (from Cockell et al. 2003a).

3.2 Radiation

Living within shocked rocks is a double-edged sword in terms of the radiation environment. On the one hand the overlying rock will reduce photosynthetically active radiation; required as an energy source by phototrophic communities, but on the other hand, it will reduce the level of damaging ultraviolet radiation.

The reduction in the photosynthetically active radiation (PAR is radiation between 400 to 700 nm used for photosynthesis) can be quantified by measuring the spectrum of light received under a defined thickness of gneiss and comparing it to the unattenuated source radiation. Using this technique PAR is reduced by approximately one order of magnitude under 0.5 mm of shocked gneiss (Cockell et al. 2002) giving organisms typical midday PAR levels of 10 μmoles $m^{-2}s^{-1}$ during July at a depth of 1 mm, corresponding to the 'compensation point' (1% of ambient light). The level of PAR reduction will vary depending on the composition of the rock and level of shock. This reduced PAR may not be a problem for phototrophic organisms. Provided it is above the minimum required for photosynthesis then the organisms can persist, as they do in Antarctic sandstones (Nienow et al. 1988). Cyanobacteria in polar regions generally succeed as a result of slow, persistent growth and tolerance of environmental extremes, not by outcompeting other microorganisms (Tang and Vincent 1999). Thus, the reduced PAR compared to living outside the rock would not necessarily cause any selective disadvantage *per se*, unless, of course, the rock becomes covered in epilithic organisms that take advantage of high light levels and completely block light penetration to the organisms beneath.

The light penetration through the shocked material is on average one order of magnitude greater than the penetration through low shocked (<10 GPa) material, showing that the increase in porosity and the volatilization of minerals during shock processing has increased the translucence of the rock in the region of the spectrum important for photosynthesis. Indeed, in low shocked rocks the light penetration is completely insufficient for photosynthetic organisms, explaining why colonization is only observed in weathering crusts in this material.

Similarly to visible light, ultraviolet radiation (wavelengths to 400 nm) is also reduced by approximately one order of magnitude under 0.5 mm of shocked gneiss. Using rock sections placed over small 1 x 2 cm dosimeters made from a monolayer of *Bacillus subtilis*, a spore forming soil bacterium, the protection afforded to organisms can be measured directly as a function of biological damage. The *B. subtilis* spores are killed by UV radiation and the percent that are killed can be quantified in the laboratory diation and the percent that are killed can be quantified in the laboratory (Horneck et al. 1986). By comparing the inactivation of spores covered in defined thicknesses of rock to control dosimeters exposed to unattenuated solar radiation the protective effects of the rock can be determined. The one order of magnitude reduction of UV damage measured under 0.5mm of shocked gneiss using this technique is consistent with the level of reduction of PAR measured using a spectrometer.

The protection afforded against ultraviolet radiation by shocked rock is also manifested in the production of UV screening compounds produced by phototrophs in response to UV exposure. When terrestrial cyanobacteria are exposed to UV radiation they synthesize carotenoids, which are used to quench reactive oxygen states generated from interaction of energetic UV light with cellular components (Quesada et al. 1999). They also synthesize scytonemin, a passive UV screening compound made from tryptophan derivatives that screens UV radiation before it has a chance to interact with cell components (Garcia-Pichel et al. 1992, Ehling-Shulz et al. 1997). This double line of defence (along with DNA and cellular repair processes) allows cyanobacteria to live under high environmental UV fluxes. Other microorganisms also possess UV screening compounds and carotenoids.

The concentrations of scytonemin and carotenoids in cyanobacteria removed from the interior of four shocked gneiss samples were 3.3 ± 0.36 and 0.93 ± 0.26 μg/cm^2. However in four samples of cyanobacteria scraped from the surface of shocked rocks, the corresponding concentrations were 27.0 ± 15.4 and 7.8 ± 4.2 μg/cm^2, respectively, just over eight times higher than the endolithic organisms (Cockell et al. 2002). Although the community compositions of these samples may not be identical, the data show that cyanobacteria within the shocked rocks are substantially protected against UV radiation and need to expend less energy in synthesizing UV-protecting compounds compared to those exposed to the unattenuated UV flux. This data is consistent with the *B. subtilis* biofilm dosimetry data.

Thus, phototrophic organisms growing on the surface of rocks get higher visible light levels required for photosynthesis, but need to expend more energy synthesizing UV protecting compounds to protect against the high UV flux than organisms within the shocked rocks. Organisms within shocked rocks get less light for photosynthesis, but they also need to expend less energy on UV-protection. Thus, a trade-off occurs depending on where an organism becomes attached and grows.

Because UV radiation is dramatically cut down by thin layers of gneiss, it follows that for non-photosynthetic organisms living within the interior of the rock at depths of just 2-3 mm and greater, UV radiation damage over the short summer season becomes negligible. For an organism at 1 mm depth the damage received over the summer season will be similar to the damage received by an organism on the surface of the rock in one day. Shocked rocks can be considered a highly effective refugium from UV radiation damage.

4
Microbial colonization

The microbial colonization of impact-shocked rocks can, for convenience, be split into two categories. First, phototrophic organisms inhabit the near-surface of the rocks. They are confined to the region of the rock where light levels exceed the minimum required for photosynthesis and growth at levels sufficient to make the communities viable. Second, non-phototrophic communities inhabit the interior of the rock and are not limited by light (but can be limited by the depth of water penetration into the rock). This latter category includes heterotrophic microorganisms that use organics as a source of carbon and/or energy and chemolithotrophic organisms that use inorganic redox couples as a source of energy. Other groups of organisms, such as fermenters, cannot be ruled out for some shocked rocks with high organic contents that would be found, for instance, in impact structures with high vegetation biomass.

The phototrophic communities within the Haughton shocked gneisses are exclusively cyanobacteria. The effect of the higher porosity of the shocked material and the increase in translucence in the photosynthetically active region of the spectrum compared to low shocked material is evident in the colonization by this group of organisms. Seventy-three percent of a sample of 30 rocks were colonized by phototrophic organisms either endolithically (within the rock substratum) or chasmolithically (growth within macroscopic cracks). In 26% of samples a coherent endolithic band > 1 cm long was observed within the substratum of the rock. In only 3% of the low-shocked material was any type of colonization observed at all and in these cases it was confined to a weathering crust (Cockell et al. 2002).

The difference in the colonization patterns is accounted for by the higher porosity and the number of micro-fractures in the shocked rocks and the greater translucence of the shocked material. The higher porosity allows microbes to gain access to the subsurface of the material and also provides surfaces for lateral growth within the rock. In the unshocked rock there is neither the access to the subsurface, nor a network of interconnected fractures and pores that allows for biofilms of microbes to penetrate and grow throughout the subsurface space.

Cyanobacteria are found to penetrate to a depth of a maximum 5 mm. At such a depth light levels are about 1000-10,000 times lower than at the surface of the rock, depending on the structure of the rock at the micro-scale. As the minimum light level for photosynthesis is ~10-100 nmoles $m^{-2}s^{-1}$ (Littler et al. 1986, Raven et al. 2000) and midday light levels in the arctic are ~1100 μmoles $m^{-2}s^{-1}$ during July, this depth would be consistent with the notion that the lower level of the phototrophic band in the shocked rocks is determined by the lack of availability of photosynthetically active radiation.

The cyanobacteria that do inhabit the subsurface space are exclusively coccoid species such as that shown in Figure 1c. We have not observed filamentous species, which are often found colonizing the surface of the rock (epilithic habitat) with coccoid species. We attribute this to the fact that the filaments are limited in their ability to penetrate and move throughout the surface space. The coccoid species that do inhabit the rock are themselves taxonomically restricted in their distribution compared to surface-dwelling coccoid species. This may be caused by morphology or colonial growth patterns that may have different abilities to attach to and fill the subsurface space. Similarly eukaryotic algae are not found inside the rocks, but are found colonizing the surface of the rocks. The surface of the rock is inhabited by a diverse coccoid cyanobacteria assemblage including *Aphanothece*, *Gloeocapsa* spp. and unicellular chlorophytes are found. However, the interior of the shocked rock is dominated apparently by one species. We attributed this species to the genus *Chroococcidiopsis* on account of its morphology and colonial growth pattern (Cockell et al. 2002). As cyanobacterial identification is difficult and has become confused with new genetic sequencing techniques, we note that in morphology and colonial growth pattern the organisms are '*Chroococcidiopsis*-like'.

It is evident that the interior of the rocks, well below the photosynthetic zone, also provides an environment for microbial growth. Using aseptically collected fragments from the interior of shocked rocks, we isolated heterotrophic bacteria using tryptic soy agar (TSA) plates and we serially streaked them to isolate species (Fike et al. 2003). Polymerase Chain Reaction (PCR) and sequencing was employed to determine their identity. A total of 27 bacteria were isolated and sequenced. Genera that were isolated were *Arthobacter*, *Planococcus*, *Bacillus*, *Pseudomonas*, *Stenotrophomonas*, *Janthinobacter*, and *Caulobacter*. The isolates that we sequenced most closely resembled bacteria that have already been found in soil, polar and marine environments.

The origin of these microorganisms is speculative, but they may have been transported into the rock by meltwater, perhaps in some cases being initially deposited on the rocks by wind. The microbes then filter into the subsurface. The species we identified are not unusual or endemic and their previous characterization in polar environments suggest that the populations within the rocks are similar to those that inhabit the soils and water in the arctic.

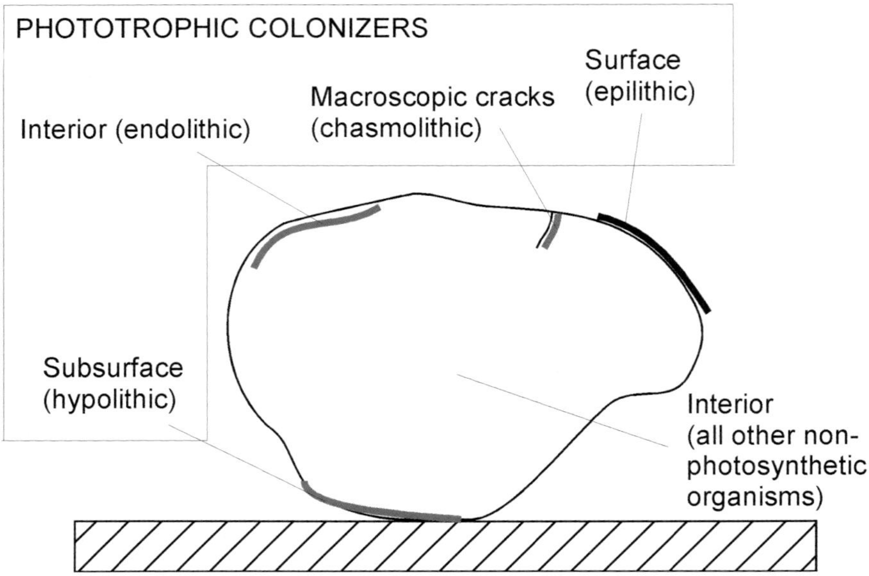

Fig. 5. The different modes of colonization of impact-shocked rocks by microorganisms.

Some of the isolates we cultured closely matched psychrophilic species that need to grow at 0 to 15°C. This is consistent with the environmental conditions to be found in the arctic. All of the species we isolated are strict aerobes and all use respiratory pathways, none were fermentative, consistent with the culture conditions used to isolate them.

We have not determined the metabolic activity of these organisms *in situ*. This preliminary examination of the heterotrophic bacteria within the shocked rocks suggests that the rocks are home to a diverse microbial assemblage. As many of the rocks become soaked by rain during the summer it is likely that there is a diversity of inactive bacteria that have leached into the rocks and become attached. However, the presence of biofilms suggests that the rock interior is also a habitat for actively growing communities.

Figure 5 shows a summarized view of the shocked rocks as a microbial habitat.

5 Astrobiological perspectives on the data

At the time of writing the question of life elsewhere in the Solar System, let alone on extrasolar planets, remains speculative, but it is worth considering the data presented here within a more 'universal' setting. The impact of asteroids and comets with rocky planetary surfaces is, presumably, a universal process as no Solar System can be expected to form 'perfectly' without leftover cometary and asteroidal material. Therefore, one might venture to say that on any solid planet on which one might wish to postulate the evolution of life, there will be impact craters (Melosh 1989). It is also probably a reasonable astrobiological statement to say that life will tend to survive and grow better in environments where extremes that are detrimental to it are minimized. In extreme deserts of the world endolithic habitats are ubiquitous refugia for life (e.g., Büdel and Wessels 1991, Friedmann 1977, 1980, Wierzchos et al. 2003). 'Extreme', when used in the context of habitats, is a loaded word because some organisms can, for example, benefit from UV radiation (such as many metazoans that use UV radiation in vision).

The amelioration of environmental extremes within impact shocked rocks might make them attractive places for primitive life on other planets. A particularly case would be the planet Mars, where the lack of recent plate tectonics and hydrological activity means that craters are well preserved. Craters on Mars have previously been recognized as potentially important sites for exobiology (Newsom 1980, Cabrol and Grin 1995, Rathbun and Squyres 2002). If Mars had more liquid water in its early history, which would have ponded in depressions and craters (McKay and Davis 1991, Scott et al. 1991, Newsom et al. 1996), impact shocked rocks would have been a favourable habitat for life.

Finally, impact craters were undoubtedly more common on early Earth when the impact flux was supposed to have been much higher than today (Chyba et al. 1994). Early Earth may have lacked an ozone screen and before atmospheric oxygen levels rose, damage to DNA on the surface of the planet may have been at worst three orders of magnitude higher than on the surface of the Earth today, if there had been no other atmospheric UV screens (Cockell and Horneck 2001). Impact-shocked rocks could have provided a refugium from this early, intense, UV radiation environment.

Not all habitats can be considered to have a universal applicability. Glaciers are not present on all planets, neither are sand dunes. As evident on Mars, lakes and ponds are not necessarily universal either. However, impact-shocked rocks can be considered one of the few microbial habitats that can be postulated with confidence on all rocky planets.

6 Conclusions

Impact-shocked gneiss in the Haughton impact structure provides a microhabitat for diverse microbial communities. As impact shocked rocks can persist for billions of years, much longer than impact lakes and hydrothermal systems, they are a long-lasting and biologically important habitat that is directly caused by the impact of asteroids and comets. Physically, the interior of the shocked rocks provides many advantages compared to living outside of the rock, particularly in an extreme polar location. UV radiation damage is reduced, moisture is retained and the temperatures are often higher than air temperatures. As UV damage, low temperatures and desiccation are three of the major environmental stressors to polar microorganisms, in the case of the Haughton structure the impact event can be seen to have ameliorated these extremes for many present-day microbial communities, providing an important example of a beneficial effect of an impact event to microorganisms.

Acknowledgments

We would like to acknowledge the reviewers and also the ESF IMPACT program for the opportunity to present this work. This work was made possible by a Leverhulme grant to study UV penetration into microhabitats (no. F/00 802/A) and British Council Academic Research Collaboration (ARC) Project No. 1210. We also acknowledge the Natural Environmental Research Council (NERC) for use of the NERC ICP Facility.

References

Bouchard MA (1989) L'histoire naturelle du Cratere du Nouveau-Quebec. Collection Environment et géologie, 7. Départment de géologie, Université de Montréal, 420 pp

Büdel B, and Wessels DCJ, 1991 Rock inhabiting blue-green algae/cyanobacteria from hot arid regions. Algological Studies 64: 385-398

Cabrol NA, Grin EA (1995) A morphological view on potential niches for exobiology on Mars. Planetary and Space Sciences 43: 179-188

Chyba CF, Owen, IP, Ip WH (1994) Impact delivery of volatiles and organic molecules to Earth. In: Gehrels (ed) Hazards due to comets and asteroids, University of Arizona Press, Arizona, pp 9-58

Cockell CS, Horneck G (2001) The history of the UV radiation climate of the Earth – Theoretical and space-based observations. Photochemistry and Photobiology 73: 447-451

Cockell CS, Lee P, Hildalgo L, Schuerger A, Stokes D, Jones J (2001) Microbiology and vegetation of micro-oases and polar desert, Haughton Impact Crater, Devon Island, Nunavut, Canada. Arctic, Alpine and Antarctic Research 33: 306-318

Cockell CS, Lee P, Osinski G, Horneck G, Broady P (2002) Impact-induced microbial endolithic habitats. Meteoritics and Planetary Sciences 37: 1287-1298

Cockell CS, McKay CP, Omelon C (2003a) Polar endoliths – an anticorrelation of climatic extremes and microbial biodiversity. International Journal of Astrobiology 1: 305-310

Cockell CS, Osinski G, Lee P (2003b) The impact crater as a habitat – effects of impact processing of target materials. Astrobiology 3: 181-191

Cremer H, Wagner B (2003) The diatom flora in the ultra-oligotrophic Lake El'gygytgyn, Chukotka. Polar Biology 26: 105-114

Ehling-Shulz M, Bilger W, Scherer S (1997) UV-B-induced synthesis of photoprotective pigments and extracellular polysaccharides in the terrestrial cyanobacterium *Nostoc commune*. Journal of Bacteriology 179: 1940-1945

Fike DA, Cockell CS, Pearce D, Lee P (2003) Heterotrophic microbial colonization of the interior of impact-shocked rocks from Haughton impact structure, Devon Island, Nunavut, Canadian High Arctic. International Journal of Astrobiology 1: 311-323

Friedmann EI (1977) Microorganisms in antarctic desert rocks from dry valleys and Dufek Massif. Antarctic Journal of the United States XII: 26-30

Friedmann EI (1980) Endolithic microbial life in hot and cold deserts. Origins of Life and Evolution of the Biosphere 10: 223-235

Frisch T, Thorsteinsson R (1978) Haughton astrobleme: a mid-Cenozoic impact crater Devon Island, Canadian Arctic Archipelago. Arctic 31: 108-124

Garcia-Pichel F, Sherry ND, Castenholz RW (1992) Evidence for an ultraviolet sunscreen role of the extracellular pigment scytonemin in the terrestrial cyanobacterium *Chlorogloeopsis* sp. Photochemistry and Photobiology 56: 17-23

Gibson RL, Reimold WU (2001) The Vredefort impact structure, South Africa. Memoir 92, Council for Geoscience, Geological Survey of South Africa, Pretoria. 111 pp

Golubic S, Friedmann I, Schneider J (1981) The lithobiontic ecological niche, with special reference to microorganisms. Journal of Sedimentary Petrology 51: 475-478

Gronlund T, Lortie G, Guilbault JP, Bouchard MA, Saanisto M (1990) Diatoms and arcellaceans from Lac du Cratere du Nouveau-Quebec, Ungava, Quebec, Canada. Canadian Journal of Botany 68: 1187-1200

Horneck G, Rettberg P, Rabbow E, Strauch W, Seckmeyer G, Facius R, Reitz G, Strauch K, Schott JU (1996) Biological UV dosimetry of solar radiation for different simulated ozone column thicknesses. Journal of Photochemistry and Photobiology B: Biology 32: 189-196

Jessberger EK (1988) ^{40}Ar-^{39}Ar dating of the Haughton impact structure. Meteoritics 23: 233-234

Littler MM, Littler DS, Blair SM, Norris JN (1986) Deep-water plant communities from an uncharted seamount off San Salvador Island, Bahamas: distribution, abundance and primary production. Deep-Sea Research 33: 881-892

McCarville P, Crossey LJ (1996) Post-impact hydrothermal alteration of the Manson impact structure. In: Koeberl C, Anderson RR (eds) The Manson Impact Structure, Iowa: Anatomy of an Impact Crater, Geological Society of America Special Paper 302, Denver, Colorado, USA, pp 347-376

McKay CP, Davis WL (1991) The duration of liquid water habitats on Mars. Icarus 90: 214-221

McKay CP, Friedmann EI (1985) The cryptoendolithic microbial environment in the Antarctic cold desert: temperature variations in nature. Polar Biology 4: 19-25

Melosh H J (1989) Impact cratering. A geologic process. Oxford University Press, 245 pp

Metzler A, Ostertag R, Redeker HJ, Stöffler D (1988) Composition of the crystalline basement and shock metamorphism of crystalline and sedimentary target

rocks at the Haughton Impact Crater, Devon Island, Canada. Meteoritics 23: 197-207
Newsom HE (1980) Hydrothermal alteration of impact melt sheets with implications for Mars. Icarus 44: 207-216
Newsom HE, Graup G, Sewards T, Keil K (1986) Fluidization and hydrothermal alteration of the suevite deposit at the Ries crater, West Germany, and implications for Mars. Journal of Geophysical Research 91: E239-E251
Newsom HE, Brittelle GE, Hibbitts CA, Crosse LJ, Kudo AM (1996) Impact crater-lakes on Mars. Journal of Geophysical Research 101: 14,951-14,955
Nienow JA, McKay CP, Friedmann EI (1988) The cryptoendolithic microbial environment in the Ross Desert of Antarctica: light in the photosynthetically active region. Microbial Ecology 16: 271-289
Osinski GR, Spray JG, Lee P (2001) Impact-induced hydrothermal activity within the Haughton impact structure, arctic Canada: generation of a transient, warm, wet oasis. Meteoritics and Planetary Sciences 36: 731-745
Osinski GR, Spray JG (2001) Impact-generated carbonate melts: evidence from the Haughton structure, Canada. Earth and Planetary Science Letters 194: 17-29
Quesada A, Vincent WF, Lean DRS (1999) Community and pigment structure of arctic cyanobacterial assemblages: the occurrence and distribution of UV-absorbing compounds. FEMS Microbial Ecology 28: 315-323
Rathbun JA, Squyres SW (2002) Hydrothermal systems associated with Martian impact craters. Icarus 157: 362-372
Raven JA, Kübler JE, Beardall J (2000) Put out the light, and then put out the light. Journal of the Marine Biology Association of the UK 80: 1-25
Redeker HJ, Stöffler D (1988). The allochthonous polymict breccia layer of the Haughton impact crater, Devon Island, Canada. Meteoritics 23: 185-196
Schoeman FR, Ashton PJ (1982) The diatom flora of the Pretoria Salt Pan, Transvaal, Republic of South Africa. Bacillaria 5: 63-99
Scott DH, Rice JW, Dohm JM (1991) Martian paleolakes and waterways: exobiological implications. Origins of Life and Evolution of the Biosphere 21: 189-198
Svoboda J, Freedman B (1981) Ecology of a high arctic lowland oasis Alexandra Fiord (78°53'N, 75°55'W), Ellesmere Island, NWT, Canada. University of Toronta, Department of Botany, Toronto, Canada. 268 p
Tang EPY, Vincent WF (1999) Strategies of thermal adaptation by high latitude cyanobacteria. New Phytologist 142: 315-323
Walker BD, Peters TW (1977) Soils of the Truelove lowland and plateau. In: Bliss LC (ed) Truelove Lowland, Devon Island, Canada: A high arctic ecosystem. Edmonton: University of Alberta Press, pp 31-62
Wierzchos J, Ascaso C, Sancho LG, and Green A (2003) Iron-rich diagenetic minerals are biomarkers of microbial activity in antarctic rocks. Geomicrobiology Journal 20: 15-24

Zurcher L, Kring DA (2004) Post-impact hydrothermal alteration in the Yaxcopoil-1 hole, Chicxulub impact structure, Mexico. Meteoritics and Planetary Science 39: 1199-1221

Bacterial Spores Survive Simulated Meteorite Impact

Gerda Horneck

German Aerospace Center DLR, Institute of Aerospace Medicine, D 51170 Koeln, Germany (gerda.horneck@dlr.de)

Abstract. Lithopanspermia, i.e., the hypothesis of viable transport of microorganisms between the terrestrial planets by means of meteorites, requires that microorganisms, embedded in rocks, have to cope with three major steps: (i) escape from the planet by impact ejection, (ii) journey through space over extended time periods, and (iii) landing on another planet. Whereas step two of the scenario, the survival in space, has been studied in depth in space experiments, there are only limited data on the survivability of microorganisms of the first step, i.e. the impact ejection. Hypervelocity impacts of large objects, such as asteroids or comets are considered as the most plausible process capable of ejecting microbe-bearing surface material from a planet into space. The shock damage of rocks induced by the ejection process is quite substantial and leads to localized melting in the ejected rocks. However, due to a spallation effect, moderately shocked, solid rock fragments from the uppermost layer of the target can be accelerated to very high velocities (e. g., > 5 km/s) as documented by the meteorites that originated from the moon or Mars. To simulate this impact scenario, in shock recovery experiments with an explosive set-up, resistant microbial test systems (bacterial endospores of *Bacillus subtilis*), sandwiched between two quartz layers, were subjected to a shock pressure of 32 GPa, which is in the upper range indicated by the Martian meteorites and which can be assumed to hold also for "Earth" meteorites. Although the spore layer showed an intense darkening after the shock treatment, up to 500 spores per sample survived, resulting in a survival rate up to 10^{-4}. The data demonstrate that a substantial fraction of spores are able to survive the severe shock pressure and temperature conditions which must be expected for collisionally produced rock fragments from a medium-sized terrestrial planet that have escape velocities of approximately 5 km/s.

1 Introduction

Since its formulation in 1903, the theory of Panspermia (Arrhenius 1903), which postulates that microscopic forms of life, e.g., spores, can be dispersed in space by the radiation pressure from the sun, thereby seeding life from one planet to another, has been subjected to severe criticism. It was argued that the theory cannot be experimentally tested, that it shunts the question of the origin of life to another celestial body, or that spores will not survive long-time exposure to the hostile environment of space (arguments reviewed by Horneck 1995). Although it will be difficult to prove whether viable microorganisms have been transported within our solar system, recent discoveries, such as the occurrence of Martian meteorites (e.g., Nyquist et al. 2001), the ubiquity of endolithic microbial communities (Friedmann 1980) and the reported long-term survival of bacterial spores, e.g., in 25-40 Ma old amber (Cano and Borucki 1995) have given new support to revisit the idea of Panspermia.

However, experiments in space have shown that isolated bacterial spores – as postulated by the Panspermia theory - will not survive a journey through the hostile regions of interplanetary space. During the Spacelab I mission, spores of *Bacillus subtilis* were exposed for the first time to the concerted action of all parameters of outer space, including the high vacuum of pressures in the range of 10^{-14} Pa, the unfiltered spectrum of the intense solar ultraviolet (UV) radiation, the mixture of protons and heavy ions of cosmic radiation of galactic and solar origin, and extremes in temperature. It was found that the spores in space were inactivated within a few seconds by several orders of magnitude (Horneck et al. 1984). This dramatic killing of isolated bacterial spores in space was mainly caused by the highly energetic wavelength ranges of solar extraterrestrial UV radiation, which do not reach the surface of the Earth because they are effectively absorbed by the Earth's atmosphere. The photobiological effects of extraterrestrial solar UV radiation were caused by the production of specific photoproducts in the DNA of the spores that are highly mutagenic and lethal. Hence, the original concept of Panspermia conceiving the transport of small particles of <1.5 μm in size, which corresponds to the size of single spores, seems not to be a feasible route to interplanetary transfer of life.

2 The scenario of lithopanspermia

The detection of the Martian meteorites has shown us another more feasible route to a viable transport of life within our solar system, i.e. the transport of microorganisms through space via meteorites (Figure 1). This process has been termed Lithopanspermia (Mileikowsky et al. 2000; Nicholson et al. 2000). Experiments in space have proven that a thin layer of less than 1 cm of meteorite material is sufficient to protect microorganisms against the harmful action of solar UV radiation (Horneck et al. 2001a). If shielded against solar UV radiation, up to 80 % of *B. subtilis* spores survived a 6 year journey in space on board the Long Duration Exposure Facility (Horneck et al. 1994). This is the longest exposure time of microorganisms in space so far investigated.

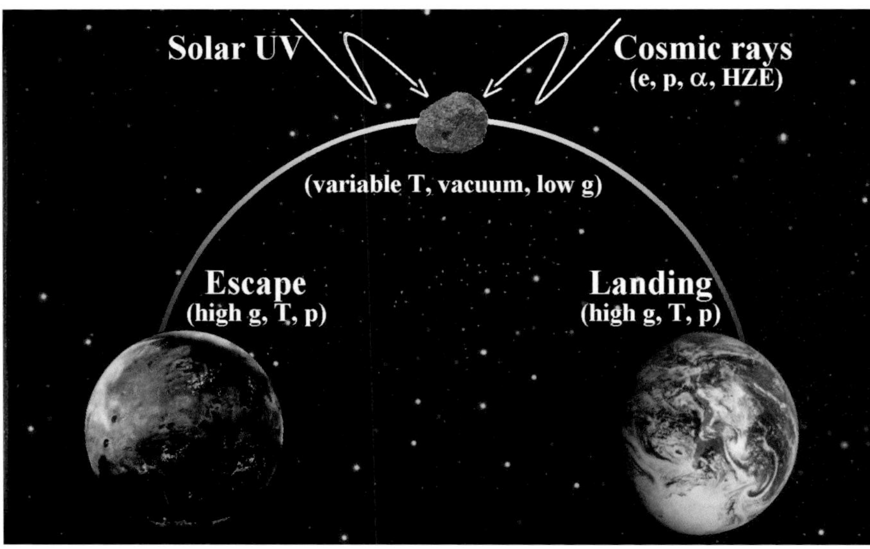

Fig. 1. Scenario of lithopanspermia, i.e. an interplanetary transfer of life in the solar system via meteorites, which requires that microorganisms survive the escape process, the journey through space as well as the entry process on another planet. (g=acceleration, T=temperature, p=pressure, e=electrons, p=protons, α=alpha particles, HZE=high atomic number charged particles).

The most plausible process capable of ejecting microbe-bearing surface material from a planet into space is the hypervelocity impact of a large object, such as an asteroid or comet. The shock damage of rocks induced by the ejection process is quite substantial and leads to localized melting in the ejected rocks as documented by the meteorites which originate from the moon or Mars (Stöffler et al. 1986, Bischoff and Stöffler 1992). The peak shock pressure estimates for the presently studied ~15 Martian meteorites range from about 20 GPa to about 45 GPa and estimates of associated post-shock temperature range from about 100 °C at 20 GPa to about 600 °C at 45 GPa (Fritz et al. 2003). As the peak shock pressure is directly proportional to the shock-induced particle velocity, i.e., the ejection velocity, it is to be expected for a planar shock wave geometry that rocks with a velocity of 5.0 km/s required to escape the gravity field of Mars, should be shock molten.

However, Melosh (1985, 1989) and Artemieva and Stöffler (2002) have shown that for spherical shock wave geometries applicable to planetary surface impacts moderately shocked, solid rock fragments from the uppermost layer of the target can be accelerated to very high velocities (e. g., > 5 km/s) due to a spallation effect. This is perfectly compatible with the observed degree of shock in Martian meteorites. Estimates suggest that within the last 4 Ga, more than 10^9 fragments of a diameter of ≥2 m and temperatures ≤100 °C were ejected from Mars, of which about 5 % arrived on Earth after a journey in space of ≤8 Ma (Mileikowsky et al. 2000). The corresponding numbers for a transfer from Earth to Mars are about 10^8 fragments ejected from the Earth with about 0.1% arriving on Mars within 8 Ma. During the preceding period of "heavy bombardment" even 10 times higher numbers are estimated. Hence, the 26 Martian meteorites, so far detected on Earth, represent probably only an infinitesimally small fraction of those imported from Mars within Earth's history.

2.1
Bacterial endospores as a model system

Viable transfer from one planet to another via lithopanspermia requires that microorganisms survive the following three steps: firstly, the escape process, i.e. ejection into space e.g., by a large impact on the parent planet, secondly, the journey through space, i.e. time scales in space comparable with those experienced by the Martian meteorites, approximately 1-20 Ma (Eugster et al. 1997; Nyquist et al. 2001); and finally, the entry process, i.e., non-destructive deposition of the biological material on another planet (Mileikowsky et al. 2000; Clark 2001).

In order to investigate whether microorganisms are capable of surviving these different steps of lithopanspermia, most studies, so far, have been performed using bacterial endospores as test systems (Figure 2). Several species of bacteria spend at least part of their life histories as dormant cellular structures known as spores. Those spores that are formed by differentiation processes within the mother cell and then released are called endospores. Most common rod-shaped soil inhabitants belonging to the genera *Bacillus* produce such endospores. The reason for selecting bacterial endospores for these studies is that they are especially formed to cope with unfavorable environmental conditions. Bacterial endospores withstand extremely hostile conditions in the dormant state. They exhibit a high degree of resistance to inactivation by various physical and chemical stresses, such as desiccation, extreme temperatures, UV and ionizing radiation, and various aggressive chemical insults including alcohol, acid and basic solutions and oxidizing agents. Hence, bacterial endospores have been recognized as the hardiest known forms of life on Earth (Nicholson et al. 2000). The high resistance of *Bacillus* endospores is mainly due to two factors: a dehydrated, highly mineralized core enclosed in a thick protective envelop, the cortex and the spore coat, and the saturation of their DNA with small, acid-soluble proteins whose binding greatly alters the chemical and enzymatic reactivity of the DNA. In the presence of appropriate nutrients spores respond rapidly by germination and outgrowth, resuming vegetative growth and cell replication. Hence, spore formation represents a strategy by which a bacterium escapes temporally and/or spatially from unfavorable conditions.

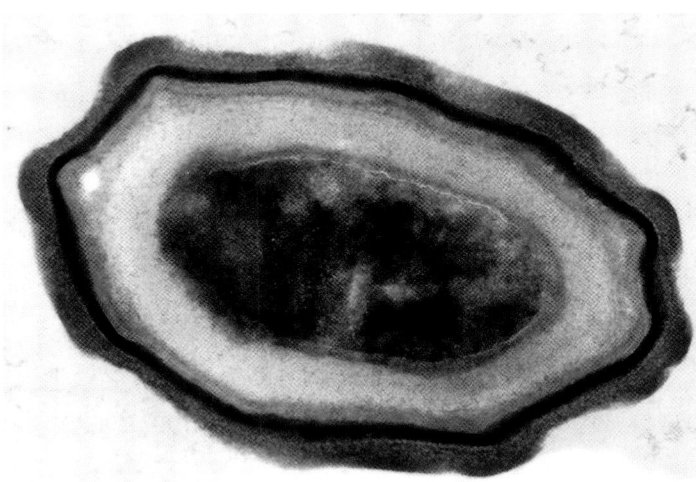

Fig. 2. Scanning electron micrograph of a spore of *B. subtilis* with the inner core (dark part) containing the DNA, surrounded by the inner spore membrane, a thick cortex (white part) and several outer spore wall layers. The long axis of the spore is 1.2 µm, the core area 0.25 µm^2 (courtesy of S. Pankratz).

Bacterial endospores exhibit incredible longevity – survival times up to 25 to 250 Ma have been reported (Cano and Borucki, 1995; Vreeland et al. 2000). Because of their microscopic size, they can be easily relocated e.g., by wind and water, over long distances to remote areas. Many microorganisms found in terrestrial soils are capable of forming spores. Soil generally accommodates microorganisms at a mean concentration of 10^6-10^8 microorganisms/g. Slightly less concentrations have been found in rocks colonized by microorganisms. However, only a small fraction of the soil microorganisms is in the dormant spore state. Up to 28 viable spores were selected recently from the interior of near-subsurface basalt rocks collected in the Sonoran desert in Texas (Benardini et al. 2003). The isolates, closely related to *B. pumilus* and *B. subtilis* were found to be substantially more resistant to UV radiation and extreme acceleration than the reference laboratory strains.

2.2 Simulation experiments

Different attempts have been made to simulate impacts comparable to those experienced by the Martian meteorites and to study the survival of microorganisms after subjecting them to such simulated impacts. Mastrapa et al. (2001) have subjected spores of *B. subtilis* to accelerations, jerks or shock waves by firing the bacteria from a rifle into a plasticene target. Such ballistic experiments provide rise times to reach maximum accelerations equivalent to those estimated for an object receiving escape velocity during an impact. It has been shown that bacterial spores as well as cells of *Deinococcus radiodurans* survived gun shots with accelerations up to 4270 km s^{-2} (436.000x g) with a rise time of <1 ms, which are equivalent to or even higher than the acceleration and jerk values calculated for an object ejected from Mars. Using a two stage gas gun, Burchell et al. (2001) shot bacteria (*Rhodococcus*)-laden projectiles into nutrient gel at typically 5 kms^{-1}. Only one of 7 shots showed survival of at least one bacterium, resulting in a survival rate in the order of 10^{-7}.

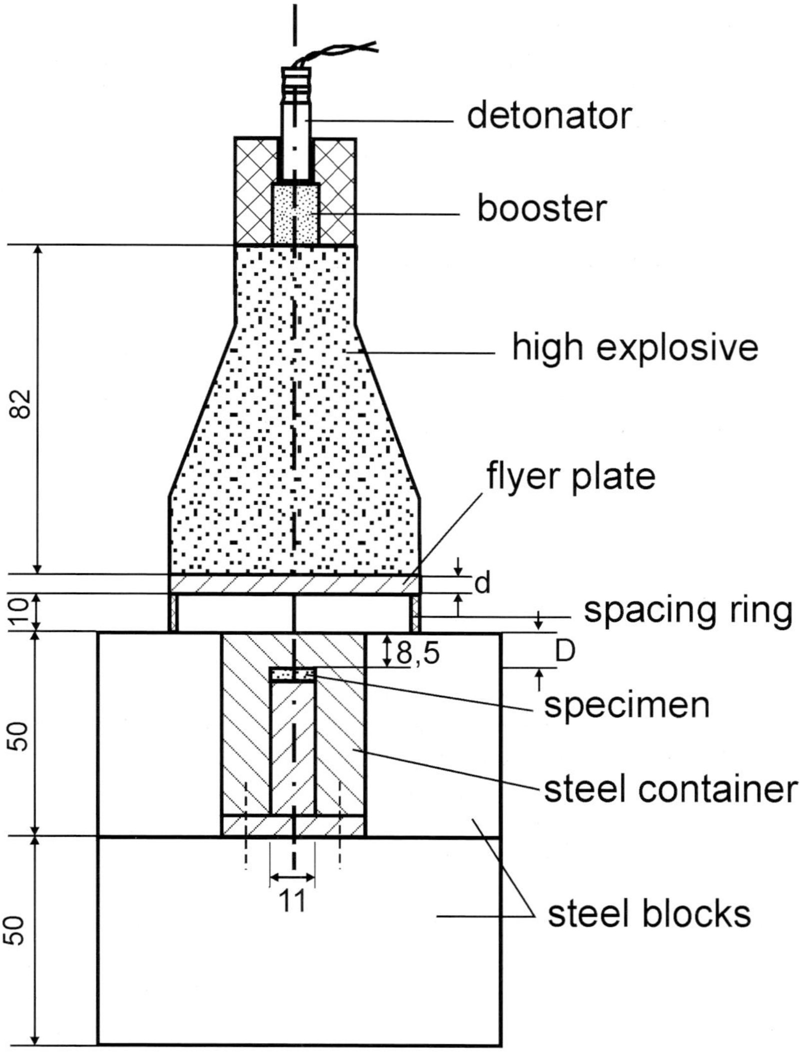

Fig. 3. Set-up of the shock recovery experiments with an explosive set-up; distances are given in mm, d = thickness of flyer plate; D = thickness of cover plate on specimen (Horneck et al. 2001b).

Another approach to simulate an meteorite impact was made in shock recovery experiments with an explosive set-up (Figure 3). The sample was positioned in a bore hole of a cylindrical container of ARMCO steel at a depth of 8.5 mm below the plane container surface. The container was surrounded by thick steel plates and the explosive device was mounted on top of the container. The explosion accelerated a flat steel plate of 3 mm thickness and 64 mm diameter striking the container surface after a free flight of 10 mm (Figure 3). From the impacted surface a plane shock wave propagated through the 8.5 mm thick ARMCO steel plate into the sample which was at room temperature. The shock pressure at the upper interface between the steel plate and the sample was calculated from free surface velocity measurements by a pin technique in separate test series. The shock wave was reflected several times at the two steel-quartz interfaces (reverberation technique) until a final peak shock pressure was reached in both the sample and the adjacent iron container. The shock pressure at the quartz sample was determined by impedance matching in multiple reflection mode using the Hugoniot data of ARMCO steel and single crystal quartz. Details of this technique and of the determination of the shock pressure are described in Müller and Hornemann (1969) and Stöffler and Langenhorst (1994).

The sample consisted of a sandwich of two discs of quartz (diameter: 11 mm) on which spores of *B. subtilis* (Figure 2) were mounted in a monolayer (Figure 4). The quartz discs were cored and cut from a single crystal of quartz, ground to a thickness of 0.54 mm and polished with diamond paste on both sides. The complete sandwiched quartz sample (thickness: 1.1 mm) was loaded in the sample container, made out of steel.

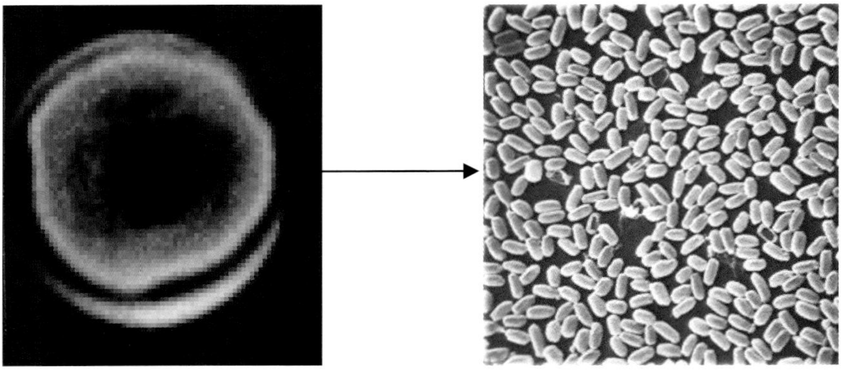

Fig. 4. Monolayer of dry spores of *B. subtilis* used in the shock recovery experiments using an explosive set-up; left side: whole monolayer about 6 mm in diameter; right side: micrograph of spore layer, each spore about 1.2 µm long (Horneck et al. 2001b).

The effects of a simulated meteorite impact on spores of *B. subtilis* were studied for a shock treatment with a peak shock pressure of 32±1 GPa and a post-shock temperature of about 250°C, which lies in the upper pressure range which some Martian meteorites have experienced according to well calibrated shock effects of their mineral constituents (Stöffler et al. 1986; Bischoff and Stöffler 1992; Fritz et al. 2003). The first pressure increase at the steel-quartz interface brought the sample to a pressure of 22 GPa. After several reverberations of the shock wave the peak shock pressure at the sample site, i.e. in the inner part of the quartz discs where the spores were mounted, reached a final value of 32 GPa after a few microseconds, whereas at the upper steel-quartz interface the shock pressure reached a final value of 42.5 GPa. The steel container was only slightly deformed at the impacted surface and the sample could be recovered completely in the original position by careful opening of the ARMCO steel container.

After the shock treatment, the two quartz discs of each sandwich were carefully separated by use of a scalpel and the spores were recovered. For this purpose, the area of the quartz or glass carrying the dry spore layer was punched out, crushed into fragments by use of a glass rod, and transferred into 1 *ml* of distilled water where the spores were resuspended by fractionated shaking with glass beads (30 min shaking, overnight storage, 15 min shaking) followed by ultrasound treatment for 15 min. The colony forming ability of the spores was determined by plating diluted spore suspensions on nutrient broth (NB, Difco) agar plates. Colonies were scored after 16 h of incubation at 37°C. Mean values and standard errors were calculated for each sample from up to six replicate plates (Figure 5). Although the spore layer showed an intense darkening after the shock treatment, up to 500 spores per sample survived, resulting in a survival rate up to 10^{-4} (Horneck et al. 2001b). This darkening was probably caused by heat polymerization of the outer layer of polysaccharides of the spores.

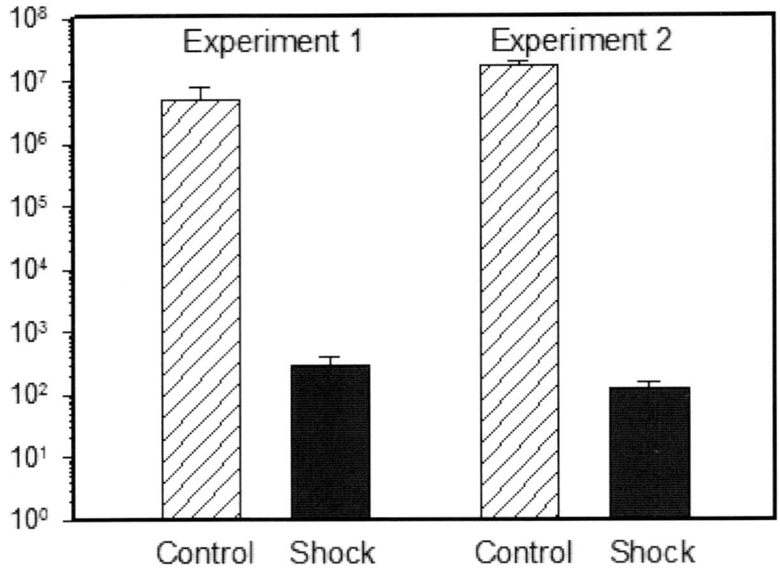

Fig. 5. Survival of spores of *B. subtilis* after treatment with a peak shock pressure of 32 GPa during a shock recovery experiment (Horneck et al. 2001b). The initial spore concentration was 2×10^7/sample.

3 Conclusions

The results of the shock recovery experiments on bacterial spores of *B. subtilis* at a peak shock pressure of 32 GPa and a post-shock temperature of about 250 °C demonstrate that a substantial fraction of the spores (up to 10^{-4}) are able to survive the severe shock pressure and temperature conditions which must be expected for collisionally produced rock fragments from a medium-sized terrestrial planet that have escape velocities of approximately 5 km/s. Assuming a mean spore density of 10^8 spores/g in e.g., desert soil or rock, a 1 kg rock would accommodate approximately 10^{11} spores, of which up to 10^7 could survive the shock pressures occurring during a meteorite impact. These findings have direct implications to the possibility of a transfer of bacterial organisms from Mars to Earth during the entire history of both planets. Since all Martian meteorites have been moderately to severely shocked to peak shock pressures ranging from about 20 GPa in the case of nakhlites (clinopyroxenites) to about 45 GPa in the case of peridotitic shergottites (feldspar-bearing ultramafic rocks), the experimental shock pressure (32 GPa) was in the medium range of the shock pressures observed in the Martian meteorites. Hence, it was related to a quite realistic scenario. However, several of the Martian meteorites were more moderately shocked. Calculations by Melosh (1985, 1988) indicate that even rock fragments shocked to less than 20 GPa and possibly as low as 0.1 GPa could be launched from the very surface and leave the gravity field of Mars. Hence, if 10^{-4} is the survival rate of spores at a shock pressure of 32 GPa corresponding to a post shock temperature of some 250 °C, the survival rate should be substantially higher in rocks exposed to 20 GPa or less. At this pressure the post-shock temperature in basaltic rocks is expected to be in the order of 100 to 150 °C (Stöffler 2000) and somewhat less in ultramafic rocks such as the nakhlites.

Summing up, the data presented here give experimental evidence that bacterial spores may be capable of surviving the first step within a hypothetical transfer of life, namely the planetary ejection process. More experiments at more moderate shock pressures are in preparation.

References

Arrhenius S (1903) Die Verbreitung des Lebens im Weltenraum. Die Umschau 7: 481-485

Artemieva NA, Stöffler D (2002) Conditions for the "launch window" of Martian meteorites: Observation and modeling [abs.]. Meteoritics and Planetary Science 37: A13

Benardini JN, Sawyer J, Venkateswaran K, Nicholson WL (2003) Spore UV and acceleration resistance of endolithic *Bacillus pumilus* and *B. subtilis* isolates obtained from Sonoran desert basalt: implications for lithopanspermia. Astrobiology 3: 709-717

Bischoff A, Stöffler D (1992) Shock metamorphism as a fundamental process in the evolution of planetary bodies: Information from meteorites. European Journal of Mineralogy 4: 707-755

Burchell MJ, Mann J, Bunch AW, Brandao PFB (2001) Survivability of bacteria in hypervelocity impact. Icarus 154: 545-547

Cano RJ, Borucki MK (1995) Revival and identification of bacterial spores in 25- to 40-million-year-old Dominican amber. Science 268: 1060-1064

Clark BC (2001) Planetary interchange of bioactive material: probability factors and implications. Origins of Life and Evolution of the Biosphere 31: 185-197

Eugster O, Weigel A, Polnau E (1997) Ejection times of Martian meteorites. Geochimica et Cosmochimica Acta 61: 2749-2757

Friedmann EI (1980) Endolithic microbial life in hot and cold deserts. Origins of Life and Evolution of the Biosphere 10: 223-235

Fritz J, Greshake A, Stöffler D (2003) Launch conditions for Martian meteorites: Plagioclase as a shock pressure barometer [abs.]. Lunar and Planetary Science XXXIV, abs. # 1335 (CD-ROM)

Horneck G (1995) Exobiology, the study of the origin, evolution and distribution of life within the context of cosmic evolution: a review. Planetary and Space Science 43: 189-217

Horneck G, Bücker H, Reitz G, Requardt H, Dose K, Martens KD, Mennigmann HD, Weber P (1984) Microorganisms in the space environment. Science 225: 226-228

Horneck G, Bücker H, Reitz G (1994) Long-term survival of bacterial spores in space. Advances in Space Research 14: (10)41-(10)45

Horneck G, Rettberg P, Reitz G, Wehner J, Eschweiler U, Strauch K, Panitz C, Starke V, Baumstark-Khan C (2001a) Protection of bacterial spores in space, a contribution to the discussion on Panspermia. Origin of Life and Evolution of the Biosphere 31: 527-547

Horneck G, Stöffler D, Eschweiler U, Hornemann U (2001b) Bacterial spores survive simulated meteorite impact. Icarus 149: 285-290

Mastrapa RMF, Glanzberg H, Head JN, Melosh HJ, Nicholson WL (2001) Survival of bacteria exposed to extreme acceleration: implications for Panspermia. Earth and Planetary Science Letters 189: 1-8

Melosh HJ (1985) Ejection of rock fragments from planetary bodies. Geology 13: 144-148

Melosh HJ (1988) The rocky road to Panspermia. Nature 332: 687-688

Melosh HJ (1989) Impact Cratering - A Geologic Process. Oxford University Press, New York, 245 pp

Mileikowsky C, Cucinotta FA, Wilson JW, Gladman B, Horneck G, Lindegren L, Melosh HJ, Rickman H, Valtonen M, Zheng JQ (2000) Natural transfer of viable microbes in space. Part 1: From Mars to Earth and Earth to Mars. Icarus 145: 391-427

Müller WF, Hornemann U (1969) Shock induced planar deformation structures in experimentally shock loaded olivines and in olivines from chondritic meteorites. Earth and Planetary Science Letters 7: 251-264

Nicholson WL, Munakata N, Horneck G, Melosh HJ, Setlow P (2000) Resistance of *Bacillus* endospores to extreme terrestrial and extraterrestrial environments. Microbial and Molecular Biology Reviews 64: 548-572

Nyquist LE, Bogard DD, Shih C-Y, Greshake A, Stöffler D, Eugster O (2001) Ages and histories of Martian meteorites. Space Science Reviews 96: 105-164

Stöffler D (2000) Maskelynite confirmed as diaplectic glass: Indication for peak shock pressures below 45 GPa in all Martian meteorites [abs.] Lunar Planetary Science XXXII, abstract # 1170 (CD-ROM)

Stöffler D, Langenhorst F (1994) Shock metamorphism of quartz in nature and experiment: I. Basic observation and theory. Meteoritics 29: 155-181

Stöffler D, Ostertag R, Jammes C, Pfannschmidt G, Sen Gupta PR, Simon SB, Papike JJ, Beauchamp RM (1986) Shock metamorphism and petrography of the Shergotty achondrite. Geochimica et Cosmochimica Acta 50: 889-903

Vreeland RH, Rosenzweig WD, Powers DW (2000) Isolation of a 250 million-year-old halotolerant bacterium from a primary salt crystal. Nature 407: 897-900

Impact-Generated Hydrothermal System – Constraints from the Large Paleoproterozoic Sudbury Crater, Canada

Doreen E. Ames [1], Ian R. Jonasson [1], Harold L. Gibson [2], Kevin O. Pope [3]

[1] Natural Resources Canada, Geological Survey of Canada, 601 Booth Street, Ottawa, ON, K1A 0E8, Canada (dames@nrcan.gc.ca)
[2] Department of Earth Sciences, Laurentian University, Ramsey Lake Road, Sudbury, P3E 2C6, Canada
[3] Geo Eco Arc Research, 16305 St Mary's Church Road, Aquasco, MD 20608, USA

Abstract. The 1848 Ma impact-generated hydrothermal system in the ~200-km-diameter Sudbury structure in Canada is exceptionally well preserved and provides the opportunity to study potential fossil ecosystems associated with impact craters. The hydrothermal alteration fingerprint at the Sudbury impact site is preserved for ~1 km below the melt sheet and ~2 km above. The system was capable of producing sufficient heat and fluid flow to form sinter deposits on the crater-floor. Fluid-rock interaction and resultant alteration mineral products record the waxing and waning phases of a complex hydrothermal system within the impact crater with temperatures in the basin ranging from 250-300°C down to ambient. Below the melt sheet fluid-rock interaction took place at <420°C. The exceptional preservation of the Sudbury impact structure including fractured and shocked basement rocks, melt sheet, impact-related crater-fill breccias, chemical sediments on the crater-floor and post impact sedimentation, yields significant new insights into the physical, chemical and potentially the biological framework of impact-generated hydrothermal systems in large craters. Significant to the development of microbial niches is defining the lower temperature regimes (<120°C) of the habitable zone. In the Sudbury basin from base to top, lies a 1.4-km-thick sequence of suevite (Onaping Formation) that has undergone extensive water-rock interaction manifested as regionally extensive semiconformable alteration zones, a thin ~ 14-m-thick exhalative-sedimentary sequence (Vermilion Formation) and in a metal-enriched hydrothermal plume extending another <1 km into the post-impact basin sediments (Onwatin Formation). The hydrothermal signature includes basin-wide semiconformable alteration zones defined by silicification, albitization, carbonate-chlorite alteration in the Onaping Formation. Also present are discordant alteration zones with focussed fluid flow which produced local higher temperature perturbations imposed on the more extensive lower temperature ($<250^{\circ}$C) alteration zones within the crater-fill sequence. The Vermilion Formation represents a subaqueous hydrothermal vent complex with a proximal hydrothermal Ca-Fe-Mg-Mn carbonate mound facies containing replacement type Zn-Pb-Cu-Fe mineralization, a distal finely laminated carbonate facies, or "carbonate-facies iron formation", buried by distal turbidite sediments. Prolonged post-mineralization diffuse fluid flow and unfocussed low temperature emanation of hydrothermal plumes and the Fe-Mn-rich distal carbonates produce favourable habitats for thermophilic micro-organisms.

1 Introduction

Terrestrial impact craters offer an accessible venue for studying impact related hydrothermal processes pertinent to other cratered planetary objects; and in particular the quest for water and physiochemical niches suitable for sustaining life. An understanding of the range of hydrothermal processes permissive on earth may provide insight into possible microbial environments favourable to the evolution and propagation of microbial life (Humphris et al. 1995). Hydrothermal environments in modern and ancient volcanic settings are well-known established niches (Baross and Hofmann 1985) with impact hydrothermal settings recently recognized as potential sites (Newsom et al. 1996; Allen et al. 1982). Key elements for the evolution of an impact hydrothermal system involve a heat source, rock permeability and availability of fluid. Significant differences in the extent, style and conditions of alteration are evident in various terrestrial craters (McCarville and Crossey 1996; Naumov 2002; Ames et al. 1998).

The Paleoproterozoic Sudbury Structure is generally regarded as the eroded remnant of a large >200 km peak ring impact structure (Stöffler et al. 1994). It hosts a major impact-induced hydrothermal system directly dated at 1848 +3.9/-1.8 Ma (Ames et al. 1998). The exceptional exposure of the Sudbury crater from basement rocks through to post-impact basin sediments provides a glimpse into the subsurface hydrothermal plumbing of large impact structures, a feature that cannot be appreciated in other large craters, such as the deeply buried Chixculub crater or the deeply eroded Vredefort crater. Impact-generated hydrothermal processes are quite evident both below and above the Sudbury melt sheet (Fig. 1). The Sudbury impact and subsequent hydrothermal systems produced anthraxolite veins in the post impact basin sediments, carbonate sinters and Zn-Pb-Cu deposits on the paleocrater-floor, large magmatic Ni-Cu (PGE) sulfide accumulations below the melt sheet, and regional subsurface alteration zones whose volume (>6000 km^3) and intensity are not recognized in suevitic rocks in other impact craters. Extensive and pervasive hydrothermal alteration in the Sudbury impact crater is noteworthy, as most craters have only minor, volumetrically insignificant alteration that occurs as open space fillings and veins (e.g., Manson, Puchez-Katunki, Popigai, Ries craters) and minor pervasive alteration (Puchez-Katunki, Kara, Kärdla and Lockne craters), (McCarville and Crossey 1996; Allen et al. 1982; Naumov 2002; Sturkell et al. 1998). The terrestrial craters noted above are all <100 km in diameter. Differences in hydrothermal alteration in the Sudbury Structure compared to other studied impact hydrothermal systems are likely a function of the large size, greater volume of melt and a subaqueous paleoenvironment.

The paper examines the large impact-induced hydrothermal system in the Sudbury crater in Canada, focusing on hydrothermal alteration and mineralization within the crater fill succession. We present data on the characteristics of: 1) regional semi-conformable alteration zones; 2) associated hydrothermal base metal showings and deposits in the suevitic crater fill; 3) chemical sediments on the paleoseafloor; and 4) an extensive hydrothermal plume in the post-impact sediments within the Sudbury basin. Detailed study of the crater-fill sequence revealed a complex hydrothermal history controlled by the structural, magmatic and stratigraphic history of the impact crater. This study provides a deeper understanding of cratering processes during the crater modification stage within one of Earths largest impact craters.

2 Geological setting, metamorphism and structure

The 1.85 Ga Sudbury Structure is one of the largest, well preserved and exposed example of a terrestrial impact crater. It is generally regarded as the deformed and eroded remnant of a 200-250-km-diameter peak-ring or possible multi-ring, impact basin (Stöffler et al. 1989, 1994; Thompson and Spray 1994). Timing of the meteorite impact event at Sudbury is interpreted as, equivalent to the age of the Sudbury Igneous Complex (SIC) magmatic event, 1850 ± 1 Ma (Krogh et al. 1984), supported by geochronology of shocked zircons (Krogh et al. 1996).

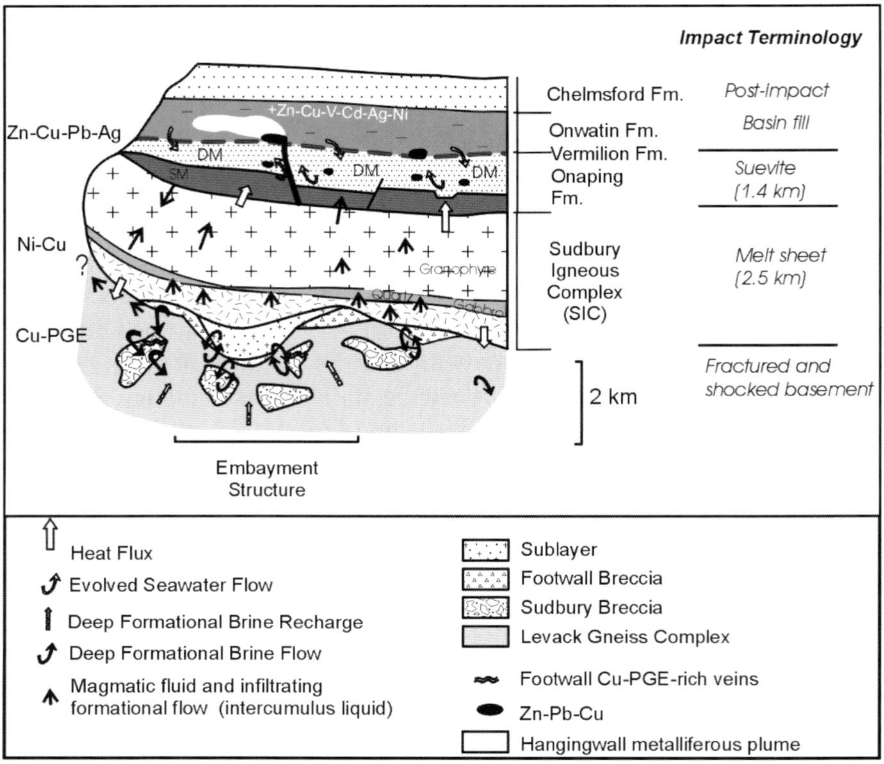

Fig. 1. Schematic diagram showing the principal components of the Sudbury impact crater and associated hydrothermal system (modified from Farrow and Watkinson 1999). Arrows indicate fluid and heat flow paths below, above and within the melt sheet. The relative positions of the magmatic and hydrothermal ore deposits are shown, as well as a metal enriched hydrothermal plume in the post-impact basin sediments.

The duration of impact events from shock metamorphism through to impact-induced hydrothermal alteration in the Sudbury Structure is within the error of U-Pb geochronology, less than 4 Ma (Ames et al. 1998). The Sudbury Structure is unparalleled in terms of its famous magmatic Ni-Cu-PGE mineral wealth that overshadows the lesser known impact-generated hydrothermal Zn-Pb-Cu-Au deposits.

The Sudbury Structure encompasses the outer Sudbury Igneous Complex (SIC) and its Ni-Cu-PGE mineralization, brecciated footwall rocks, and the overlying Whitewater Group, of four formations, from oldest to youngest, the Onaping with its Zn-Pb-Cu mineralization (Table 1), Vermilion, Onwatin and Chelmsford informally referred to as the Sudbury basin (Pye et al. 1984),

(Figs. 1 - 3). Prior to the 1850 Ma Sudbury impact event, the northern footwall rocks were affected by several igneous, deformation and metamorphic episodes during the Archean and early Proterozoic.

The Proterozoic southern footwall rocks record two major tectono-metamorphic events, the Blezardian and Penokean (Card 1978; Card and Innes 1981; Wodicka 1997; Dressler 1984; Riller et al. 1996; Rousell et al. 1997). Coeval with the Sudbury Event are a series of faults, fractures and zones occupied by (1) Onaping Formation melt (Ames et al. 2002a), (2) offset dykes (Grant and Bite 1984; Corfu and Lightfoot 1997), (3) pseudotachylite generally known as Sudbury breccia (Dressler 1984; Thompson and Spray 1994),(4) footwall breccia (Farrow 1994) and (5) Cu-Ni-PGE mineralization (Pye et. al. 1984). Shock metamorphic features in the Sudbury Structure include shatter cones (Dietz 1964; Gibson and Spray 1997) up to 20 km from the SIC, planar deformation features in zircon from the Levack gneiss and Onaping Formation (Krogh et al. 1984; 1996), impact diamonds in the Onaping Formation (Masaitis et al. 1997), planar deformation features in quartz and feldspar, and kink bands in biotite up to 8-10 km from the SIC (Dressler 1984).

Regional metamorphic grade increases from greenschist grade in the north to amphibolite grade in the south across the Sudbury Structure (Card et al. 1984; Card 1978). The SIC and Onaping Formation in the North Range are largely undeformed, affected only by broad open folds, although deformation in the South Range is intense and attributed to the Penokean Orogeny (Cowan and Schwertdner 1994; Shanks and Schwertdner 1991; Card and Jackson 1995). Major folding occurs about northeast- and northwest-trending axes (Rousell 1975; Card 1978; Cowan and Schwertdner 1994). The result has been an overall northwest-southeast compression of the SIC and Sudbury Basin resulting in the prominent elliptical shape with the East Range perturbation controlled by pre-existing basement faults (Morris 1999). Thrust faulting affects mainly the southern half of the structure including the 5 x 50 km, ductile, South Range shear zone that strongly deformed the footwall to the Errington and Vermilion Zn-Cu-Pb deposits in the Onaping Formation (Fig. 2) (Shanks and Schwerdtner 1991). High resolution seismic data (Milkereit et al. 1992; Wu et al. 1994) shows the strong asymmetric shape of the Sudbury Structure at depth which is due to northwest directed thrusting during the Penokean orogeny. Interpretation of the geophysical data suggests a basin-shaped SIC, 6 km deep, with the Onaping Formation extending to ~4 km depth (Card and Jackson 1995; Wodicka 1997). Alternatively, a parabolic shape was reconstructed with a large accumulation of Onaping Formation at depth (Wu et al. 1994). Impact at Sudbury likely took place during the peak influences and effects of the Penokean Orogeny (Card et al. 1984).

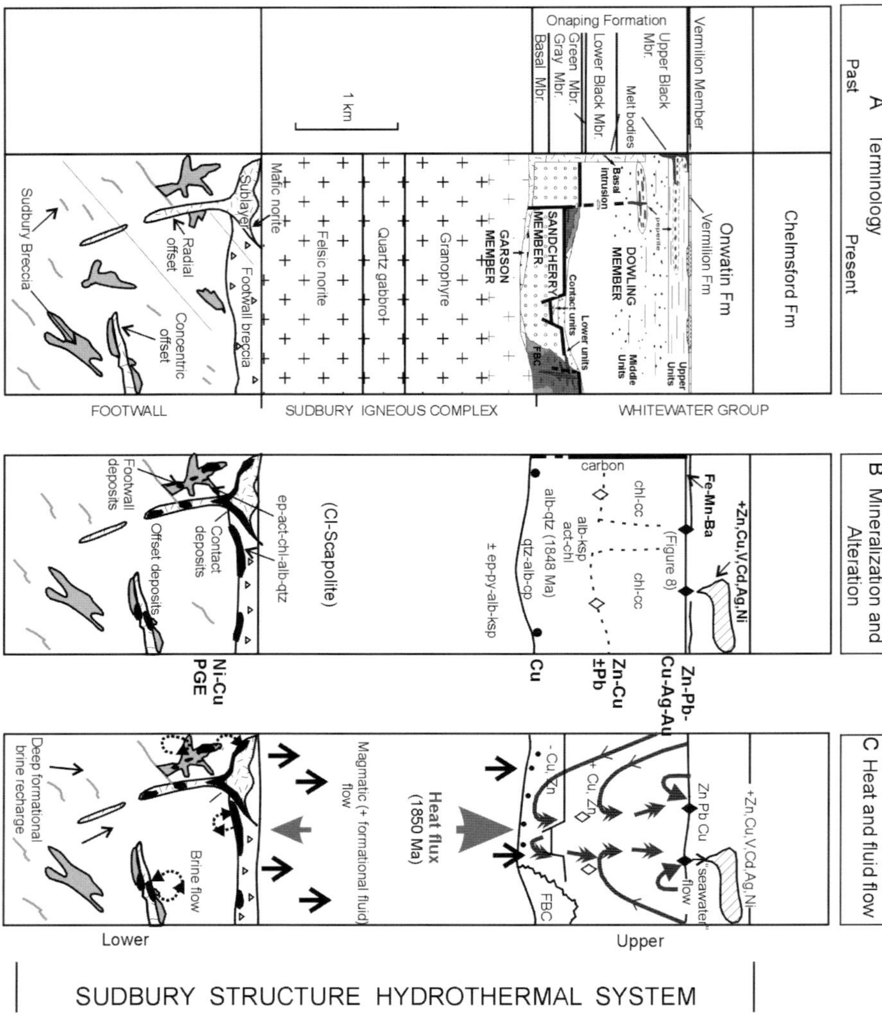

Fig. 2. Schematic sections through the Sudbury Structure showing terminology, mineralization, hydrothermal alteration, fluid flow (black arrows) and heat flow (gray arrows) from the Sudbury Igneous Complex. A) The terminology in the Onaping Formation is changed as a carbon boundary that delimits obsolete "Gray and Black Members" irregularly transects stratigraphic contacts (Fig. 3B). Modified from Grieve et al. (1991). B) The distribution of hydrothermal alteration relative to the various ore types in the Sudbury Structure. See text for discussion on hydrothermal mineral assemblages. C) Section showing heat and fluid flow fueled by the 1850 Ma SIC. Fluids include seawater, evolved seawater and "magmatic" fluid flow through the crater-fill, magmatic and formational fluid flow within the SIC and in the fractured basement rocks, deep formational brine flow of groundwaters (Ames et al. 2002a).

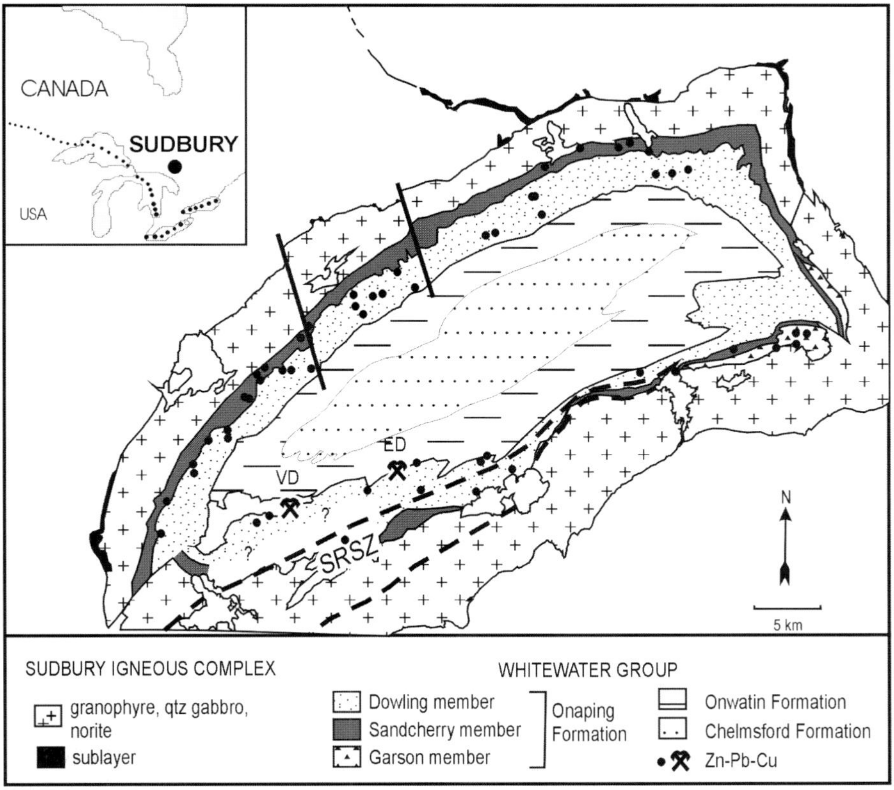

Fig. 3. General geology map of the Sudbury Structure showing the distribution of >50 hydrothermal Zn-Pb-Cu showings and deposits in the crater-fill Onaping Formation. The Vermilion Formation, not shown, is a thin ~14 m thick unit at the contact between the Onaping and Onwatin Formations. The Sandcherry and Dowling members are regionally extensive but the Garson member outcrops in the southeastern sector of the Sudbury Stucture only. SRSZ=South Range shear zone. VD, ED = Vermilion and Errington Zn-Pb-Cu (Ag) deposits. (Data sources: Falconbridge Ltd.; Gray 1995; Gibbins 1994; Pye et al. 1984; Ames et al. 1998; Ames 1999; Shanks and Schwerdtner 1991).

Table 1. Production and reserve history of the carbonate-hosted Errington and Vermilion mines (data from Martin 1957; Young 1996).

Mine	TONS	Zn wt%	Cu wt%	Pb wt %	Au oz/t	Ag oz/t	
ERRINGTON							
	Production 1926-1931, tonnage through mill						
1926-1928	32,708	5.75	1.02	1.12	0.029	1.79	Treadwell Yukon Co. Ltd.
1928-1929	89,221	4.49	1.04	0.99	0.028	1.64	"
1929-1930	64,859	-	-	-	-	-	"
total tons [1]	186,788	-	-	-	-	-	[1](Young 1996)
total tons [2]	186,172	4.60	1.07	1.10	0.030	1.70	[2](Martin 1957)
	Reserves						
1929	833,000	5.47	2.44	1.20	0.027	1.84	Sudbury Basin Mines Ltd.
1954	7,513,007	3.24	1.02	0.75	0.017	1.49	Consolidated Sudbury Basin Mines Ltd.
1957-Errington #2	10,432,47	3.82	1.10	0.97	0.021	1.58	Ontario Pyrites Co. Ltd.
1957-Errington #3	1,437,500	4.79	1.05	1.96	0.017	2.22	"
1991	6,900,000	4.21	1.22	1.09	-	-	Falconbridge Ltd.
VERMILION							
1954	2,819,220	4.56	1.43	1.10	0.020	1.78	Consolidated Sudbury Basin Mines Ltd.
1957-Vermilion #4	5,077,779	3.92	1.26	0.97	0.025	1.40	Ontario Pyrites Co. Ltd.
1957-Vermilion North	862,500	3.73	0.34	1.30	0.020	1.64	"
1991	2,700,000	5.11	1.49	1.37	0.032	1.93	Falconbridge Ltd

3 Geology of the crater-fill sequence

The Sudbury Structure records a coherent history of crater-fill emplacement, crater-collapse and development of crater floor-fractures that are critical controls on melt and hydrothermal fluid flow. The Onaping Formation is a 1.4-km-thick hydrothermally altered sequence of vitric-rich fragmental rocks with minor intrusive rocks (Figs. 3 and 4). The three regionally mappable members of the Onaping Formation (from base to top: Garson, Sandcherry and Dowling) have different vitric morphologies and abundance, percentage of matrix and lithic fragments, and depositional characteristics (Ames et al. 2002a; Ames 1999; Gibbins 1994), (Fig. 3). Vitric clast size and percentage variations are defined using Fishers (1966) classification for volcaniclastic rocks are used in a non-generic sense. This in no way infers a volcanic derivation for the Onaping Formation.

The laterally discontinuous ~25 km long Garson member (0-500 m thick) (Fig. 3) consists of quartzite block-rich megabreccia to breccia units with <90% basement fragments in a melt-rich matrix.

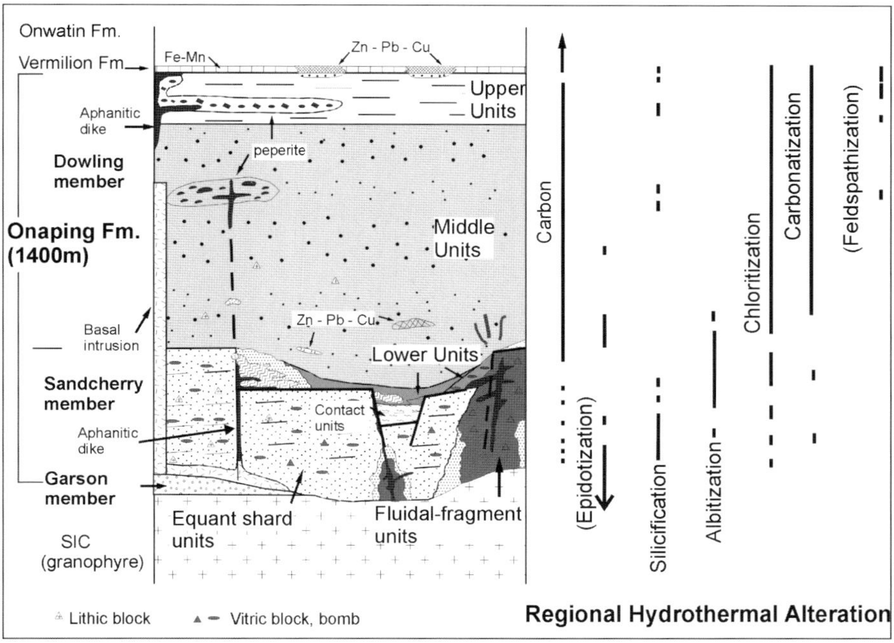

Fig. 4. Simplified stratigraphic section of the crater-fill showing the general distribution of carbon and regional semiconformable alteration zones within the Onaping Formation. Carbonatization refers to calcite dominant alteration. Brackets denote alteration types with poorly constrained regional distribution patterns.

The conformably overlying Sandcherry member (~300-500 m thick) is characterized by >60 vol% altered vitric fragments that are equant or fluidal in morphology and contain 5-15 percent basement blocks in a fine matrix that may or may not contain carbon. The Dowling member (>1000 m thick) units are characteristically matrix-rich (~60 vol%), dominantly carbon-bearing (0.4 wt% C), and have fewer vitric fragments (~ 25-40%) that are lenticular, cuspate and amygdaloidal and commonly chloritized. The well defined stratigraphic contacts are cut by a carbon boundary that separates carbon from non-carbon bearing strata (Figs. 4 and 5). The carbon boundary was historically used to define the former "Gray and Black Members" (Muir 1984; Ames et al. 1998), (Figs. 4 and 5). The presence or absence of carbon in the Onaping Formation is irrelevant in terms of stratigraphic correlation and nomenclature, but important to the discussion of favourable sites for sustaining microbial colonies.

Facies analysis, architecture of the fragmental pile and structural analysis were used to define a two-stage process for the deposition of the suevitic crater-fill Onaping Formation (Ames 1999; Ames et al. 2002a; Ames and Gibson 2004 a-e). The establishment of the least altered glass composition in the crater-fill, Onaping Formation, led to the suggestion that this early andesitic melt composition represents the original Sudbury Structure impact melt or shock-melt (Ames et al. 2002a) that evolved along various paths during its cooling history.

3.1
Emplacement of the Garson and Sandcherry members, lower Onaping Formation

The Sandcherry and Garson members of the lower Onaping Formation were deposited in a very dynamic and structurally unstable environment. The Garson member is restricted to the southeastern quadrant of the Sudbury Structure and consists of breccia and mega-breccia units dominated by basement Huronian quartzite blocks in an andesitic melt matrix (Fig. 3). The Garson member is interpreted to have formed during slumping immediately after impact possibly from a quartzite-dominated (basement Huronian sediments) peak ring structure (Ames et al. 2002a). This is consistent with the position of the reconstructed Sudbury crater (Golightly 1994). The Garson member was followed by the deposition of the conformably overlying Sandcherry membe. It is composed dominantly of equant shard units that are laterally continuous, block and bomb-rich, non-bedded and vitric-rich strata which are interpreted as fall back breccia (Peredery 1972; Peredery and Morrison 1984; Ames 1999), however this interpretation has been contested (Gibbins 1994). These units were injected by andesitic impact melt from below to form discordant and concordant, intrusive/extrusive fluidal-breccia complexes comprising tabular fluidal fragment units, discrete aphanitic dykes and autobrecciated amoeboid dykes (Fig. 4), (Ames and Gibson 2004 a-e).

3.2 Emplacement of the Dowling member (1000 m), upper Onaping Formation

The Dowling member, which comprises the upper 1000 m of the Onaping Formation, includes the Contact units, lower middle and upper units (Figs. 2A, 4, 5). These strata are distinguished by a lenticular, cuspate shard morphology and a much higher percentage of matrix component (~60vol%) than the Sandcherry member (Gibbins 1994). The two units at the base of the Dowling member are, (a) the laterally discontinuous Contact units (formerly the Green member, Avermann 1997; Stöffler et al. 1994) and, (b) the Lower units (Figs. 4 and 5a). Contact units are 30-300 m thick, laterally extensive yet discontinuous units, with local incipient welding. They have a block- and bomb-rich base containing fragments of the underlying Sandcherry member and are interpreted as mass flow deposits (Ames 1999; Gibbins 1994) . The Lower units are 20-30 m thick, laterally extensive crudely stratified deposits or small (20 x 150 m) channel-shaped debris flow deposits controlled by syn-depositional faults commonly infilled with vitric andesitic melt dykes. Both of these units record a change in the vitric morphology from blocky at the base to lenticular at the top reflecting differences in the glass fragmentation mechanism (e.g., Wohletz 1983) between the Sandcherry and Dowling members. These units at the base of the Dowling member record a major period of mass wasting and crater-fill deposition likely triggered by crater collapse (Ames et al. 2002a).

The ~600 m thick, Middle units are large volume (>3600 km^3) deposits that are block and bomb-poor, laterally continuous tuffaceous units with minor fragments and rare beds of ash and lapillistone (Figs. 4, 5a). The contact with the underlying Lower units is sharp. The Middle units have detailed characteristics similar to pyroclastic ash-flow deposits and therefore were likely deposited by collapse of the impact vapour plume (Ames 1999). A plume origin is also supported by the presence of anomalous Ir within the Middle and Upper units of the Dowling member (Mungall et al. 2002). The contact of the Middle units with the overlying Upper units is gradational over a few metres, and marked by an increase in the fine tuff component (vitric < 2 mm). The 140-220 m thick Upper units are characterized by >70vol% fine tuff component (Figs. 4, 5a). The Upper units mark the first appearance of significant sedimentation in the Onaping Formation producing thin, normally to non-graded bedded units in an aquagene environment.

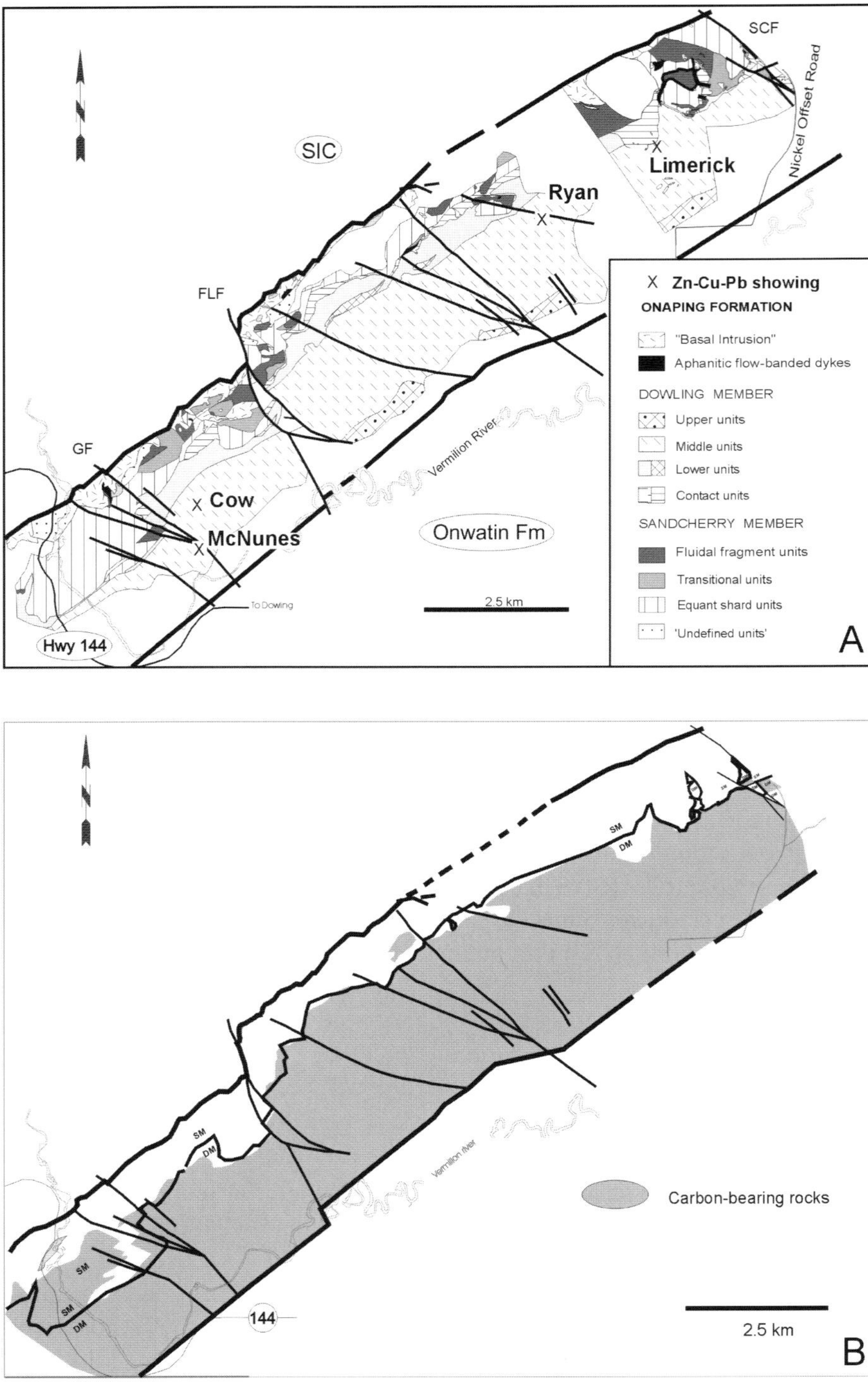
SCF
Nickel Offset Road
SIC
Limerick
Ryan
FLF
GF
Cow
McNunes
Vermilion River
Onwatin Fm
To Dowling
2.5 km
Hwy 144
X Zn-Cu-Pb showing
ONAPING FORMATION
"Basal Intrusion"
Aphanitic flow-banded dykes
DOWLING MEMBER
Upper units
Middle units
Lower units
Contact units
SANDCHERRY MEMBER
Fluidal fragment units
Transitional units
Equant shard units
'Undefined units'
A
SM
DM
Vermilion river
Carbon-bearing rocks
144
2.5 km
B

Fig. 5. a) Simplified geology map of >15 km wide segment of the crater-fill Onaping Formation, North Range, Sudbury Structure. Locations of Zn-Cu-Pb showings described in this study. Note the dominance of fluidal-fragment units around the long-lived Fecunis Lake (FLF) and Sandcherry Creek (SCF) faults. Mapping from Ames 1999; Gibbins 1994; Ames and Gibson 2003). Note the undefined units (Gibbins 1994) are silicified Sandcherry member units. **b)** Distribution of carbon within the Onaping Formation (shaded area is carbon-bearing rocks). The carbon front is diachronous across the Sandcherry member (SM)- Dowling member (DM) stratigraphic contact. Note the isolated patches of carbon in the lower Onaping Formation, particularly the large area of carbon-bearing Sandcherry member strata along Highway 144 in the High Falls area.

3.3 Intraformational dykes

Andesitic, aphanitic dykes extend throughout the Onaping Formation and are compositionally similar from base to top and to vitric bombs and blocks in the Sandcherry member strata (Ames et al. 2002a). Dyke contacts are commonly peperitic, and extensive zones of peperite are present in the Dowling member Middle and Upper units indicating contemporaneous melt injection into the unconsolidated, wet host strata. This indicates an extended period of melt injection into a water-saturated crater-fill succession (Fig. 4) (Ames et al., 2002a). The syn-crater fault set, (310/70 and 015/65; Ames et al. 2002a) also hosts igneous textured, xenolithic, intraformational dykes and pods (Basal intrusion) and focused hydrothermal fluid flow. These faults, interpreted as crater floor-fractures similar to those in large planetary impact craters (e.g. Isabella on Venus or Ritter and Sabine on the Moon), form a distinct set whose orientations were controlled by the regional strain field imposed by the local Penokean Orogeny in the Sudbury area.

3.4 Post-impact basin sediment

The impactites are overlain by > 2000 m of Paleoproterozoic sedimentary rocks (Figs. 1-3). The thin, 5 to 50 m thick, Vermilion Formation is composed of carbonate, grey argillite and chert and conformably overlies the Onaping Formation (Martin 1957; Stoness 1994; Gray 1995). The Vermilion Formation consists of (1) The Lower Carbonate Member (LCM), subdivided into a proximal hydrothermal carbonate mound facies and a distal, finely laminated carbonate facies, (2) the Grey Argillite Member composed of distal turbidites; and (3) the Upper Carbonate Member which contains aerially restricted, concretionary carbonate units (Fig. 6) (Stoness 1994; Gray 1995). The mineralogically complex Ca-Mg-Fe-Mn carbonates have been interpreted to be the result of subaqueous, fumerolic hot spring activity (Martin 1957; Card and Hutchinson 1972; Whitehead et al. 1990; Stoness 1994), but Arengi (1977) attributed them to clastic sedimentation.

The 600-m to 1000-m-thick Onwatin Formation is a poorly exposed sequence of conductive, carbonaceous, pyritic argillite with minor greywacke turbidite units (Figs.1-3), (Rousell 1984a). The massive to laminated, black carbonaceous mudstone and siltstone average 3.5 wt.% of organic carbon. These pelagic sediments are interpreted to have been deposited in a restricted basin under anoxic conditions (Rousell 1984a). They grade upward into the 600 to 850 m thick Chelmsford Formation (Cantin and Walker 1972) that consists of greywacke and minor siltstone. It is interpreted to be a proximal turbidite succession (Cantin and Walker 1972; Rousell 1984a) (Figs. 1-3).

4
Impact-generated hydrothermal system

A major impact-induced convective hydrothermal system in the Sudbury crater was instigated and sustained largely by heat produced from the cooling of the superheated melt sheet (SIC). This resulted in extensively alteration of the rocks both above and below (Figs. 1, 2). Above the melt sheet the Sudbury hydrothermal system produced: (1) large-scale semiconformable alteration zones defined around the entire circumference of the exposed structure within the suevitic Onaping Formation; (2) >50 hydrothermal Cu-Zn-Pb showings within the Onaping Formation; (3) carbonate sinters on the paleoseafloor (Vermilion Formation); (4) the replacement-type Zn-Pb-Cu Errington and Vermilion massive sulfide deposits; and (5) hangingwall hydrothermal metalliferous alteration zone in the post-impact basinal sediments (Onwatin Formation mudstones), (Ames et al. 1998) (Figs. 1, 2). The heat flux from the crystallizing melt sheet also initiated convective hydrothermal fluid flow beneath the SIC by heating footwall groundwaters with recharge from deep formational brines that altered and remobilized metals from the Ni-Cu-PGE deposits (Farrow and Watkinson 1992, 1999; Marshall et al. 1999) (Fig. 1). A study of the magmatic-hydrothermal history of the SIC shows evidence of volatile-rich alteration (high F, Cl) (Ames et al. 2001; Ames 2002).

4.1
Hydrothermal processes in rocks beneath the SIC: Lower hydrothermal component

The complex fluid history recorded in the basement to the melt sheet indicates a protracted hydrothermal evolution of the Sudbury region from pre-1850 Ma to ~ 5-15 Ma (Marshall et al.1999; Molnar et al. 2001). Hydrothermal alteration, recognized in the basement rocks below the SIC, occurs in large zones associated with impact brecciated footwall rocks, Sudbury breccia and Cu-PGE mineralization (Watkinson 1990; Farrow and Watkinson 1992, 1999; Farrow 1994; Watkinson 1994; Molnar et al. 1997, 1999, 2001; Hanley and Mungall 2003). The regional distribution of alteration in the basement rocks has not been mapped despite and its century-old recognition. Detailed fluid inclusion and isotopic studies have recently defined multiple generations of fluids due to emplacement of the SIC, the Penokean Orogeny and later neotectonic events (Farrow 1994; Molnar et al. 2001; Marshall et al. 1999). Mineralogy, mineral chemistry, fluid inclusion petrography and microthermometry, and stable and radiogenic isotope studies have focused on understanding the lower hydrothermal system and its possible relationship to footwall Cu-PGE mineralization which are PGE, Au and Ag enriched (see Farrow and Watkinson 1999; Watkinson 1999; Molnar et al. 1997, 1999 and references therein). At 1850 Ma, deep formational Cl-rich, high temperature fluids (400-420°C) dominate the hydrothermal activity below the SIC, whereas high-Cl aqueous fluids rich in CO_2 and CH_4 or magmatic fluids migrated upwards through the crystallizing SIC melt (Farrow and Watkinson 1999).

Analogies for the hydrothermal system below the melt in the impact crater environment can be made to mafic sill emplacement into wet or aquifer-bearing country rocks and these provide comparisons for (1) fluid sources - infiltrating country-rock fluids, exsolved high-temperature volatile-rich fluids and intercumulous fluids (Boudreau and Meurer 1994; Mathez 1999); (2) patterns of fluid and heat flow (Cathles et al. 1997; Delaney 1982); and (3) the duration of a hydrothermal system due to a single intrusive event (Cathles et al. 1997).

4.2
Hydrothermal processes in rocks above the SIC: Regional alteration zones

Regionally extensive detailed mapping of the crater-fill Onaping Formation along with mineralogical, geochemical and isotopic studies have established a solid geologic framework for the hydrothermal system directly related to the impact event ca 1850 Ma (Ames et al. 1998 2002a, b; Ames 1999 Ames and Gibson 2004 a-e). A major hydrothermal circulation system altered the Onaping Formation around the entire circumference of the exposed Sudbury Structure (>6000 km^3) (Ames et al. 1998). The vitric-rich nature of the suevite provides readily reactive zones to fluid-rock interaction. Vertically stacked, basin-wide, semiconformable alteration zones consist of an upper calcite-chlorite zone, transition zone, actinolite-chlorite zone, albite zone and lower silicified zone (Ames and Gibson 1995; Ames et al. 1998; Ames 1999). Alteration minerals are generally pseudomorphous after glass fragments and also pervasively alter the matrix. Most of the zones have a minor, yet significant, discordant fracture-controlled component. The volumetrically minor discordant zones are significant for determining alteration paragenesis.

The upper calcite zone is pervasive in the upper 800-1000 m of the Onaping Formation and extends around the circumference of the Sudbury basin (Ames et al. 1998; Ames 1999; Ames and Gibson 2004 a-e). It mostly affects the Dowling member Middle and Upper units but has a patchy distribution in the lower part of the Dowling member. Calcite also occurs intermittently the lower part of the Onaping Formation and adjacent to the granophyre of the SIC. Calcite alteration is locally absent (usually <10-20 m wide haloe) around Zn-Cu-Pb sulfide showings and local silicified zones have ~ 200 m wide discordant calcite-free corridors through the semi-conformable calcite alteration zone (Ames and Gibson 2004 a,d,e). Although the upper 1 km of stratigraphy is also carbonaceous, there are discrete carbon-only and calcite-only zones distributed throughout the crater-fill succession (Figs. 4,5B) (Ames 1997; 1998).

The semiconformable calcite zone has vitric shards totally to partially replaced by calcite, calcite in amygdales and locally the matrix. Most vitric fragments (shards) are only partially replaced by calcite thereby revealing the relict textures. Relict palagonite and perlitic textures are retained by calcite particularly in the lower 200-500 m of the calcite alteration zone. Above this, most of the regional calcite zone generally has (1) abundant calcite that totally to partially replaces small 0.4 mm shards; (2) minor calcite in larger, 1-2 mm devitrified feldspathic shards and (3) calcite in trace amounts in the 1-2 mm chloritic shards. Shards that are totally replace by calcite locally resemble fragments however, they contain relict amygdales. The mineral assemblage consists of chlorite-calcite, albite ± K-feldspar-quartz and minor to trace carbon-titanite-pyrrhotite-chalcopyrite-sphalerite ± epidote-prehnite, actinolite, pyrite and magnetite. Field, textural, stable and radiogenic isotope data for the regional carbonate zone are compatible with the interpretation of the source of carbonate carbon a mixture of magmatic CO_2 degassed from melt deeper in the system and diluted by Proterozoic seawater (Ames et al. 1999; Ames 1999).

The transition zone is situated between the upper chlorite-calcite and underlying actinolite-chlorite and/or albite alteration zones and ranges from 0-500 m wide although it is typically 150 m wide. This weak alteration affects the carbonaceous debris flow units, Lower units and the base of the Middle units, Dowling member (Fig. 4). The lower units consist of 20-30 vol% lenticular chloritic shards with or without a fine bluish rim, and 10-25 vol% characteristic bluish, lapilli-sized altered glass fragments. Altered vitric fragments are mineralogically heterogeneous along strike but dominated by those with distinctive albitic rims with minor chlorite-epidote and occasional prehnite. Altered glass shards are texturally complex, variably devitrified and compositionally variable. They consist of varying amounts of chlorite- albite-K-feldspar-titanite-quartz ± prehnite-epidote-actinolite-pyrrhotite-sphalerite-chalcopyrite-pyrite. Bluish shards are a complex mixture dominated by (10-80 vol%) feldspar-quartz in radiating, globular, spherulitic, sector crystallized and poorly defined masses with disseminated chlorite (5-15 vol%), titanite (<5 vol%) and variable amounts of radiating, reniform masses of prehnite (5-15 vol%). Textures pseudomorphous after palagonite are evident in replacement feldspar and locally chlorite along shard margins with micron-scale layered textures outlined by thin Fe-Ti oxides or Leisegang rings. Shard cores are composed of spherical chlorite, titanite and fibro-radial spherulites of feldspar and trace prehnite-epidote. Perlitic crack outlines, decorated with titanite and Fe-Ti oxides, attest to the former glassy nature of the shards. Prehnite, typically a minor component but locally up to10 vol% of the rock, lines amygdales and shards. It partially to totally replaces shards that contain amygdales and forms 0.4 mm bowtie masses finely disseminated within chloritic shards or coarse crystals in the core of devitrified albite-chlorite shards. The heterogeneous and complex devitrification textures in the shards, the lack of distinctive mineralogical zonation and minimal mass change (Ames 1999) suggest only weak alteration.

The chlorite alteration zone regionally mimics the distribution of the discontinuous Contact units Dowling member (Ames and Gibson 2004a). Regionally, the contact unit outcrops as discontinuous areas up to 10 km long and 5 to 150 m wide, averaging 30 m for a strike length of 50 km. An abrupt termination of the chlorite altered Contact units against albitized Sandcherry member units in Morgan Township (Ames and Gibson 2004a) indicates that faulting post-dates alteration of the contact units. Chloritization of shards is evident throughout the entire Onaping Formation, although the proportion of chlorite changes (Fig. 4). Variation in the amount of chlorite is dependent on the regional alteration type, its intensity and abundance of vitric shards. The abundance of chlorite, as a replacement of shards and vesicle fillings, increases in the weakly albitized Sandcherry member towards the contact with the Dowling member. The percentage of glass shards in the contact units is higher than that of the other Dowling member units. The mineral assemblage in the chlorite alteration zone consists of actinolite-chlorite-albite-orthoclase-titanite-prehnite-pyrrhotite-chalcopyrite-sphalerite. Fe-chlorite (pycnochlorite-ripidolite) compositions are present throughout the altered formation except near base metal mineralization, the zone of silicification and zones of peperite, where more Mg varieties of chlorite occur (Ames 1999).

A prominent feature of the contact units is 20-45 vol% dark green altered shards in a bluish to greenish gray matrix in which alteration was used to define a former stratigraphic unit ie: Green Member (Avermann 1994), or chlorite shard horizon; Muir and Peredery (1984). The "chloritic" domains have been interpreted as collapsed chlorite-filled cavities or void spaces between fragments (Deutsch et al. 1995). The presence of vesicles, relict devitrification and palagonitic textures and vesicles truncated across glass boundaries clearly indicate that the "chloritized" domains in the contact unit are altered vesicular vitric fragments or altered glass shards (Gibbins 1994; Ames 1999).

The chlorite alteration is overprinted and/or coeval with albitization at the base of the alteration zone and is overprinted by more calcic alteration resulting in actinolite, pyrrhotite-chalcopyrite-sphalerite, prehnite and finally calcite. The structural control on the distribution of the weakly chloritized Contact units in Morgan Township suggests that chlorite-amphibole alteration predates or is synchronous with block faulting and pre-dates dyke injection. The Contact units were preferentially altered as a consequence of their hot, subaerial to shallow water emplacement during a major period of faulting and introduction to water. These faults were also local conduits for fluids responsible for albitization of the contact units as albite alteration extends <1 m into the contact units adjacent to the trace of the faults.

The regional albite zone affects the Sandcherry member strata but is also present within the Contact and Lower units of the Dowling member. The ~300 m wide sulfide-poor zone of albitization is situated 1 km below the top of the Onaping Formation (paleoseafloor) and 0-280 m above the SIC (Fig. 4). Albitization is spatially associated with the emplacement of flow-banded aphanitic andesitic sills and dykes that are local feeders to the Sandcherry member fluidal fragment units. Intense replacement of shards by albite within 1-m-wide envelopes around isolated dyke apophyses attest to the syndepositional timing of this alteration. Above sills, intense alteration is areally extensive and the density of sill and dykes of andesitic melt correlates with the intensity of albite alteration.

Albitization is characterized by the following mineral assemblages: (1) weak alteration, albite-K-feldspar-chlorite-quartz-prehnite-pumpellyite; 2) moderate alteration, albite-K-feldspar-chlorite-actinolite; and 3) intense alteration, albite-actinolite-quartz (Ames and Gibson 2004c). Albite alteration replaces the vitric fragments and matrix, and overprints the basement lithic fragments giving a pinkish beige hue to the rocks. Pervasive albite alteration is overprinted by more discrete fracture-controlled albitization which forms linear replacement zones within the Sandcherry member and basal intrusion units. Coarse euhedral titanite associated with this fracture controlled replacement zones was dated at 1848 Ma (Ames et al. 1998).

Semiconformable, discontinuous silicified zones up to 1300 m long by 280 m thick have been traced for more than 30 km at the exposed base of the Onaping Formation (Fig. 4). Silicification preferentially replaces the groundmass, overprints the regional albite zone and is characterized by the higher temperature assemblage, quartz-epidote-clinopyroxene-pyrrhotite-chalcopyrite-sphalerite.

5 Intra-Onaping hydrothermal base metal mineralization

The intra-Onaping Zn-Cu-Pb showings are generally situated in the Lower units and at the base of the Middle units in the Dowling Member, whereas the larger deposits are situated at the top of the Onaping Formation, hosted dominantly by the carbonates of the Vermilion Formation (Figs. 1, 3A, 5 and 6). Base metal showings within the Onaping Formation occur as replacement of lapillistone beds and fragments, sulfide veins, stringer zones, disseminations and sulfide fragments and are most abundant in the carbonaceous upper part of the Onaping Formation (Rousell 1984b; Gibbins 1994; Ames 1997). The stratigraphy, alteration characteristics and structural setting of the intra-Onaping prospects were defined through study of the larger showings - Limerick, Ryan, Cow, McNunes and Simmons (Fig. 3A), as well as a detailed mineralogical study of the proximal alteration and sulfidic mineralization of the Errington deposit (Fig. 5) (Ames 1999). Characteristics of the base metal showings are listed in Table 2.

Table 2. Summary of hydrothermal base metal showings in the Onaping Formation.

(BI=basal intrusion, DM=Dowling Member, SM=Sandcherry Member, act-actinolite, alb-albite, cc-calcite, chl-chlorite, cpy-chalcopyrite, ga-galena, ilm-ilmenite, Ksp-K-feldspar, pre-prehnite, po-pyrrhotite, py-pyrite, sph-sphalerite, tnt-titanite)

Showing	Location	Host unit	Style	Mineralization	Alteration
Simmons	NR SW Dowling T.	DM-contact unit	veins	ga-sph-cpy-po,py	regional: chl-cc
		near SM/	5 veins in 2 m	py replaces po	proximal: shards-
		DM contact (Ames and Gibson 2004d)	167°/80	ga-cpy cores	alb-Ksp (70-80%)
				secondary anglesite sph (12-14 mole% FeS)	chl (15-20), pre-tnt
					distal: (~10 m)
					alb-Ksp (50-60%)
					act-Hb (30-40%)
					cc, Mn-ilm, tnt
					cpy-ga in wallrocks
		DM-Lower unit	dissemi nated gossans	sph-cpy-po	cc-absent local zones
		-spatially with very amygdaloidal	20x40 cm		some silicification.
		BI pods and dykes	stringers	cpy-py cut diss'd	
Limerick	NR Morgan T.	base of DM-middle units	stringer	cpy, po, sph	regional: calcite-absent zone at
		above topographic		vein gangue: barian-Ksp,	base of chl-cc
		high and close to 020°		quartz, muscovite, ankerite	100x300 m gossan
		syndepositiona l fault		calcite, siderite	Mg-rich chlorite assoc. with po
		mineralized "fault"		sph (10-13 mole% FeS)	and/or muscovite)
		(Ames and Gibson 2004a)			Fe-rich chlorite associated with
				phengitic muscovite, low	ankerite and cpy
				Ba and F	Fe-rich epidote
				calcite, ankeritic dolomite	Muscovite: phengitic
				and late siderite	low Ba and F
Ryan	NR	DM-middle units	replace ment of	po-py (2-15%)	In a 100 m^2 calcite-absent

	SW Morgan T.	above topographic high formed by proximal fluidal breccia complex along trajectory of 305° (Ames and Gibson 2004a)	lapilli-stone beds	cpy (<3%), sph (<3%), trace galena grab samples 6200 ppm Cu, 1900 ppm Pb, 3980 ppm Zn Early po-red sph-cpy overprinted by py, marcasite, honey sphalerite. Second phase colloform textured fills open spaces with prehnite/epidote, tr. mica	zone 600 m above base of regional chl-cc zone relict shards are micro-spherulitic and/or chlorite altered
Cow Lake	NR NE Dowling T.	DM-middle units 500 m above SM-DM contact unit (Ames and Gibson 2004e)	veins, gossans N-trending joint control (Map 4B inset; Ames 1999)	gossans: 10-20% po-cpy -sph-ga impregnated matrix and cut chloritic shards Po-cpy-ga calcite veins Secondary py rims po and is colloform in open spaces Tertiary, coarse py euhedra.	Local calcite absent zone within regional calcite-chlorite zone veins are filled with calcite
McNunes	NR NE Dowling T.	DM-middle units 295° syndepositional fault control lithic-rich lapilli tuff.	gossans and replacement of coarser beds Gossans zones 330° and 060° (Ames and Gibson 2004e)	20% po-cpy-sph. 1.9% Zn and 0.2% Cu grab samples (Gibbins et al. 1989)	Localized calcite absent zone 20 x 60 m, 140 m above an offset base of the regional chl-cc zone (Ames and Gibson 2004e)

The major intra-Onaping Formation base metal showings occur in decreasing order of significance: (1) at or near the base of the regional semi-conformable calcite zone within carbonaceous strata of the Dowling member; (2) in localized calcite-free zones with very minor sulfide; and (3) in the basal silicification zone near the granophyre/SIC contact (Ames 1999). Showings within the regional semi-conformable calcite zone form localized calcite-absent pockets and discordant corridors with carbonate-sulfide veinlets or replacement zones at Ryan, McNunes and in eastern Morgan Township (Fig. 2B, 5A) (Ames and Gibson 2004 a, e).

Important geologic controls on the distribution of base metal showings are syndepositional faults (e.g., crater-floor fractures) and paleotopographic highs in the footwall Sandcherry member units. Both the N/NE and NW orientations of faults in the Onaping Formation focused hydrothermal fluid flow. The Limerick occurrence overlies a 020° syndepositional fault structure characterized by the abrupt termination of depositional units, melt or aphanitic dyke injection into the Dowling member, and intense discordant albitization. This zone contains anomalous disseminated chalcopyrite, pyrrhotite and lesser sphalerite and calcite. In the immediate footwall of the Limerick showing, aphanitic dykes and pods are coarsely spherulitic and have peperitic contacts with the host Dowling member Contact units. Similarly, the Ryan mineralization is situated above an apparent paleotopographic high and fault, and the McNunes showing is along strike of a 295° trending syndepositional fault that is responsible for a complex geometry between units due to apparent sinistral movement. Small pods of basal intrusion are aligned along this trend, as well as a steeply east- dipping, flow banded aphanitic dyke, again illustrating the control of melt by faults (Ames 1999). Amygdales in small pods of basal intrusion are filled with pyrrhotite, chalcopyrite and calcite.Sulfide assemblages of the intra-Onaping base metal showings are dominated by sphalerite, chalcopyrite, pyrrhotite, minor galena, arsenopyrite and later pyrite and anglesite. Alteration minerals include, calcite, ankerite, siderite and trace Fe- or Mg- chlorite, phengite, barian K-feldspar and epidote (Table 3). The mineral-chemistry and textural details are documented in Ames (1999). Sphalerite contains 6-14 mole% FeS (Table 2) suggesting that the hydrothermal fluids were buffered close to the pyrite-pyrrhotite fO_2 stability boundary (Hannington et al. 1995). Secondary pyrite replaces pyrrhotite and is nickeliferous (Ames 1999; Desborough and Larson 1970). Barium anomalies typically associated with some intra-Onaping base metal occurrences and the Errington and Vermilion deposits (Rousell 1984b; Paakki 1992; Gray 1995; Gibbins 1994) are due to the presence of barian K-feldspar and phengite in mineralized veinlets, ore gangue and proximal alteration haloes (Table 2).

6
Crater-floor hydrothermal ore deposits, carbonate sinters and hanging wall metalliferous plumes

6.1
Carbonate sinters, Vermilion Formation

The Vermilion Formation is thin (~14 m) and composed of the Lower Carbonate (LCM), Grey Argillite (GAM) and Upper Carbonate (UCM) members (Fig. 6). Iron-rich chemical sedimentary rocks or "exhalites" commonly have interbedded volcaniclastic-detrital and chemical sedimentary components (Spry et al. 2000). The LCM characterized by 1-2 mm laminations and minor colloform textures is discontinuous yet aereally extensive (Stoness 1994). The proximal facies is grey to pink and contains fine laminae composed of Ca-Mg (<Fe, Mn) carbonates, phengite, hyalophane, celsian, stilpnomelane and minor organic carbon. The laterally extensive distal facies LCM has typically beige brown laminations of alternating Mn-Fe carbonates, Ba-feldspar, carbon (average 1.24 wt% $C_{non\text{-}carbonate}$), quartz and trace phengite. Geochemically it is a carbonate-facies iron formation (Ames et al. 2002b). The hydrothermal signature of the LCM coupled with their mound-like morphology are compatible with the interpretation of proximal carbonate hydrothermal vents (Stoness 1994) similar to those on the modern seafloor (Sagalevich et al. 1992; Gamo et al. 1991; Chertkova and Stunzhas 1990). The Grey Argillite Member is a 10-20 m thick succession of distal turbidite that extends around the basin (Stoness 1994). The geochemical signature suggests a significant terrigenous component (Stoness 1994). The detrital component in the chemical sediments in the Sudbury crater corresponds to the initial influx of extra-basinal sedimentation likely as the crater rim breached. These extra-basinal sourced siliciclastic sediments capped the low temperature vents of the LCM. Local sites of Upper Carbonate Member deposition are restricted to interpreted hydrothermal vent sites in the Errington and Vermilion mine areas and as rare pods up to 50 cm thick on the North range. The UCM contains ~80 vol% carbonate nodules, 0.05 to 20 mm in diameter, composed of complex intergrowths of calcite-ankerite-dolomite-siderite-kutnohorite. They are cored by carbon and set in a matrix of carbonaceous mudstone (Fig. 6C). Deposition of the Upper Carbonate Member marks the initiation of an extended period of quiescence and deposition of fine carbon-rich hemipelagic mudstones that continued during deposition of the overlying Onwatin Formation. Geologic, petrographic and isotopic data suggest that the fluid evolution responsible for

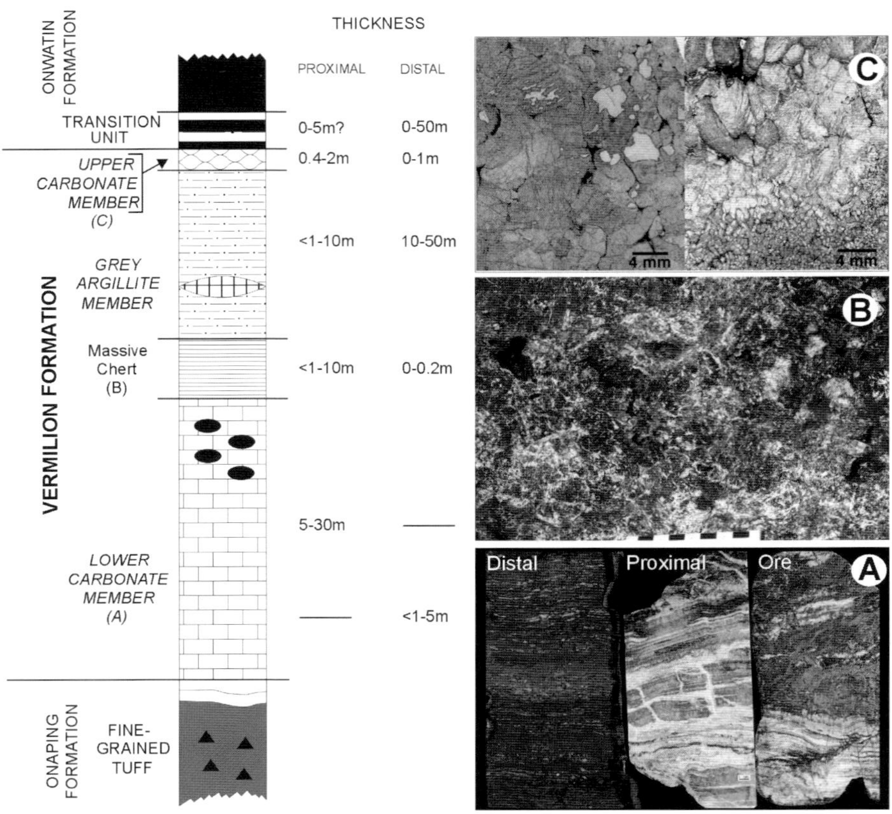

Fig. 6. Schematic stratigraphic section through the Vermilion Formation (modified from Gray 1995). A) Polished slabs of the Lower Carbonate Member: brown distal facies, pink and white proximal facies and replacement ore. Note the laminated carbonate bands as relicts of replacement in the sulfide ore. B) Outcrop photograph of silicified Lower Carbonate Member, "cherty breccia". Tourgney trench, South Range. C) Thin section photographs of concretionary carbonates of the Upper Carbonate Member.

carbonate deposition in the Onaping and Vermilion Formations progressed from early seawater introduction into the Onaping Formation to low temperature interaction and subsequent development of pore and evolved ore fluids. The Vermilion Formation is interpreted to have precipitated from the bicarbonate-rich fluid that initially formed the regional calcite zone in the Onaping Formation and ascended to the paleocraterfloor via syndepositional faults to precipitate aerially restricted carbonate mounds of the Lower

Carbonate Member (Ames 1999). Diffuse fluid flow and low temperature dispersal of hydrothermal plumes resulted in the deposition of the Fe-Mn-rich distal facies. The development of euxinic conditions in late Onaping to Onwatin time (Rousell 1984a,b) was likely triggered by subsidence, widespread diffuse hydrothermal input of Fe^{2+}, H_2S, SO_2 and CH_4 due to degassing of the SIC melt sheet, methane generation from organic matter in the Onaping Formation and subsequent bacterial bloom in a restricted and deepening Sudbury crater. This was followed by deposition of turbidites of the Grey Argillite Member with diffuse hydrothermal emissions precipitated as the Upper Carbonate Member local to the vents. The distal low temperature facies of the LCM and silicified zones are most prospective for biomarkers in the impact-related chemical sedimentary rocks.

6.2 Errington and Vermilion Zn-Pb-Cu deposits

The Errington and Vermilion hydrothermal base metal deposits together contain 6.4 Mt grading 4.36 wt% Zn, 1.37 wt% Cu, 1.15 wt% Pb, 55 g/t Ag and minor Au (Table 1) (Jonasson and Gibson 1993). The Zn-Pb-Cu deposits occur in the southwestern part of the Sudbury basin as multiple sulfide lenses at the top of the Onaping Formation and within carbonates of the Vermilion Formation (Figs. 1-3). The construction of the trans-Canada railway was responsible for many of the mineral finds in northern Ontario during the 1800s and discoveries in the Sudbury basin were the result of a Canadian Pacific Railway line cut across the basin in 1883. Zn-Pb-Cu sulfide mineralization was first discovered in 1884 near Stobie Falls on the Vermilion River by James Stobie with the Errington and Vermilion mines found later in the 1800s. An historical account of discovery and development is summarized in Ames (1997). Metal deposition in the Errington-Vermilion deposits is largely within carbonate mounds of the Vermilion Formation and rooted in a footwall sequence of hydrothermally altered Onaping Formation. It is overlain by carbonaceous argillites of the Onwatin Formation (Martin1957; Paakki 1992; Gray 1995). The origin of the massive sulfide deposits has been ascribed to syn-tectonic replacement or syngenetic exhalation (Card and Hutchinson 1972; Whitehead et al. 1990; Rousell 1984b) and as the sub-seafloor replacement of a carbonate host - the currently favoured model (Martin 1957; Paakki 1992; Gray 1995; Jonasson and Gibson 1993).

The Errington and Vermilion deposits are at a distinct stratigraphic level (Vermilion Formation) underlain by a laterally extensive sodium depletion zone extending over 10 km east of the Errington deposit and 5 km to the west of the Vermilion deposit (Fig. 7). The deposits are characterized by a marked increase in Ba, Si and Zn near the orebody and a zone of Mn depletion within the Vermilion Formation (Paakki 1992; Gray 1995). Deformation within the area of these deposits is manifested by tight isoclinal folding and numerous axial planar and thrust faults. The ore lenses occur within tight inclined and locally overturned folds. Ore lenses occur within 3 parallel anticlines, with the ore concentrated in the crest of each anticline (Gray 1995; Paakki 1992). The Vermilion deposit contains discontinuous, lensoid, gold-rich zones about 15 x 30 m in size within the main ore zone with high-grade linear copper zones hosted by silicified rocks. The 030° trending linear copper-rich zones are interpreted as syn-depositional structures in the Vermilion area (Gray 1995), which is consistent with the orientation of syn-depositional faults defined in the footwall Onaping Formation (Ames et al. 2002a).

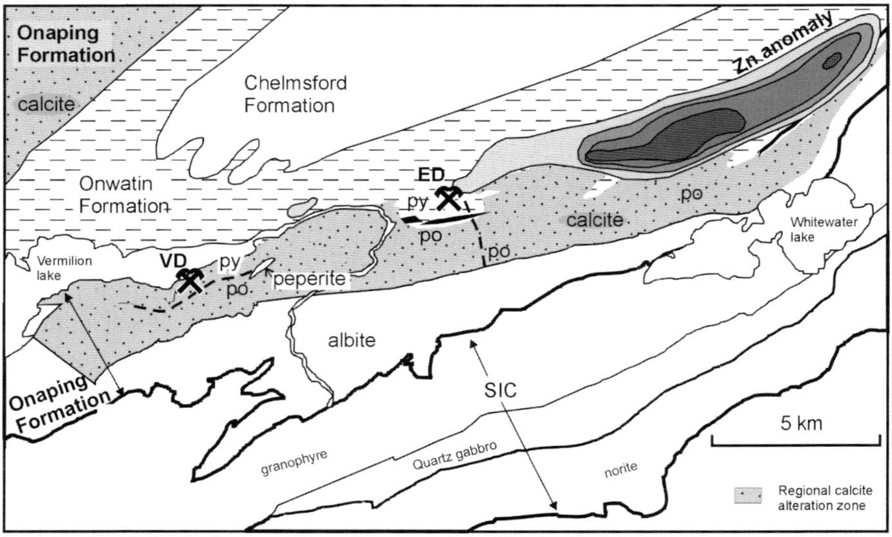

Fig. 7. Distribution of regional semiconformable and deposit proximal, footwall alteration zones in the Onaping Formation and metal enrichment halos in the hangingwall carbonaceous argillite, Onwatin Formation. Vermilion Deposit=VD, Errington Deposit=ED. Zn anomaly in hangingwall Onwatin Formation, 1000 ppm increments from dark >5000 to light 1000-2000 ppm. (Ames 1999; data compiled from Rogers 1995, Paakki 1992, and Gray 1995).

6.3 Hydrothermal alteration and sulfide mineral assemblages

The 2.44 Mt Vermilion deposit (Table 1) has a lower discordant chlorite alteration zone, an upper discordant silicification zone, and a stratiform Ca-Mg carbonate chalcopyrite-rich stockwork zone (Gray 1995). In contrast, the 4.36 Mt Errington deposit does not contain a chlorite-rich footwall alteration zone, nor a stockwork zone which are common features in many proximal VMS deposits (Paakki 1992). The variation between the two deposits is likely related to the relative proximity to the hydrothermal upflow zones (i.e. Vermilion deposit closer and hotter).

The Errington deposit is mineralogically simple, containing pyrite, chalcopyrite, sphalerite, galena, marcasite, arsenopyrite with only trace amounts of pyrrhotite, freibergite, and argentite (Table 3). Trace amounts of cuprite, bornite, covellite, pentlandite, hematite, magnetite, native gold and silver have been reported in the Vermilion deposit along with distinct chlorite-rich discordant alteration that is absent at the Errington deposit (Gray 1995). The gangue is dominated by various types and generations of Fe-Mg-Mn-Ca carbonate, quartz, hyalophane, celsian, barian muscovite, pyrobitumen and traces of barian stilpnomelane and rare Fe-chlorite (Table 3). Carbonates are dominantly ankerite, dolomite and late siderite at the base of the deposit, but calcite dominates in the carbonate-hosted mineralization. Barium within the Errington deposit is distributed between feldspar and muscovite in the Basal Pyrite Zone, Onaping Formation, in muscovite in the lower part of the carbonate hosted mineralization in the LCM, Vermilion Formation and in minor amounts within stilpnomelane and feldspar in the upper part of the orebody.

In general, the deposits include from base to top: 1) conductive andesite tuff in the footwall Onaping Formation with local carbonate concretions; 2) basal pyrite zone, hosted in Onaping Formation; 3) carbonate-hosted mineralization, the main ore zone within the Lower Carbonate Member, Vermilion Formation; 4) silicified, cherty zones referred to as chert breccia at top of the orebody within the proximal facies of the Lower Carbonate Member (locally at the base) and 5) an upper pisolitic carbonate unit defining the Upper Carbonate Member of the Vermilion Formation (Paakki 1992; Stoness 1994) (Fig. 8).

The basal pyrite zone replaces the top of the Onaping Formation and comprises the lowermost 2.5-6 m of the massive sulfide deposits and averages 2 wt% Cu, 1.34 wt% Zn and 0.31 wt% Pb in the Errington deposit (Paakki 1992). The lower contact of the basal pyrite zone is defined by a marked increase in the abundance of 5 mm spherical pyrite grains (up to 70%) and this zone likely developed on or near the paleocrater floor. The basal pyrite zone is characterized by the assemblage of framboidal pyrite, sphalerite, quartz, ankerite-dolomite, Ba-K-feldspar, chalcopyrite, Ba-muscovite (phengite), arsenopyrite, and galena, and a later stage of calcite, ankerite, and euhedral pyrite.

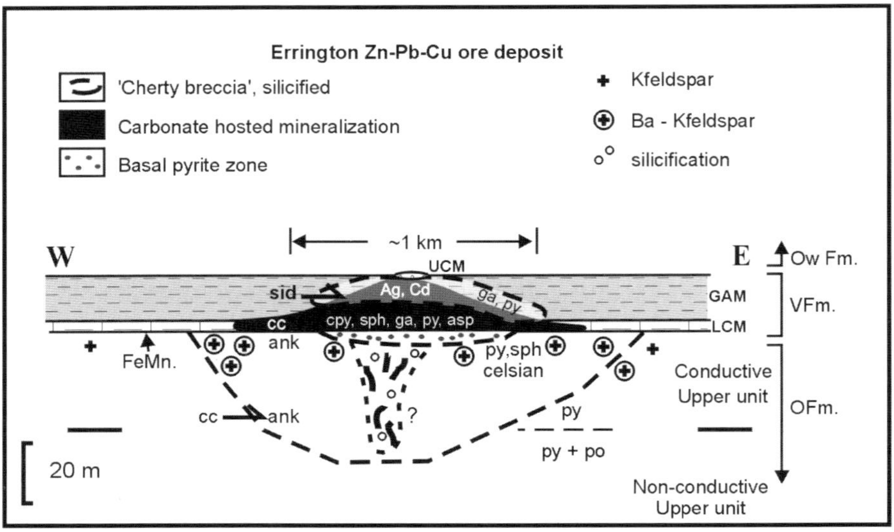

Fig. 8. Schematic diagram of the proximal alteration, sulfide mineralogy and trace element variations, Errington Zn-Pb-Cu deposit (modified from Paakki 1992).

The main sulfide precipitation stage is characterized by 500 μm spheres of pyrite-sphalerite-chalcopyrite-ankerite and a later crosscutting phase of Fe-rich chlorite-chalcopyrite-ankerite-siderite-pyrrhotite veins. Ankerite is the dominant carbonate in the basal pyrite zone coeval with sulfide. Barian K-feldspar (<18 wt% Ba; maximum 38.1 wt%) occurs as <300 μm, skeletal, anhedral grains associated with quartz in the matrix. Celsian (Ba K-feldspar) is an early alteration phase that is overprinted by higher temperature assemblages. Barian muscovite partially replaces celsian at the top of the Onaping Formation directly below the basal pyrite zone.

The carbonate-hosted mineralization at the Errington deposit forms a zone with an apparent thickness of 35 m above the basal pyrite zone. Sphalerite, chalcopyrite and galena are the principal ore minerals associated with pyrite, along with accessory arsenopyrite, freibergite (Ag-rich tetrahedrite), argentite and pyrrhotite. Common gangue minerals are calcite, minor ankerite-dolomite and locally quartz, trace siderite, stilpnomelane and barian muscovite. Sphalerite and chalcopyrite are the major sulfide phases. High Cd and Hg levels in the sphalerites are noted near the top of the ore zone compared to those at the base and middle (<0.5 wt%; Ames 1999).

Element correlations of the Vermilion ore linked Cd-Zn and Hg-Zn reflecting the trace element characteristics of the sphalerite (Gray 1995). Chalcopyrite occurs as veins and pods in carbonate and is commonly associated with galena which occurs as inclusions in the chalcopyrite and as isolated grains. Compositional differences in chalcopyrite are confined to variations in Ag and As (Ames 1999). The beginning and progressive replacement of chalcopyrite by argentite, is visible initially as bluish-red tarnish on chalcopyrite and finally resulting in black, porous aggregates of acicular argentite. Freibergite ($Ag_5 (Cu Zn)_2(Sb,As)_4S_3$) with up to 30 wt% Ag and 25 wt% Sb occurs as fairly large (<145 x 50 μm) grains associated with chalcopyrite, arsenopyrite, pyrite and galena near the top of the orebody. Locally, tiny 1-2 μm Ag-sulfides extend along fractures that cut the freibergite-chalcopyrite grain boundary.

Carbon inclusions also occur within some chalcopyrite grains and along grain boundaries. Stilpnomelane may overgrow chalcopyrite. The carbonate minerals display a complex paragenesis. The dominant carbonate is Mn-poor calcite which has replaced an earlier minor phase of Mn-bearing ankerite (Ames 1999). The deposit is generally zoned with respect to carbonates with an ankeritic base (basal pyrite zone), calcitic core (carbonate-hosted mineralization) and sideritic top (chert breccia and Upper Carbonate Member), whereas Mn-rich minerals predominate lateral to the deposits (Fig. 8).

Barian muscovite persists into the main zone of mineralization above the basal pyrite zone as subhedral to euhedral inclusions in massive chalcopyrite. It also occurs in the matrix at the top of the orebody where it comprises up to 15% of the calcite-quartz-dolomite matrix to minor disseminated sulfides. In the main part of the orebody, the Ba content of the muscovite is high and variable (3.74-6.86 wt%) locally with Ba-rich cores. These micas are phengitic, and compositionally form a tight cluster with relatively constant Fe/(Fe+Mg) at ~0.32. This zone of mineralization contains micas with the highest Ba and F contents in the crater-fill hydrothermal system with 0.74-1.48 wt.% F (Ames 1999).

Table 3. Mineralogical summary for mineralization in the Sudbury Basin.

ERRINGTON Zn-Cu-Pb DEPOSIT	
Carbonate-hosted mineralization	
Pyrite	Ankerite-dolomite
Arsenopyrite	Siderite (replaces sulphide)
Argentite	Graphitic carbon
Galena	Muscovite (Ba)
Tetrahedrite (Freibergite; native silver)	Stilpnomelane
Sphalerite (Low-Fe 9 mol%)	Calcite
Chalcopyrite	
Basal Pyrite Zone	
Pyrite	Ankerite-dolomite
Sphalerite (High-Fe 12 mol%)	Ba-K-feldspar
Chalcopyrite	(+- Phengite)
Pyrrhotite	(+- Chlorite [high Fe])
Galena	
Arsenopyrite	
INTRA-ONAPING MINERALIZATION	
Limerick Cu -Zn showing	
Chalcopyrite	Ankerite-dolomite
Pyrrhotite	Siderite
Sphalerite (High-Fe 13 mol%)	Ba-K-feldspar
	Epidote
	Phengite (Ba)
Simmons Pb-Zn-Cu showing	
Galena	Dolomite
Sphalerite (High-Fe 14 mol%)	
LW showing	
Arsenopyrite	Ankerite-siderite
Pyrite	
Sphalerite (Low-Fe 6 mol%)	

Silicification in the upper part of the Errington deposit has produced a cherty, sulfide carapace at the top of the deposit known as the cherty breccia (Paakki 1992) (Fig. 8). It consists of a very fine matrix with subequal amounts of pyrite (with trace galena) and quartz that is cut by stringers, pods and veins. Fine (1-2 μm), euhedral to subhedral pyrite and trace galena occur in 10-20 μm aggregates in fine microcrystalline quartz. Veins are composed of pyrite, chalcopyrite, galena, arsenopyrite and carbon in calcite, siderite and stilpnomelane gangue. Galena has the highest Ag content (0.7 wt%, Ames 1999). Se, in galena, is elevated relative to the underlying carbonate-hosted mineralization but lower than that in the basal pyrite zone (~ 0.3, 0.1, 0.9 wt% respectively, Ames 1999). Sphalerite shows a decrease in the mole% of FeS in this upper zone as well as relatively high Cd levels (0.5 wt%, Ames 1999). All of the carbonates in the silicified upper zone of the orebody are manganiferous in contrast to Mn-poor carbonates analyzed below this zone. Calcite, the dominant carbonate contains zoned manganoan ankerite with dolomitic cores and may contain subhedral to euhedral grains of inhomogeneous siderite. Siderite also forms rims on sulfide minerals and locally contains inclusions of calcite, quartz and pyrite. Stilpnomelane is associated with the Fe-rich minerals in the silicified zone such as sulfides and siderite. Trace amounts of barian K-feldspar (4-6.9 wt% Ba) occur as inclusions in chalcopyrite and trace barian phengite, disseminated in the silicified cap rocks, has relatively high F and Ba contents (Ames 1999). The low temperature mineral assemblages containing Ba, Mn, and Pb occur in the upper part of the orebody and distal to the deposits reflecting the exhalative chert precipitation (Fig. 8)

6.4 Hangingwall hydrothermal plume

The carbonaceous mudstones of the Onwatin Formation have anomalous metal values relative to average values in shale and in particular have very high Ag, Zn, and Cu values and high Co, Pb, Ni, and Cr (Rousell 1984b). Pyrite forms bedding-parallel disseminations and forms massive lenses 1-3 cm and locally 20 cm thick. Rogers et al. (1995) did a systematic study of the spatial distribution of lithogeochemical anomalies relative to the Errington and Vermilion base metal deposits. An alteration halo, enriched in Zn, Cu, Fe, Ba As, V and S/C_{org}, is widespread, extending into the overlying carbonaceous mudstones of the Onwatin Formation covering an area 25 km^2 around the base metal deposits (Rogers et al. 1995) (Fig. 7). This metalliferous halo in the post-impact basin sediments is interpreted to be the result of continued hydrothermal fluid emanation into the water column during deposition of the basin shales (Rogers et al. 1995). This signifies an extended period of post-impact hydrothermal activity within the Sudbury impact crater.

7 Summary of Sudbury hydrothermal system

Although hydrothermal alteration extends below the melt sheet (~1 km) into permeable Sudbury breccia and forms haloes around Cu-PGE mineralization (~150 m) it is most extensive (>2 km), and of lower temperature, above the melt sheet. Regional seawater ingress, convection and resultant subseafloor fluid/rock interaction throughout the 1.4 km thick Onaping Formation mobilized base metals from the devitrified andesitic pile and precipitated them as massive sulfide deposits on the paleocrater floor. The variation in alteration mineral assemblages down section shows evidence for thermal gradients in excess of 100°C/km (Ames 1999). Mineralogical mapping identified subsurface regional alteration zones that broadly define paleoisotherms within the crater. The hydrothermal system was low temperature (<260°C) and developed regional scale zones of (1) silicification (quartz-albite-epidote), (2) albite (albite-actinolite, albite-K-feldspar-actinolite-chlorite, albite-K-feldspar-chlorite-prehnite), (3) chlorite (actinolite -chlorite) and (4) carbonate (calcite-chlorite) generally showing a decrease in temperature upwards. Mineralogic data indicate that the lower silicification zone adjacent to the SIC at the exposed base of the pile formed at temperatures greater than 340°C (Ames 1999). Lower temperature albite and actinolite-rich assemblages, restricted to below 1000 m in the Onaping Formation, are typical for such subseafloor depths, with lower temperature (250-300°C) seawater-rock interaction (Alt 1996; Schiffman and Fridleifsson 1991). The albite-rich alteration zones in the lower Onaping Formation are depleted in Cu, Zn, and S, indicating conditions in the lower part of the hydrothermal system favourable for stripping and removing metals (Fig. 9). Although it has been stated that the entire Onaping Formation contains abundant sulfide (Muir and Peredery 1984), in general the albitized Sandcherry member shows depletion in Cu, Zn, and S in a zone between 1100 to 1400 m below the the top of the Onaping Formation. This metal-depleted alteration was dated with U-Pb zircon geochronology at ca. 1848 Ma (Ames et al. 1998) and is overprinted by the pyrrhotite-chalcopyrite-amphibole rich assemblage in the silicification zones. The higher temperature pyrrhotite-chalcopyrite-Fe-chlorite veinlets cutting the Errington deposits may be due to the release of deeper, hotter fluids from the base of the sequence. The lower temperature calcite and likely metamorphosed clay-rich alteration assemblages in the upper 1000 m typically form at temperatures less than 180°C (Alt et al. 1986).

The presence of disseminated sulfides and small sulfide stockwork zones in the lower Dowling member above paleotopographic highs and/or syndepositional faults indicates pathways for hydrothermal fluid that remobilized metals upsection. Such faults are well defined in the crater-fill sequence through mapping of melt dyke and fluidal breccia complex orientations, unit offsets, channel-shaped debris flows, extensive zones of peperite, linear distribution of Lower Carbonate Member sites, thickness variations in the Grey Argillite Member and the localization of hotspring deposits possibly in depressions along the paleo-crater floor. Isotopic compositions of the calcite from the regional calcite zone and Vermilion Formations show that the carbonate is likely the product of degassing from the SIC (Ames et al. 1999) (Fig. 1) whereby the melt sheet served not only as a heat source but also contributed volatiles. Textural and paragenetic observations are compatible with the proposed replacement origin of the Errington and Vermilion Zn-Cu-Pb deposits (Martin 1957; Gray 1995; Gibson et al. 1996). Early dolomite of the hydrothermal carbonate mounds is replaced by calcite-sphalerite-chalcopyrite-pyrite assemblages.

Bulk compositions of modern seafloor deposits are the product of the rock types in the subsurface and accordingly the Pb-Ba-Zn and Cu in the Errington and Vermilion deposits are the result of leaching from the abundant andesitic glass in the Onaping Formation and subsequent deposition due changes in pH upon interaction and replacement of carbonate mounds of the Vermilion Formation. Mineral assemblages at the Errington deposit are dominated by low temperature phases such as barian minerals, argentiferous tetrahedrite, galena and arsenopyrite (Barnes 1979). The SIC likely contributed volatiles Hg, As, Zn and Co to the massive sulphide deposits. In general, the deposit has a Ba-K-Zn-rich base and a Si-Ag-As-Sb-Cd-Pb-Mn -rich top (Fig. 8). The Ba-rich nature of proximal alteration is attributed to the composition of the feldspathic footwall rocks, from which barium was leached during fluid-rock interaction (Von Damm et al. 1985). However, the preservation of barian K-feldspar as early phases in the basal pyrite zone and in the siliceous cap rock at Errington, is compatible with low-temperature (<200°C) deposition (Barnes 1979). The small size of the Errington and Vermilion deposits is likely due to a combination of (1) the relatively short-lived and low temperature nature of the impact-generated hydrothermal system compared to tectonically and magmatically active volcanic settings (Barrie and Hannington 1999) and (2) perhaps more importantly, the broad scale dispersion and loss of metalliferous fluids through the permeable Dowling member and subsequently, precipitation of disseminated sulfides as intra-Onaping disseminations and showings. Development of relatively late structures coupled with hydrothermal sealing of Dowling member contact units (welded) in the cratering process are needed to focus hydrothermal fluids to produce hydrothermal ore deposits in the impact crater environment.

Defining the timing of the various regional and discordant alteration phases is critical to understanding the evolution of fluids, temperatures, ore deposits and subsequent prospectivity for microbial activity in the crater. Recognition of high and low temperature alteration assemblages, their controls, style of occurrence (open space filling versus replacement of impact glass) and architecture within the crater are important. Alteration vectoring towards upflow zones is evident in the lower metal-depleted albite alteration, discordant carbonate-free zones in the regional calcite zone, and the carbonate sinters that immediately preceeded and now host Zn-Pb-Cu mineralization. The composition of the sulfide and carbonate deposits is a function of the composition of the target rocks and melt sheet. The bulk composition of the hydrothermal ore deposit will be controlled by the composition of the suevite in the subsurface. The dominant metals leached and redeposited in an impact environment are a function of the target rock composition. Convective hydrothermal circulation of fluids in the Sudbury crater allowed for the effective dissipation of impact-generated heat at 1848 Ma (Ames et al. 1998). The following features are significant; (1) the relatively late timing of the metal-bearing semi-conformable silicification overprinting the metal-leached regional zone of albitization; (2) discordant mineralized silicification zones that cut the regional calcite alteration zone, and represent upflow zones; (3) intra-Onaping base metal showings have local carbonate-free zones indicating overprinting of the regional calcite alteration; (4) significant base metal showings occur near the base of the regional calcite zone; (5) similar alteration assemblages are present in the ore deposits and peperite zones on the South Range. Field and petrographic observations clearly show that the intra-Onaping showings and ore deposits overprint the regional zone of calcite-chlorite alteration in the upper 1 km of the Onaping Formation.

Basin deepening, development of water column anoxicity and deposition of the fine carbonaceous upper units of the Dowling member was interrupted by local minor faulting and dyke injection into the water saturated sediments (Fig. 9). These vitric andesitic dykes intrude up the entire Onaping Formation and are the same composition as vitric rinds on bombs and as vitric fragments in the lower suevite (i.e., melt; Ames et al. 2002a). The occurrence of fault-controlled peperite bodies below the Errington and Vermilion Zn-Pb-Cu deposits indicates the presence of a thermal and structural anomaly along which were focused ascending hydrothermal fluids that reached the seafloor. The spatial association of peperite with the main ore zones in the southwestern sector of the Sudbury basin is not entirely fortuitous. Peperite bodies in the footwall to ore deposits have been noted in the Iberian Pyrite Belt and Guaymas Basin (Boulter 1993; Gieskes et al. 1982). The intrusion of magma close to the paleoseafloor is known to trigger short-lived hydrothermal activity and displace large volumes of pore-water, super-heated seawater and hydrothermal fluids, from deep reservoirs along pre-existing but sealed fault systems to the paleoseafloor and overlying seawater (Baker et al. 1987, 1989; Embley and Chadwick 1994). The intrusive activity associated with peperite formation in the Onaping Formation is an unlikely heat source for productive hydrothermal convection, owing to the small volume of melt and shallow level of emplacement in the entire hydrothermal system. However, heat supplied by the melt could induce shallow lithification of the sediments and thus alter convective fluid flow patterns and focus fluid flow to the paleoseafloor and/or favourable traps in the subseafloor. The most extensive sealant is likely the reworked Upper units of the Dowling member which are carbonaceous, bedded sediments. These provide a near surface seal to diffuse fluids convecting in the permeable Onaping formation, and cause ponding and eventually overpressuring. Typically, low temperature siliceous-carbonic fluids (<200°C) are expelled onto the seafloor due to shallow sill injection with the higher temperature systems resulting from a deeper level heat source (Gieskes et al. 1982). Silica alteration of upper Onaping strata about these peperite bodies, and the early silicification of the mounds at the Errington vent site are the likely result of similar processes that operated in the Sudbury crater. It was the early precipitation of carbonate on the Sudbury crater floor as sinter mounds that allowed the later trapping of hot, metal-rich hydrothermal fluids before all their metals could be dispersed as vent plumes during Vermilion Formation time and precipitated in metalliferous haloes in the Onwatin Formation (Figs. 1 and 7). Seismic activity and melt intrusion throughout the deposition of the crater-fill sequence focused fluid flow through discordant structures and produced local higher temperature perturbations in the broad-scale lower temperature (<250°C) hydrothermal system.

8 Implications

The Sudbury crater hydrothermal system is but one variant (end member) within impact craters that span the range from little hydrothermal alteration to geothermal-like systems reported in numerous craters (Puchez-Katunki, Manson, Ries, Kara; Naumov 2002) to robust hydrothermal systems (Sudbury). In order to sustain fluid convection and focus hydrothermal discharge the impact crater must have fluid(s), a heat source/melt sheet, a thick crater-fill sequence and, a prolonged history of crater adjustment to produce syndepositional faults.

Alteration assemblages record the initiation and cooling of an impact hydrothermal system that, when combined with the structural, stratigraphic and magmatic history of the crater, allow a glimpse into the subsurface plumbing system in large impact craters. The change of crater morphology with increasing crater diameter is well documented (Melosh 1989) however, crater collapse and subsequent related hydrothermal processes have not been successfully modelled especially for large craters (Melosh and Ivanov 1999).

Systematic mapping has defined a distinctive and consistent stratigraphic succession in the Sudbury crater-fill showing that impact breccias are not simply chaotic fallback but record the fragmentation, deposition and emplacement mechanisms of distinctive units overprinted by a succession of hydrothermal alteration events. Crater faults control melt emplacement in the Sudbury Structure in the outer part of the crater as radial and concentric offset dykes including the South Range Breccia Belt, embayment structures at the edge of the transient crater and, in the craters interior as crater floor-fractures (Ames et al. 2002a). Fluids in the crater environment include seawater, "magmatic" and intra-formational fluids, and their evolved mixtures. The deposition of the crater infill succession in an initially shallow submarine environment, a heat source, the permeable and highly reactive nature of the vitric-rich Onaping strata, the presence of syn-depositional faulting and the persistence of intra-crater magmatic activity are basic conditions important for the generation of hydrothermal ore deposits in impact craters.

Significant to the development of life microbial habitats is recognizing the lower temperature regimes (<120°C) that define favourable surface and subsurface zones. Hydrothermal systems typically result in the rapid, low-temperature crystallization of carbonates and silicates that can fossilize microbial life and their chemical signatures (Walter and Des Marais 1993; Farmer and Des Marais 1994; Cady and Farmer 1996; Mojzsis and Arrhenius 1998). Temperatures in the Sudbury hydrothermal system in the basement below the melt sheet are high (<420°C) however, upon cooling this system could reach habitable temperatures with precipitation of low temperature hydrothermal minerals likely in open space vugs and veins.

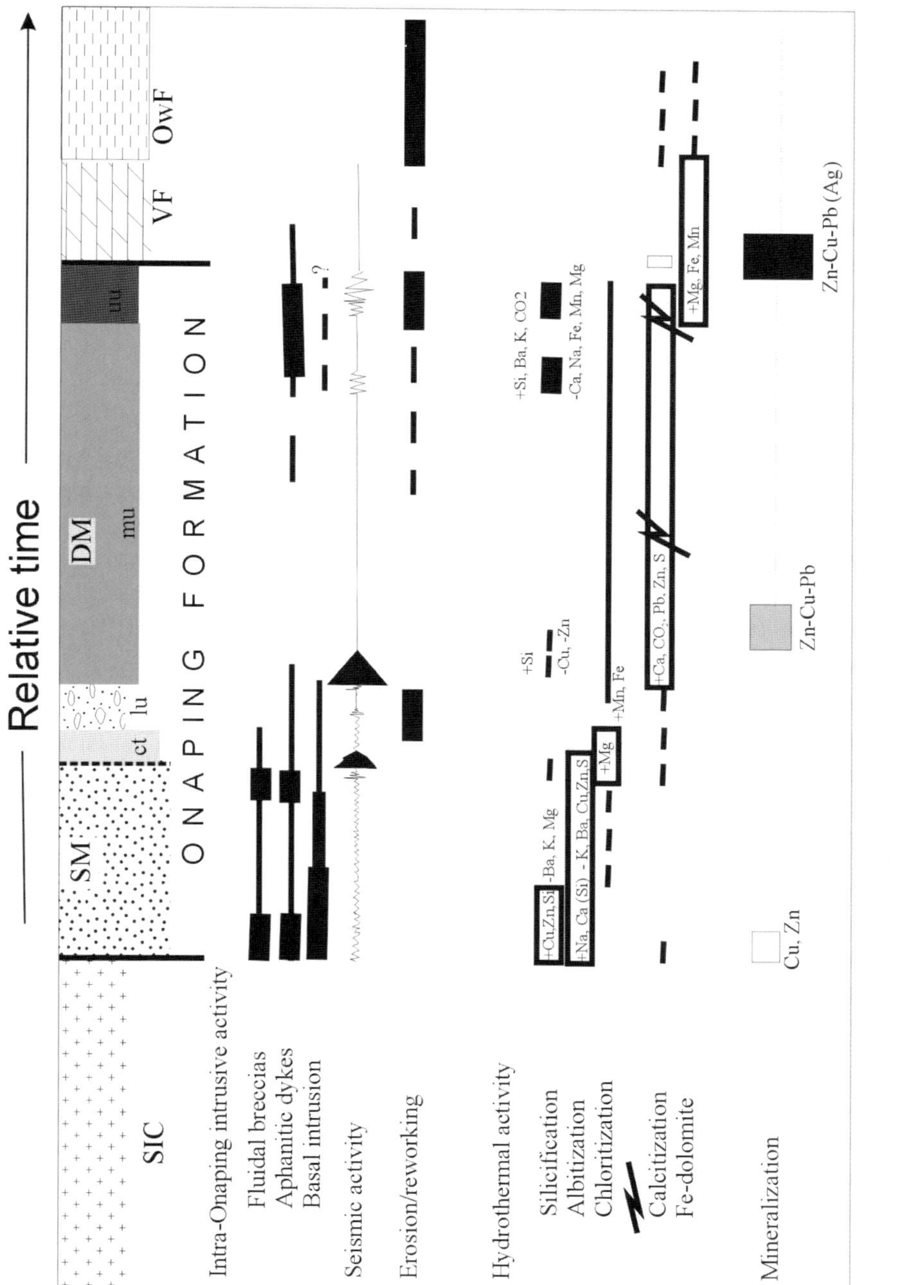

Fig. 9. Relative time-space chart for events in the Sudbury Basin (Ames 1999).

Impact hydrothermal systems of sufficient duration may generate and convect nutrients to create environments suitable for proliferation of microbes, likely preserved in the extensive low temperature zones in the subsurface at Sudbury as regional semiconformable calcite and carbon zones, distal carbonate-facies iron formation on the crater floor and low temperature hydrothermal plumes. It is the remarkably large extent of these zones in the upper part of the Sudbury hydrothermal system that make them highly exploitable.

Acknowledgements

This work was supported by Natural Resources Canada, Geological Survey of Canada and in part by NASA Exobiology program (Grant NAG5-10744) awarded to KOP and DEA. We would like to thank AG Galley and K Kirsimae for critical comments that helped improve this manuscript. This is Geological Survey of Canada contribution 2003306.

References

Allen CC, Gooding JL, Keil K (1982) Hydrothermally altered impact melt rock and breccia: Contributions to the soil of Mars. Journal of Geophysical Research 87: 10083-10101

Alt JC (1996) Subseafloor processes in mid-ocean ridge hydrothermal systems. In: Humphris S, Lupton L, Mullineaux L, Zierenberg R (eds) Seafloor hydrothermal systems: physical, chemical and biological interactions, Geophysical Monograph 91: 85-114

Alt JC, Honnorez J, Laverne C, Emmermann R (1986) Hydrothermal alteration of a 1 km section through the upper oceanic crust, Deep Sea Drilling Project Hole 504B: Mineralogy, chemistry, and evolution of seawater-basalt interactions. Journal of Geophysical Research 91: 10309-10355

Ames DE (1997) Sudbury Structure: Whitewater Group; Hydrothermal alteration and mineral deposits. In: Ames DE, Bleeker W, Heather KB, Wodicka, N (eds) Timmins to Sudbury transect: New insights into the regional geology and the setting of ore deposits. Geological Association of Canada Fieldtrip Guidebook B6: 94-122

Ames DE (1999) Geology and regional hydrothermal alteration of the crater-fill Onaping Formation: Association with Zn-Pb-Cu mineralization, Sudbury Structure Canada. Unpublished Ph.D. thesis, Carleton University, Ottawa, Canada 460 pp

Ames DE (2002) Sudbury Targeted Geoscience Initiative (TGI: 2000-2003): Overview and Update. Summary of Field Work and Other Activites 2002 Ontario Geological Survey Open File Report 6100: 17-1 to 17-10

Ames DE, Gibson HL (1995) Controls on and geological setting of regional hydrothermal alteration within the Onaping Formation, footwall to the Errington and Vermilion base metal deposits, Sudbury Structure, Ontario. Current research 1995-E. Geological Survey of Canada: 161-173

Ames DE, Gibson HL (2004a) Geology, alteration and mineralization of the Onaping Formation, Morgan Township, Sudbury Structure. 2 sheets, 8 figures descriptive notes, 1:5000, Geological Survey of Canada Open File 3717

Ames DE, Gibson HL (2004b) Geology, alteration and mineralization of the Onaping Formation, Rockcut Lake area, Norman Township, Sudbury, Ontario. 1 sheet with 4 figures and 1 table, scale 1:2000, Geological Survey of Canada Open File 4565

Ames DE, Gibson HL (2004c) Geology, alteration and mineralization of the Onaping Formation, Joe Lake area, Wisner Township, Sudbury, Ontario. 1 sheet, 5 figures and 1 table, scale 1:2000, Geological Survey of Canada Open File 4566

Ames DE, Gibson HL (2004d) Geology, alteration and mineralization of the Onaping Formation, Simmons Lake area, Dowling Township, Sudbury, Ontario. 1 sheet with 4 figures and 1 table, scale 1:2000, Geological Survey of Canada Open File 4567

Ames DE, Gibson HL (2004e) Geology, alteration and mineralization of the Onaping Formation, Cow Lake area, Dowling Township, Sudbury, Ontario. 2 sheets with 7 figures and 1 table, scale 1:2000, Geological Survey of Canada Open File 4568

Ames DE, Watkinson DH, Parrish RR (1998) Dating of a hydrothermal system induced by the 1850Ma Sudbury impact event. Geology 26: 447-450

Ames DE, Gibson HL, Watkinson DH (1999) Constraints on the origin of regional semiconformable alteration zones, Onaping Formation, Sudbury Structure. Geological Association of Canada Abstract 24: 1

Ames DE, Galley AG, Legault D, LaFrance B, Dubois A, Parker J, Benn K, Kjarsgaard IM, Zierenberg R (2001) Sudbury Targeted Geoscience Initiative (TGI): role of volatiles, structure and host rocks in the evolution of Sudbury ores. In: Summary of Field Work and Other Activities 2001 Ontario Geological Survey Open File Report 6070: 25-1 to 25-9

Ames DE, Golightly JP, Lightfoot PC, Gibson HL (2002a) Vitric compositions in the Onaping Formation and their relationship to the Sudbury Igneous Complex, Sudbury Structure. Economic Geology 97: 1541-1562

Ames DE, Pope KO, Jonasson IR, Hofmann B (2002b) Chemical sediments associated with the impact-generated hydrothermal system, Vermilion Formation, Sudbury Structure [abs.]. Geological Society of America 34(6): 544

Arengi JT (1977) Sedimentary evolution of the Sudbury Basin: unpublished MSc. thesis, University of Toronto, Toronto, Ontario, 141 pp

Avermann ME (1997) Origin of the polymict, allochthonous breccias of the Onaping Formation, Sudbury Structure, Ontario, Canada. In: Dressler BO, Grieve RAF, Sharpton VL (eds) Large meteorite impacts and planetary evolution. Geological Society of America Special Paper 293: 265-274

Baker ET, Massoth GJ, Feely RA (1987) Cataclysmic venting on the Juan de Fuca Ridge. Nature 329: 49-51
Baker ET, Lavelle JW, Feely RA, Massoth GJ, Walker SL (1989) Episodic venting of hydrothermal fluids from the Juan de Fuca Ridge. Journal of Geophysical Research 94: 9237-9250
Barnes HL (1979) Solubilities of Ore Minerals. In: Barnes HL (ed.) Geochemistry of Hydrothermal Ore Deposits, second edition, Wiley, New York: 404-459
Baross JA, Hoffman SE (1985) Submarine hydrothermal vents and associated gradient environments as sites for the origin and evolution of life. Origin s of Life and Evolution of the Biosphere 15: 327-345
Barrie CT, Hannington MD (1999) Chapter 1. Classification of VMS deposits based on host rock compositions. In: Barrie CT, Hannington MD (eds) Volcanic-Associated Massive Sulfide Deposits: Processes and Examples in Modern and Ancient Settings: Reviews in Economic Geology, 8: 1-13
Boudreau AE, Meurer WP (1994) Supersolidus alteration processes and stratabound platinum-group-element deposits in layered intrusions. In Lentz DR (ed) Alteration and Alteration Processes associated with ore-forming systems. Geological Association of Canada Short Course Notes 11: 213-236
Boulter CA (1993) Comparison of Rio Tinto, Spain, and Guaymas Basin, Gulf of California: An explanation of a supergiant massive sulfide deposit in an ancient sill-sediment complex. Geology 21: 801-804
Cady SL, Farmer JD (1996) Fossilization processes in siliceous thermal springs: Trends in preservation along thermal gradients. In: Evolution of hydrothermal ecosystems on Earth (and Mars?) Ciba Foundation Symposium 202: 150-173, Wiley, Chichester
Cantin R, Walker RG (1972) Was the Sudbury Basin circular during deposition of the Chelmsford Formation? In: Guy-Bray JV (ed) New developments in Sudbury geology. Geological Association of Canada Special Paper 10: 93-101
Card KD (1978) Geology of the Sudbury-Manitoulin area, Districts of Sudbury and Manitoulin; Ontario. Geological Survey, Report 166: 239 pp
Card KD, Hutchinson RW (1972) The Sudbury Structure: its regional geological setting. In: Guy-Bray JV (ed) New Developments in Sudbury Geology, Geological Association of Canada, Special Paper 10: 67-68
Card KD, Innes DG (1981) Geology of the Benny area, District of Sudbury. Ontario Geological Survey, Report 206, 117 pp
Card KD, Jackson SL (1995) Tectonics and metallogeny of the Early Proterozoic Huronian fold belt and the Sudbury Structure of the Canadian Shield. Geological Survey of Canada, Open File 3139: 55 pp
Card KD, Gupta VK, McGrath PH, Grant FS (1984) The Sudbury Structure: Its regional geological and geophysical setting. In: Pye EG, Naldrett AJ, Giblin PE (eds) The geology and ore deposits of the Sudbury structure, Ontario Geological Survey Special Volume 1: 25-45
Cathles LM, Erendi AHJ, Barrie T (1997) How long can a hydrothermal system be sustained by a single intrusive event? Economic Geology 92: 763-765

Chertkova LV, Stunzhas PA (1990) Geochemistry of gases from Paramushir submarine spring, Kuril island arc. Volcanology and Seismology 3: 36-50 (in Russian)

Corfu F, Lightfoot PC (1997) U-Pb geochronology of the sublayer environment, Sudbury Igneous Complex, Ontario. Economic Geology 91: 1263-1269

Cowan EJ, Schwerdtner WM (1994) Fold origin of the Sudbury Basin. In: Lightfoot, PC, Naldrett AJ (eds) Proceedings of the Sudbury - Noril'sk Symposium, Ontario Geological Survey, Special Volume 5: 45-55

Delaney PT (1982) Rapid intrusion of magma into wet rock: Groundwater flow due to pore pressure increases. Journal of Geophysical Research B87: 7739-56

Desborough GA, Larson RR (1970) Nickel-bearing iron sulfides in the Onaping Formation, Sudbury Basin Ontario. Economic Geology 65: 728-730

Deutsch A, Grieve RAF, Avermann M, Bischoff L, Brockmeyer P, Buhl D, Lakomy R. Muller-Mohl V, Ostermann M and Stöffler D (1995) The Sudbury Structure (Ontario, Canada): a tectonically deformed multi-ring impact basin. Geologische Rundschau 84: 697-709

Dietz RS (1964) Sudbury structure as an astrobleme. Journal of Geology 72: 412-434

Dressler BO (1984) General geology of the Sudbury area. In: Pye EG, Naldrett AJ, Giblin PE (eds) The geology and ore deposits of the Sudbury Structure. Ontario Geological Survey Special Volume 1: 57-82

Embley RW, Chadwick,WW (1994) Volcanic and hydrothermal processes associated with a recent phase of seafloor spreading at the northern Cleft Segment, Juan de Fuca Ridge. Journal of Geophysical Research 99: 4741-4760

Farmer JD, Des Marais DJ (1994) Exploring for a record of ancient Martian life. Journal of Geophysical Research 104: 26977-26995

Farrow CEG (1994) Geology, alteration and the role of fluids in Cu-Ni-PGE mineralization of the footwall rocks to the Sudbury Igneous Complex, Levack and Morgan Townships, Sudbury District, Ontario. Unpublished Ph.D thesis, Carleton University, Ottawa, Canada, 373 pp

Farrow CEG, Watkinson DH (1992) Alteration and the role of fluids in Ni, Cu and Platinum-group element deposition, Sudbury Igneous Complex contact, Onaping-Levack area, Ontario. Mineralogy and Petrology 46: 67-83

Farrow CEG, Watkinson DH (1999) An evaluation of the role of fluids in Ni-Cu-PGE bearing, mafic-ultramafic systems. In: Keays RR, Lesher CM, Lightfoot PC, Farrow CEG (eds) Dynamic Processes in magmatic ore deposits and their application in mineral exploration. Geological Association of Canada, Short Course Notes 13: 31-67

Fisher RV (1966) Rocks composed of volcanic fragments and their classification. Earth Science Reviews 1: 287-298

Gamo T, Sakai H, Ishibashi J, Oomori T, Chiba H, Shitashima K, Nakashima K, Tanaka Y, Masuda H (1991) Growth mechanism of the hydrothermal mounds at the CLAM site, Mid-Okinawa Trough, inferred from their morphological, mineralogical, and chemical characteristics. Proceedings of JAMSTEC (Japan Marine Science and Technology Center) Symposium, Deep Sea Research 7: 163-184

Gibbins SFM (1994) Geology, geochemistry, stratigraphy and mechanisms of emplacement of the Onaping Formation, Dowling Area, Sudbury Structure, Ontario Canada. Unpublished MSc thesis, Laurentian University, Sudbury, Ontario, 314 pp

Gibson HM, Spray JG (1997) Diagnostic criteria for the recognition of shatter cones: In: Conference on Large Meteorite Impacts and Planetary Evolution (Sudbury 1997), LPI contribution No. 922, Lunar and Planetary Institute, Houston, p 16

Gieskes JM, Kastner M, Einsele G, Kelts K, Niemitz J (1982) Hydrothermal activity in the Guaymas Basin, Gulf of California: A synthesis. In: Curray, JR , et al. (eds) Initial Reports of the Deep Sea Drilling Project, 64: 1159-1168, Washington, D.C , U.S. Government Printing Office

Golightly JP (1994) The Sudbury structure as a meteorite crater: Where is its center and what should we expect there? [abs] In: Drury MJ (ed) Scientific drilling; the Sudbury Structure; proceedings of a workshop. Canadian Continental Drilling Program Report 88-2: 12

Grant RW, Bite A (1984) Sudbury quartz diorite offset dikes. In: Pye EG, Naldrett AJ, Giblin PE (eds),The geology and ore deposits of the Sudbury structure. Ontario Geological Survey Special Volume 1: 275-300

Gray MJ (1995) The Geological Setting of the Vermillion Zn-Cu-Pb-Ag-Au Massive Sulfide Deposit, Sudbury Basin, Canada. Unpublished MSc thesis, Laurentian University, Sudbury, Canada, 244 pp

Grieve RAF, Stöffler D, Deutsch A (1991) The Sudbury structure: controversial or misunderstood? Journal of Geophysical Research 96: 22753-22764

Hanley J, Mungall JE (2003) Chlorine enrichement and hydrous alteration of the Sudbury breccia hosting footwall Cu-Ni- PGE mineralization at the Fraser Mine, Sudbury, Ontario Canada. Canadian Mineralogist 41:857-881

Hannington M, Jonasson IR, Herzig PM, Petersen S (1995) Physical and chemical processes of seafloor mineralization at mid-Ocean ridges. In: Lupton J, Mullineaux L, Zierenberg R (eds), Physical, Chemical, Biological and Geological Interactions within Submarine Hydrothermal Systems, American geophysical Union, Monograph 91: 115-157

Humphris SE, Zierenberg RA, Mullineaux LS, Thomson RE (1995) Seafloor Hydrothermal Systems: Physical, Chemical, Biological, and Geological Interactions. American Geophysical Union Geophysical Monograph 91: 466 pp

Jonasson IR, Gibson HL (1993) Origin of the Onaping Formation and Zn-Cu-Pb massive sulfide deposits of the Proterozoic Sudbury Structure. Northern Ontario Development Agreement Summary Report 1992-1993: 93-96

Krogh TE, Davis DW, Corfu F (1984) Precise U-Pb zircon and baddeleyite ages for the Sudbury area. In: Pye EG, Naldrett AJ, Giblin PE (eds) The Geology and Ore Deposits of the Sudbury Structure, Ontario Geological Survey, Special Volume 1: 431-446

Krogh TE, Kamo SL, Bohor BF (1996) Shock metamorphosed zircons with correlated U-Pb discordance and melt rocks with concordant protolith ages indicate an impact origin for the Sudbury Structure. In: Basu A, Hart S (eds) Earth Processes: Reading the Isotopic Code. American Geophysical Union Geophysical Monograph 95: 343-353

Marshall DD, Watkinson DH, Farrow CEG, Molnar F, Fouillac A-M (1999) Multiple fluid generations in the Sudbury Igneous Complex; fluid inclusion, Ar, O, H, Rb and Sr evidence. Chemical Geology 154: 1-19

Martin WG (1957) Errington and Vermilion mines. In: Gilbert, G (ed) Structural Geology of Canadian Ore Deposits, 6th Commonwealth Mining and Metallurgical Congress, Montreal, Canadian Institute of Mining and Metallurgy 2: 363-376

Masaitis VL, Shafranovsky GI, Grieve RAF, Langenhorst F, Peredery WV, Balmoasov EL, Fedorova IG, Therriault A (1997) Discovery of impact diamonds at the Sudbury Structure. In: Conference on Large Meteorite Impacts and Planetary Evolution (Sudbury 1997), LPI contribution No. 922, Lunar and Planetary Institute, Houston: 33

Mathez EA (1999) On factors controlling the concentrations of platinum group elements in layered intrusions and chromitites. In: Keays RR, Lesher CM, Lightfoot PC, Farrow, CEG (eds) Dynamic Processes in magmatic ore deposits and their application in mineral exploration. Geological Association of Canada, Short Course Notes, 13: 251-286

McCarville P, Crossey LJ (1996) Post-impact hydrothermal alteration of the Manson impact structure: In: Koeberl C, Anderson R (eds) Geological Society of America, Special Paper 302: 347-376

Melosh HJ (1989) Impact Cratering: A Geologic Process. Oxford University Press, New York: 245 pp

Melosh HJ, Ivanov BA (1999) Impact crater collapse. Annual Reviews in Earth Planetary Science 27: 385-415

Milkereit B, Green A, Berrer E, Boerner C, Broome J, Cosec M, Cowan J, Davidson A, Dressler B, Feuten F, Grieve R, James R, Krause B, McGrath P, Meyer W, Moon W, Morris W, Morrison G, Naldrett AJ, Peredery W, Rousell D, Salisbury M, Schwerdtner W, Snajdr P, Thomas M, Watts A (1992) Deep geometry of the Sudbury structure from seismic reflection profiling. Geology 20: 807-811

Mojzsis SJ, Arrhenius G (1998) Phosphates and carbon on Mars: Exobiological implications and sample return considerations. Journal of Geophysical Research 103: 28495-28511

Molnar F, Watkinson DH, Everest J (1999) Fluid-inclusion characteristics of hydrothermal Cu-Ni-PGE veins in granitic and metavolcanic rocks at the contact of the Little Stobie Deposit, Sudbury Canada. Chemical Geology 154: 279-301

Molnar F, Watkinson DH, Jones PC (2001) Multiple hydrothermal processes in footwall units of the North Range, Sudbury Igneous Complex, Canada, and implications for the genesis of vein-type Cu-Ni-PGE deposits. Economic Geology 96: 1645-1670

Molnar F, Watkinson DH, Jones PC, Gatter I (1997) Fluid inclusion evidence for hydrothermal enrichment of magmatic ore at the Contact Zone of the Ni-Cu-platinum-group element 4b deposit, Lindsley Mine, Sudbury, Canada. Economic Geology 92: 674-685

Morris WA (1999) The Sudbury Structure, Ontario, Canada: A circular impact crater. Geological Association of Canada Abstract Volume 24: 86

Muir TL (1984) Sudbury Structure; Considerations and models for an endogenic origin. In: Pye EG, Naldrett AJ, Giblin PE (eds) The geology and ore deposits of the Sudbury Structure. Ontario Geological Survey Special Volume 1: 449-490

Muir TL and Peredery W (1984) The Onaping Formation. In: Pye EG, Naldrett AJ Giblin PE (eds) The geology and ore deposits of the Sudbury Structure. Ontario Geological Survey Special Volume 1: 139-210

Mungall JE, Ames DE , Hanley JJ (2002) Evidence for a chondritic impactor at Sudbury? Highly siderophile elements in terrestrial and meteoritic samples (workshop): implications for planetary differentiation and igneous processes [abs] Nancy, France, Aug 26-28 2002

Naumov M (2002) Impact-generated hydrothermal systems: Data from Popigai, Kara, and Puchezh-Katunki impact structures. In: Plado J, Pesonen LJ (eds) Impacts in Precambrian Shields. Impact Studies Volume 2, Springer-Verlag Berlin Heidelberg: 117-172

Newsom HE, Brittelle GE, Hibbitts CA, Crossey LJ, Kudo AM (1996) Impact crater lakes on Mars. Journal of Geophysical Research E6 101: 14951-14955

Paakki JJ (1992) The Errington Zn-Cu-Pb massive sulfide deposit, Sudbury, Ontario: Its structural and stratigraphic setting and footwall alteration. Unpublished MSc. thesis, Laurentian University, Sudbury, Ontario, Canada, 92 pp

Peredery WV (1972) The origin of rocks at the base of the Onaping Formation, Sudbury, Ontario. Unpublished Ph.D. thesis, University of Toronto, Toronto, Ontario, 366 pp

Peredery WV, Morrison G (1984) Discussion of the origin of the Sudbury Structure. In: Pye EG, Naldrett AJ, Giblin PE (eds) The geology and ore deposits of the Sudbury Structure, Ontario Geological Survey, Special Volume 1: 491-512

Pye EG, Naldrett AJ, Giblin PE eds (1984) The geology and ore deposits of the Sudbury Structure: Ontario Geological Survey, Special Volume 1, 603 pp

Riller U, Cruden AR, Schwerdtner WM (1996) Magnetic fabric and microstructural evidence for a tectono-thermal overprint of the Murray pluton, central Ontario, Canada. Journal of Structural Geology 18: 1005-1016

Rogers DF, Gibson HL, Whitehead R , Checetto J, Jonasson IR (1995) Structure and metal enrichment in the hanging wall carbonaceous argillites of the Paleoproterozoic Vermilion and Errington Zn-Cu-Pb masive sulfide deposits, Sudbury, Canada [abs.]. Geological Association of Canada- Mineralogical Association of Canada, Program with Abstracts 20: A90

Rousell DH (1975) The origin of foliation and lineation in the Onaping Formation and the deformation of the Sudbury Basin. Canadian Journal of Earth Sciences 12: 1379-1395

Rousell DH (1984a) Onwatin and Chelmsford Formations. In: Pye EG, Naldrett AJ, Giblin PE (eds) The geology and ore deposits of the Sudbury Structure: Ontario Geological Survey, Special Volume 1: 211-218

Rousell DH (1984b) Mineralization in the Whitewater Group. In: Pye EG, Naldrett AJ, Giblin PE (eds) The geology and ore deposits of the Sudbury Structure, Ontario Geological Survey, Special Volume 1: 219-233

Rousell DH, Gibson HL Jonasson IR (1997 The tectonic, magmatic and mineralization history of the Sudbury Structure. Exploration and Mining Geology 6: 1-22
Sagalevich AM, Torokhov PV, Matveyenkov VV, Galkin SV, Moskalev LI (1992) Hydrothermal manifestations at Pyip submarine volcano, Bering Sea. International Geology Review 34: 1200-1209
Schiffman P, Fridleifsson GO (1991) The smectite-chlorite transition in Drill Hole NJ-15, Nesjavellir geothermal field, Iceland: XRD, BSE and electron microprobe investigations. Journal of Metamorphic Geology 9: 679-696
Shanks WS, Schwerdtner WM (1991) Structural analysis of the central and southwestern Sudbury Structure, Southern Province, Canadian Shield. Canadian Journal of Earth Sciences 28: 411-430
Spry PG, Peter JM, Slack JF (2000) Metaexhalites as exploration guides to ore. In: Spry PG, Marshall B, Vokes FM (eds) Metamorphosed and metamorphogenic ore deposits. Reviews in Economic Geology 11:163- 201
Stöffler D, Avermann M, Bischoff L, Brockmeyer P, Deutsch A, Dressler BO, Lakomy R, Müller-Mohr V (1989) Sudbury, Canada: Remnant of the only multi-ring (?) impact basin on earth [abs]. Meteoritics 24: 328
Stöffler D, Deutsch A, Avermann M, Bischoff L, Brockmeyer P, Buhl D, Lakomy R, Müller-Mohr V (1994) The formation of the Sudbury Structure, Canada: toward a unified impact model. In: Dressler BO, Grieve RAF, Sharpton VL (eds) Large meteorite impacts and planetary evolution. Geological Society of America Special Paper 293: 303-318
Stoness JA (1994) The stratigraphy, geochemistry and depositional environment of the Proterozoic Vermilion Formation, host to the Errington and Vermilion Zn-Cu-Pb massive sulfide deposits, Sudbury, Ontario. Unpublished MSc thesis, Laurentian University, Sudbury, Ontario 205 pp
Sturkell EFF, Broman C, Forsberg P, Torssander P (1998) Impact-related hydrothermal activity in the Lockne impact structure, Jamtland, Sweden. European Journal of Mineralogy 10: 589-606
Thompson LM, Spray JG (1994) Pseudotachylitic rock distribution and genesis within the Sudbury impact structure. In: Dressler BO, Grieve RAF, Sharpton VL (eds) Large meteorite impacts and planetary evolution, Geological Society of America Special Paper 293: 275-287
Von Damm KL, Edmond JM, Grant B, Measures CI (1985) Chemistry of submarine hydrothermal solutions at 21°N, East Pacific Rise. Geochimica et Cosmochimica Acta 49: 2197-2220
Walter MR, Des Marais DJ (1993) Preservation of biological information in thermal spring deposits: Developing a strategy for the search for a fossil record on Mars. Icarus 101: 129-143
Watkinson DH (1990) Platinum-group minerals in Cu-rich sulfide from some North American mafic-ultramafic rock complexes. Mineralogical Society of Korea: 6: 37-51
Watkinson DH (1994) Fluid-rock interaction at contact of Lindsley 4b Ni-Cu-PGE orebody and enclosing granitic rocks, Sudbury, Canada. Transactions of Mining and Metallurgy 103: B121-128

Watkinson DH (1999) Platinum-group-element enrichment in Cu-Ni-rich sulfides from footwall deposits, Sudbury Igneous Complex, Canada. Chronique de la Recherche Miniere 535: 29-43

Whitehead RES, Davies JF, Goodfellow WD (1990) Isotopic evidence for hydrothermal discharge into anoxic seawater, Sudbury Basin, Ontario. Chemical Geology 86: 49-63

Wodicka N (1997) Sudbury Structure: Northern footwall rocks and Sudbury Igneous Complex. In: Ames, DE, Bleeker, W, Heather, KB, Wodicka, N (eds) Timmins to Sudbury transect: New insights into the regional geology and the setting of ore deposits, Geological Association of Canada Fieldtrip Guidebook B6: 73-93

Wohletz KH (1983) Mechanisms of hydrovolcanic pyroclast formation: Grain size, scanning electron microscopy and experimental studies. Journal of Volcanological and Geophysical Research 17: 31-63

Wu J, Milkereit B, Boerner DE (1994) Timing constraints on deformation history of the Sudbury impact structure. Canadian Journal of Earth Science 31: 1654-1660

Young RA (1996) Chelmsford, its forgotten mines: Unpublished report, Chelmsford, Ontario, 85 pp

Comparison of Bosumtwi Impact Crater (Ghana) and Crater Lake Volcanic Caldera (Oregon, USA): Implications for Biotic Recovery after Catastrophic Events

Michael R. Rampino[1] and Christian Koeberl[2]

[1]Earth and Environmental Science Program, New York University,
100 Washington Square East, New York, NY 10012, USA (mrr1@nyu.edu)
[2]Department of Geological Sciences, University of Vienna, Althanstrasse 14,
A-1090 Vienna, Austria (christian.koeberl@univie.ac.at)

Abstract. Impact craters and volcanic collapse calderas involve catastrophic processes that cause destruction of plant and animal communities in the surrounding areas, leading to new habitats that are initially barren. These events are followed by variable periods of recovery of the biota. Impact craters and calderas are rather similar in form, structure and internal geometry. Furthermore, they are commonly sites of crater lakes. In order to shed light on the processes involved in the destruction and the phases of recovery of biota after a catastrophic impact, we compared the similar-sized (~10 km diameter) Crater Lake (Oregon) caldera and the Bosumtwi (Ghana) impact crater lake. Crater Lake was produced by a major caldera-forming explosive eruption 6,845±50 radiocarbon years BP. A lake was established within about 150 years after the caldera collapse, during an early stage of hydrothermal activity. Palynological studies of lake sediments show that forests had become re-established on the higher ground around Crater Lake by about 300 years after the eruption. Bosumtwi Crater was formed by the impact of an object roughly 0.5 km in diameter that instantaneously released about 4 x 10^{19} J of energy. Impact-induced hydrothermal systems may have been active for thousands of years. Volcanic hydrothermal systems may be much longer lasting, and recurrent volcanic activity may continue to disturb the environment around calderas for extended periods of time. These factors could have led to different re-colonization patterns at impact craters and volcanic calderas.

1
Introduction

Volcanic calderas may provide good analogs for the recovery processes that take place after large impacts. Calderas such as Crater Lake, Oregon (USA), are formed by collapse along circular ring fractures after paroxysmal eruptions, and are floored by a mixture of collapse breccias and intracaldera pyroclastic material (Nelson et al. 1988, 1994). By contrast, impact craters are produced by shock-induced explosion and the subsequent collapse of the transient crater cavity, and are floored by impact breccias and impact melt (Melosh 1989).

Many impact craters and calderas subsequently become filled or partly filled with water to form lakes, such as Lake Bosumtwi, Ghana, and Crater Lake, Oregon, in which similar kinds of lacustrine sediments may be deposited. The earliest post-event lake sediments may hold a record of the environmental effects of the impact or eruption and the nature of the initial biologic recovery in the surrounding area (Cockell and Lee 2002). Several historic explosive and caldera-forming eruptions caused destruction for which there are good records of the stages of biological recovery of plants and animals in the surrounding areas (e.g., Krakatoa; Thornton 1996, 2000; Mt St Helens; Del Moral and Ward 1988; Shiro and Roger 1995).

Calderas have hydrothermal systems localized along fractures and may suffer subsequent volcanism. The hydrothermal systems produced at impact craters are related to the cooling of buried melt bodies, and may be very long-lived (Gurov 1996; Osinski et al. 2001). The nature of these hydrothermal systems may be relevant to origin of life scenarios on the Earth, Mars and other planets.

Impact craters and volcanic collapse calderas have similarities in form, structure and internal geometry. They are produced rapidly by catastrophic processes that destroy the biota in the surrounding areas, followed by extended periods of biotic recovery. In order to better ascertain the environmental effects and the processes involved in the recovery of the local biota after an asteroid impact, we compare the Bosumtwi (Ghana) impact crater with the similar-sized well-studied Crater Lake, Oregon (USA) volcanic caldera (Table 1).

Table 1. Comparison of Crater Lake, Oregon (USA) and Bosumtwi Impact Crater (Ghana)

	Crater Lake	**Bosumtwi**
Diameter (km) (rim-rim)	10	10.5
Age (Ma)	0.006	1.07
Lake Depth (m)	622	80
Structure	Caldera formed by collapse along ring fractures. Renewed activity formed volcanic constructs on caldera floor.	Complex impact structure. Crater formed by collapse of transient cavity. Central uplift (ca. 1.5 km in diameter) buried beneath lake sediments
Deposits	Floored by collapse breccia and intra-caldera tuff. Lake sediments only 20 to 40 m thick. Pyroclastic flow deposits in surrounding area up to 60 km from vent.	Lake sediments (up to ca. 300 m thick) underlain by polymict impact breccia and (possibly) impact melt rocks to ca. 800 m depth.
Hydrothermal Systems	Along ring fractures for at least a few hundred years. Eruptions up to 2,500 years after main eruption	Possibly for several thousand years after the impact event. No present-day systems.
Lake Conditions	Thermally stratified and phosphate-rich conditions in early lake.	Chemically stratified lake with anoxic below ca. 5 m water depth. Occasional turnover of anoxic bottom waters.

2 Bosumtwi Impact Crater (Ghana)

Bosumtwi is one of only 19 currently confirmed African impact craters (Koeberl 1994; Master and Reimold 2000), and one of only four known impact craters associated with tektite strewn fields (Koeberl et al. 1997a). The 1.07 Ma Bosumtwi crater (centered at 06°30'N and 01°25'W) is situated in the Ashanti Region of Ghana, West Africa, about 32 km from Kumasi, the regional capital. It is a well-preserved complex impact structure that displays a pronounced rim, and is almost completely filled by the 8 km diameter Lake Bosumtwi (Fig. 1). The crater is excavated in 2 Ga old metamorphosed and crystalline rocks of the Birimian Supergroup (Junner 1937). Bosumtwi has a rim-to-rim diameter of about 10.5 km. The crater is surrounded by a slight near-circular depression, and an outer ring of minor topographic highs with a diameter about 20 km (Jones et al.

1981; Garvin and Schnetzler 1994; Reimold et al. 1998; Wagner et al. 2002).

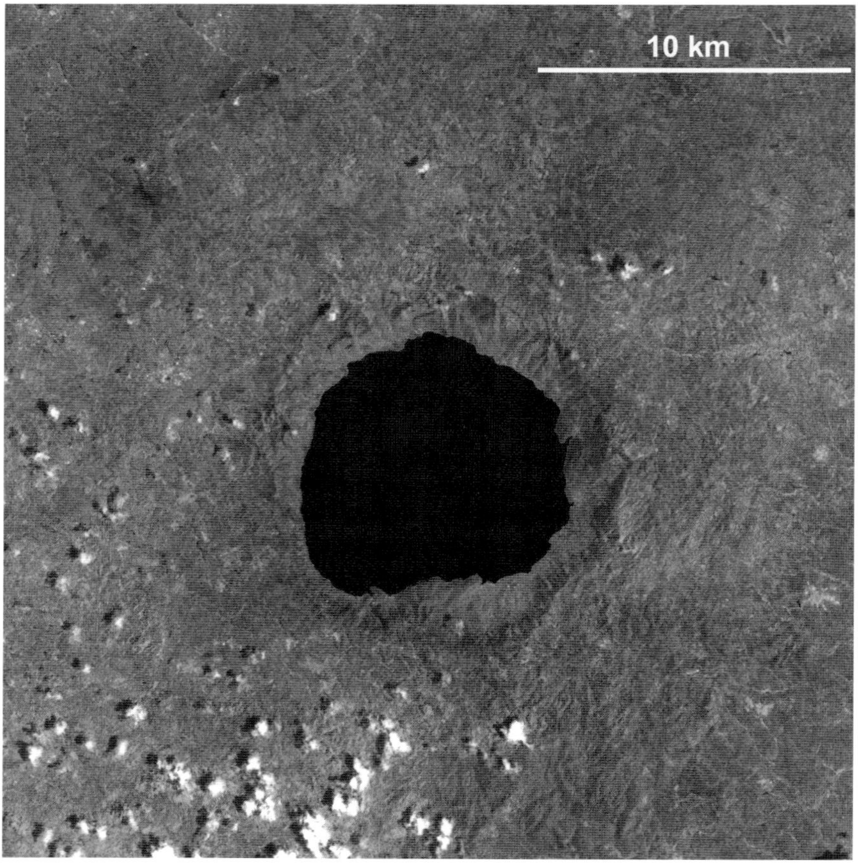

Fig. 1. Aster satellite image (a contrast-enhanced version of bands 7, 3, and 1 multispectral combination) of Bosumtwi Crater (Ghana) showing Lake Bosumtwi. Clouds show up in white, with shadows. North is up.

The origin of the Bosumtwi crater was for a long time controversial (e.g., Junner 1937; Jones 1985). Early geologists favored a volcanic origin for the crater, most likely because impact craters were not fully appreciated at that time. Renewed interest in the 1960s led to the discovery of the high-pressure silica polymorph coesite (Littler et al. 1961), Ni-rich iron spherules, and baddeleyite (a high-temperature decomposition product of zircon) within vesicular glass in suevite taken from the crater rim (El

Goresy 1966; El Goresy et al. 1968), and shocked quartz (Chao 1968). These findings provided support for the impact hypothesis. Recently, further evidence of shock effects in the clasts within the Bosumtwi suevites has been described (Koeberl et al. 1998).

Detailed geological studies and mapping of the region around Lake Bosumtwi have been carried out since the 1930s (Junner 1937; Woodfield 1966; Moon and Mason 1967; Jones et al. 1981). More recent geological studies were done along a section across the western crater rim and on exposures in the sector around the northern and northeastern parts of the crater (Reimold et al. 1998). Dense, tropical rainforest and shrubs largely cover the region around Bosumtwi. Thus, only studies of rare exposures along streams and road cuts are possible.

Target rocks at the Bosumtwi impact crater comprise mainly lower greenschist facies metasediments of the 2.1-2.2 Ga Birimian Supergroup (cf. Wright et al. 1985; Leube et al. 1990). These rocks encompass interbedded phyllites and meta-tuffs together with meta-graywackes, quartzitic graywackes, shales and slates. Birimian metavolcanic rocks (altered basic intrusives with some intercalated metasediments) reach out to the southeast of the crater. Tarkwaian clastic sedimentary rocks, which are regarded as the detritus of Birimian rocks (Leube et al. 1990), occur to the east and southeast of the crater. Relatively recent strata include the Bosumtwi lakebeds, as well as soils and breccias associated with the formation of the crater. Massive suevite deposits have been observed just outside the northern and southwestern crater rim (e.g., Boamah and Koeberl 2003, and references therein).

Substantial interest in the Bosumtwi crater has also come from studies of the Ivory Coast tektites, which occur in an area of about 40 km radius in Cote d'Ivoire, West Africa. In addition to the tektites found on land, related microtektites are found in deep-sea cores off the coast of West Africa (Glass 1968, 1969). A variety of arguments have been used to conclude that the Bosumtwi impact crater is the source of the tektites (Kolbe et al. 1967; Saul 1969; Jones 1985; Koeberl et al. 1997a, 1998). These include the similarities in chemical and isotopic compositions, and identical ages for tektites and Bosumtwi impact glasses.

Research has intensified recently with regard to a number of aspects of the Bosumtwi crater. These studies include the petrology and geochemistry of the target rocks (Koeberl et al. 1998), structural analysis of the crater rim (Reimold et al. 1998), high-resolution aero-geophysical surveys including measurements of the total magnetic field, electromagnetic field, and gamma radiation across the structure (Koeberl et al. 1997b; Ojamo et al. 1997; Pesonen et al. 1998, 1999, 2003; Plado et al. 2000).

New studies also involve the geochemistry of soils from the Bosumtwi structure (Boamah and Koeberl 2002) and of suevites from shallow drill cores outside the north side of the crater rim (Boamah and Koeberl 2003), and a large-scale remote-sensing investigation (Wagner et al. 2002). U.S. and German research teams recently conducted detailed land-based and lake-based geophysical studies in the area (e.g., Scholz et al. 2002).

The high-resolution, low altitude (~70 m) airborne geophysical survey mentioned above also yielded magnetic data (e.g., Pesonen et al. 1998, 2003), which show a circumferential magnetic halo outside the lakeshore, ~12 km in diameter. The central-north part of the lake reveals a central negative magnetic anomaly with smaller positive side-anomalies N and S of it, which is typical for magnetized bodies at shallow latitudes. A few weaker negative magnetic anomalies exist in the eastern and western part of the lake. Together with the northern one they seem to encircle a central uplift. The model of Plado et al. (2000) shows that the magnetic anomaly of the structure is presumably produced by one or more relatively strongly remanently magnetized impact melt rock or melt-rich suevite bodies.

The multichannel seismic reflection (MCS) data that were acquired by Karp et al. (2002) and Scholz et al. (2002) reveal a well-defined central uplift near the NW-central part of the lake, and a maximum lacustrine sediment thickness of ~310 m (Fig. 2). They also found that the central uplift structure has a diameter of 1.9 km and a maximum height of 130 m above the annular moat inside the crater, and has undergone faulting, probably during the latter stages of transient crater collapse and during the subsequent lacustrine phase of the structure. An intermediate velocity layer (3200 m/s) beneath the lacustrine sediment was interpreted by these authors as fallback breccia or a breccia-melt horizon, and the internal seismic velocity structure was determined from the wide-angle experiment. Scholz et al. (2002) derived an apparent depth of the crater (d_a) of 500 m, implying a slightly different diameter to depth ratio for the structure than predicted from classical scaling laws.

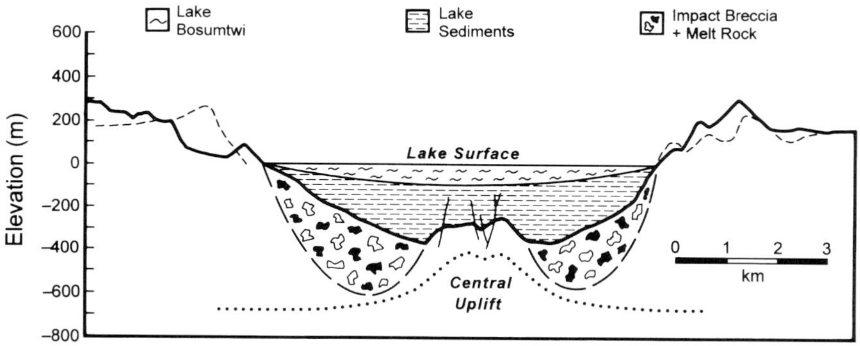

Fig. 2. East-west cross-section of the Bosumtwi impact structure, Ghana. Solid lines indicate actual topography; dashed line indicates alternative topographic cross-section from the north-west (right) to the south-east (left), with the Obuom mountain range on the left side. Vertical exaggeration 4x. The extent and location of the central uplift is derived from seismic data of Karp et al. (2002) and Scholz et al. (2002). The distribution of the crater fill breccia is estimated from the seismic data, and the crater floor location is extrapolated from seismic velocity data together with expectations from other craters. Actual drilling showed that the section designated as 'Impact Breccia and Melt Rock' consists of both *in situ* brecciated rocks as well as fallback breccia.

Lake Bosumtwi is hydrologically closed at the present time, with water balance dominated by rainfall on the lake surface and direct evaporation. Groundwater sources are thought to be negligible. The highest lake sediments occur about 110 m above the present lake level. The lowest point on the crater rim has an elevation of 210 m, about 110 m above the present lake level, which is the elevation at which the lake will overflow (Turner et al. 1996). Low salinity of about 1 per mil suggests that the dissolved material was removed by lake overflow in the relatively recent geologic past. Talbot and Delibrias (1977) and Talbot and Johannessen (1992) showed that lake-level variations correlated well with rainfall in the Sahel region. Short cores show that sediments in the deep basin of the lake are typically varved and contain sapropels. Turner et al. (1996) suggested that rapid increases in lake level might trigger episodes of sapropel deposition as a result of the rapid drowning of forests and introduction of lignin-rich biomass to the deep lake basin.

Bosumtwi is the target of an international deep drilling project, which took place from June to October of 2004, and recovered more than 2 km of core material, which will provide data for the study of impact-related information of the crater, paleoclimatic information over the past million years, and also cover geophysical and astrobiological aspects.

3
Crater Lake, Oregon (USA)

Crater Lake, Oregon (U.S.A.) (Fig. 3a), occupies a 1200-meter-deep and 8 x 10 km diameter caldera that was formed by the collapse of Mount Mazama volcano during an eruption that occurred a few thousand years ago (Fig. 3b). The present Crater Lake is 622 m deep (i.e., significantly deeper than Lake Bosumtwi at the present time). Mount Mazama itself was a composite volcano made up of several overlapping shield and stratovolcanoes composed of andesite and dacite lava flows and pyroclastic deposits (Williams 1942; Bacon 1983). The main cone-building phase of the volcanism took place over the long period between about 420,000 and 50,000 years ago (Nelson et al. 1994).

The major caldera-forming explosive eruption (~55 km^3 dense-rock equivalent magma erupted) is dated at 6845±50 radiocarbon years BP. The eruption began with a single-vent phase, during which about half of the total amount of magma released was erupted as airfall pumice and ash (Bacon 1983). The pumice and ash were deposited over at least one million km^2, mostly to the north and northeast of the volcano. The eruption column eventually collapsed and generated at least four pyroclastic flow units that cooled together to form the Wineglass Welded Tuff.

As the eruption continued, the magma chamber was partially evacuated, which led to caldera collapse along a system of circular or arcuate ring fractures. During this phase of the eruption, pyroclastic flows deposited pumiceous ignimbrite (and coarse lithic breccia in and near the caldera) (Druitt and Bacon 1986), and distal facies ignimbrite as far as 60 km from the caldera (Nelson et al. 1988).

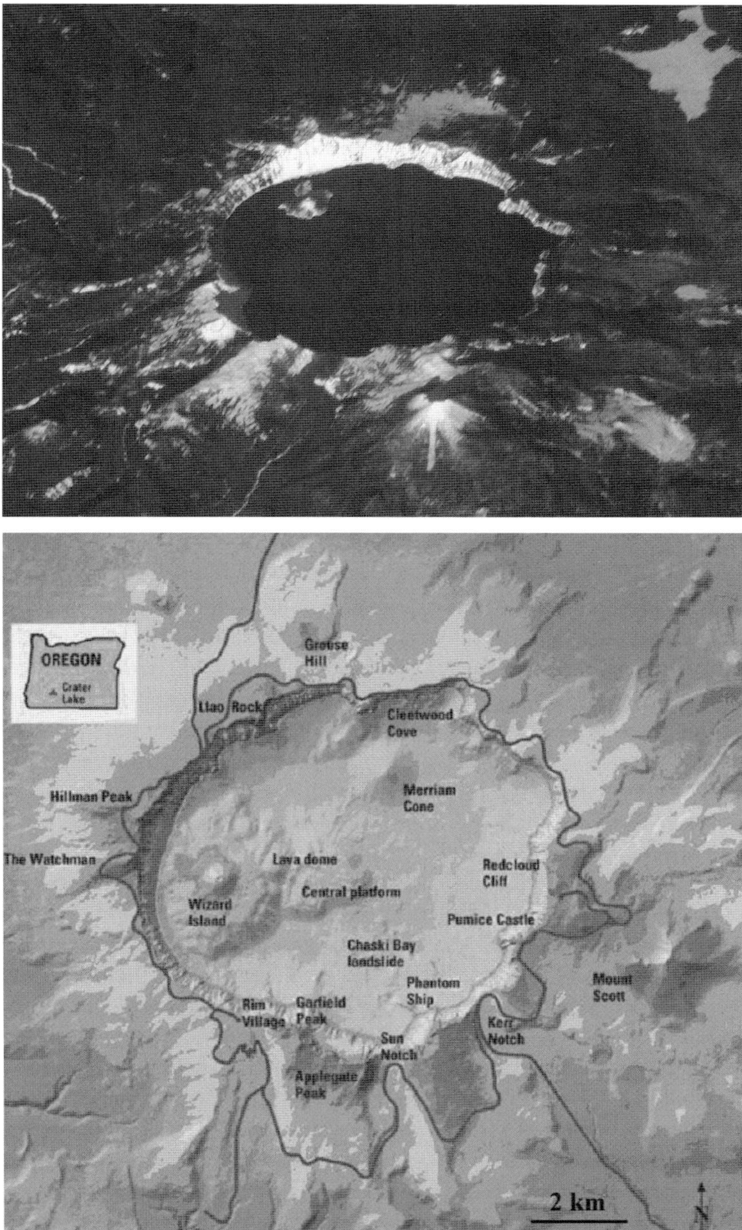

Fig. 3. a) View of Crater Lake from space (International Space Station photograph). b) Map of Crater Lake, Oregon, showing major morphologic features and extent of the caldera lake (Klimasauskas et al. 2002).

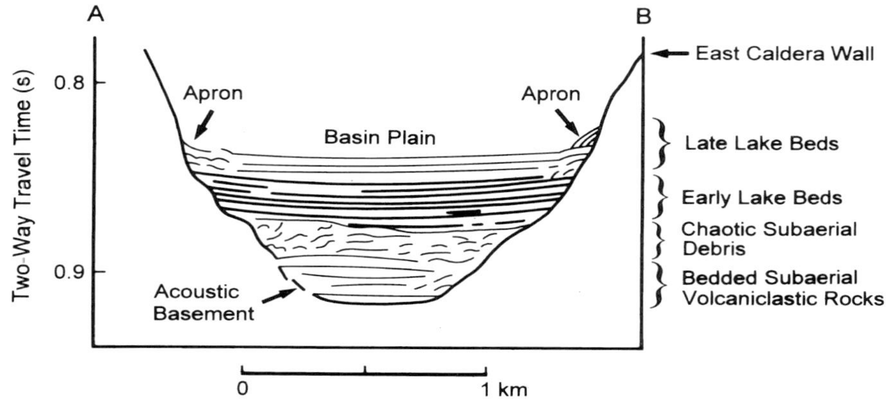

Fig. 4. Interpreted airgun (16.4 cm^3) profile oriented east to west across the east basin of Crater Lake, Oregon (after Nelson et al. 1994).

This ring-vent major eruptive phase was followed by a period of explosive phreatic eruptions within the intra-caldera debris. Renewed volcanic activity formed the central platform, Wizard Island and the Merriam Cone, which are the three volcanic edifices that rise from the floor of the caldera (Nelson et al. 1988). The 5-km-diameter elliptical ring-fracture along which the volcano collapsed is outlined by the phreatic craters, sites with observed hydrothermal activity and high heat flow, and the summit craters of Wizard Island and the Merriam Cone (Fig. 3b) (Nelson et al. 1994).

Within a short time span (probably only a few years), subaerial deposits filled the phreatic craters and then formed a wedge of mass-flow deposits shed from the caldera walls. Intense volcanic activity (phreatic explosions, subaerial lava flows, and hydrothermal activity) occurred during the early post-caldera stage, and a central platform of subaerial andesite flows and scoria formed on the caldera floor. Intra-caldera tuff and interbedded landslide deposits are about 2 km thick in the subsided portion of the caldera as can be seen in seismic reflection profiles (Fig. 4) (Bacon and Lanphere 1990). The caldera has been widened by partial collapse of the walls from an original collapse feature about 5 km in diameter to the present 10 km by 8 km topographic depression (Fig. 5). Nearly all of the collapse debris was apparently deposited by the end of the catastrophic caldera-forming eruption (Nelson et al. 1988, 1994).

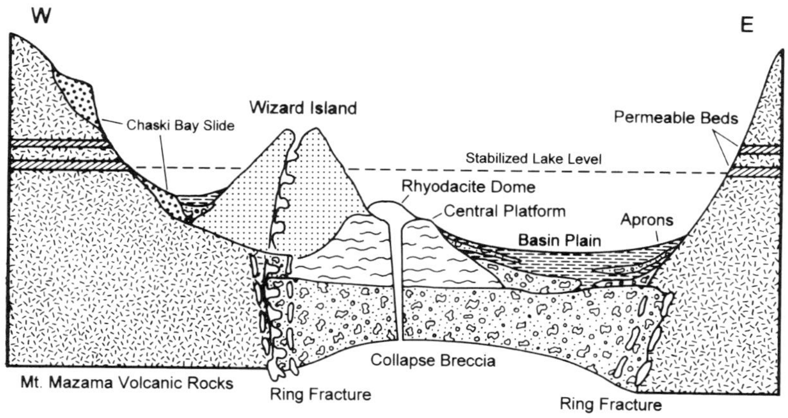

Fig. 5. Schematic cross section across the caldera floor of Crater Lake, Oregon (oriented east-west) (not to scale) (after Nelson et al. 1994).

Radiocarbon ages from short cores (< 2 m long) taken on the elevated central platform in the suggest that the deposition of open lacustrine sediment began on the central platform of the lake floor about 150 years after the caldera collapse (Nelson et al. 1994), which is also the minimum time estimated for the lake to fill halfway with water (Phillips and Van Denburgh 1968). Coarse-grained and thick-bedded turbidites within the early lakebeds in the deeper basins of the lake were deposited during the period of post-collapse volcanism when the sub-aqueous domes and cones were formed.

The most recent known post-caldera volcanic activity produced a subaqueous ash layer and an associated extrusive lava dome about 4240 radiocarbon years BP. During the volcanically quiescent period of the last 4,000 years, lakebeds with base-of-slope aprons and thin fine-grained basin-plain turbidites have been deposited (Nelson et al. 1994).

4 Impacts and Volcanic Eruptions Compared

The energy involved in the Bosumtwi impact is estimated to have been about 10^4 megatons of TNT (1 Mt TNT = 4.2 x10^{15} Joules), which is equal to about 4.2 x 10^{19} Joules (Jones et al. 1981). The volume of magma released in the Crater Lake eruption is 5.5 x 10^{10} m^3. This gives the eruption a Volcanic Explosivity Index (VEI) value of about 6.5 (Pyle 1995). The thermal energy released in the Crater Lake eruption has been estimated at between 10^{19} and 10^{20} Joules (Pyle 2000). Thus, the total amount of energy released in the impact and in the eruption is of the same order of magnitude. The major difference, of course, is that the impact releases its energy instantaneously, and the energy is a mixture of thermal and mechanical energy, whereas the eruption released its (mainly thermal) energy over hours to at most days.

The present depth/diameter ratio for Bosumtwi is ~0.048 (0.5 km/10.5 km), whereas the theoretical depth/diameter ratio for complex craters < 11 km in diameter is given by depth = 0.196 $D^{1.01}$, i.e., 2.1 km/10.5 km = 0.2 (Melosh 1989). The depth/diameter ratio of Crater Lake caldera is considerably greater than that observed for Bosumtwi, about 0.1 (1.2 km/10 km), and this may indicate an important genetic difference between impact craters and collapse calderas. Diverse subsidence geometries and collapse processes in calderas reflect the varying sizes and depths of the source magma chambers, and the geometry and materials of the roof of the chamber. Regional volcanic and tectonic conditions are also important factors.

For impacts, the area of lethal destruction corresponds to $A = 100Y^{2/3}$, where Y is the yield in Mt TNT and A is the area in km^2 (Chapman and Morrison 1994). For Bosumtwi, the area of lethal damage to plants and animals would have been about 40,000 km^2, or a circle with a radius of about 110 km. For volcanic eruptions, lateral blasts (as at Mt. St. Helens in 1980) and pyroclastic flows may reach up to several tens of kilometers from the crater (Nakada 2000). For the Crater Lake eruption, pyroclastic flows in some directions traveled more than 60 km from the vents (Wood and Baldridge 1990). A maximum area of destruction for an eruption like the Mazama event would have been about 10,000 km^2.

High-speed ground-hugging ejecta are formed in both volcanic eruptions and impacts. Field and petrographic data from the large Late Triassic Manicouagan impact crater (estimates of the size range from 65 to ~100 km diameter) in Canada suggested that the emplacement of the impact melt sheet was accompanied by erosion of the surrounding area at a rate of ~2600 kg $m^{-2}s^{-1}$. The average depth of erosion at Manicouagan was ~39 m.

Volcanic eruptions can also strip away the surface material, but on a smaller scale. Data for the erosion rates from pyroclastic flows range from about 14 to 21 $kgm^{-2}s^{-1}$. For the much smaller Mt. St. Helens lateral blast on May 18, 1980, the depth of erosion is estimated to have been about 0.3 to 0.5 m, with deeper furrows showing up to 1.2 m of erosion. Thus, the blast stripped away forest and soil down to tree root levels. A later pyroclastic flow at Mt. St. Helens eroded volcanic deposits to a depth of about 2 m (Simonds and Kieffer 1993). Volcanic eruptions (Smith 1991) and impacts (Bootsman et al. 1999) can also disrupt regional drainage and alter landscapes over large areas.

5 Recovery

Cockell and Lee (2002) proposed three stages of biological recovery following impacts: 1) The Phase of Thermal Biology in which the crater represents a thermal anomaly that commonly drives a vigorous hydrothermal circulation. For the Haughton impact structure (24 km in diameter), the cooling of the crater basement is estimated to have taken several thousand years (Osinski et al. 2001). Hydrothermal precipitates have been observed in impact crater lakes such as Boltysh, Ukraine (Gurov 1996). The biota during this stage may be largely limited to thermophilic microbes and other simple organisms that can exist in very hot acidic environments. Hydrothermally induced upwelling of lake waters could lead to blooms of microorganisms. 2) The Phase of Impact Succession and Climax, marked by a series of succession events (lacustrine and terrestrial) culminating in a climax ecology; and 3) The Phase of Ecological Assimilation, in which the infilling and erosion of the crater leads to disappearance of the crater lake and loss of ecologic distinctiveness of the crater area.

We can compare these stages with the documented recovery periods after volcanic eruptions. Large caldera-forming eruptions, such as the Mazama (Crater Lake) event, produce dry, barren, and in some cases, chemically alkaline environments. There is large-scale loss of vegetation and soils, and the ground is covered by fresh pyroclastic material that is extremely porous and has no organic material. These areas are subject to rapid erosion by water and wind. Soil production depends upon weathering, influx of nutrients and fungi, and disturbance by burrowing animals.

Diatoms in the Crater Lake sediments indicate that the early lake contained some comparatively alkaline and high salinity environments (Nelson et al. 1994). This suggests a local contribution from mineralized springs associated with terminal phases of the volcanism. Iron-rich precipitates at the base of the lake sediment section also suggest active hydrothermal systems for at least a few hundred years after the main eruption (Dymond et al. 1988; Nelson et al. 1994). The dominance of the planktonic diatom *Stephanodiscus* at the beginning of Crater Lake's history suggests a significant source of phosphorus. This may have been a result of thermal spring water mixing with cooler meteoric water, a situation that would have produced lake water with a low pH, thus favoring high dissolved phosphate concentrations (Stauffer and Thompson 1978).

Study of pollen and diatoms from cores taken through the lake sediments on the central platform and on the basin plain of Crater Lake allows a reconstruction of the vegetative and lacustrine history of the post-eruption period. The most common pollen types found at the base of the lake sediment sequence (*Pinus* and *Tsuga*, and the plant families *Taxodiacea, Cupressaceae* and *Taxaceae*) are wind-dispersed pollen, and probably represent transport of pollen into the lake from distant forests. The increase in *Abies* (true fir) pollen to values > 5% during the first ~300 years following the eruption suggests that *Abies* had become re-established on the higher ground around Crater Lake by that time (Nelson et al. 1994).

One of the best examples of the recovery of climax vegetation after an explosive volcanic eruption is the 1883 eruption of Krakatoa (or Krakatau) in Indonesia. Krakatoa was a volcanic island in the Sunda Straits between Java and Sumatra that erupted explosively in August 1883. After the cataclysmal eruption, three smaller islands remained, and all were covered with thick, hot pyroclastic flow deposits that destroyed essentially all life.

The recovery of vegetation in areas covered by the pyroclastic flows at Krakatoa is well documented (Thornton 1996, 2000). The reestablishment of plant and animal communities on the three islands required the transfer of biota from the coasts of Java and Sumatra more than 40 kilometers away. In the first three years after the 1883 eruption, plants that were waterborne or windborne made up the only vegetation. Fruiting fig trees (seeds carried by birds) were established within 14 years of the eruption, and animals arrived by rafting during the same period. Climax forests were established in some parts of the Krakatoan islands by the 1930s, about 40 years after the eruption (Ward and Thornton 2000). In the case of Krakatoa, the climate was tropical. A more comparable temperate climate case was the destructive Taupo (New Zealand) eruption of 1800 years BP, with a volume of 35 km^3. Studies suggest that full vegetation cover was reestablished in 15 to 20 years after Taupo (Smith 1991).

6 Summary and Conclusions

The Bosumtwi impact crater in Ghana, West Africa, and the Crater Lake, Oregon, USA, volcanic caldera are of similar size and both filled by lakes. A comparison between the two structures allows us to make some observations and suggestions regarding the biological consequences of natural catastrophic events of different origin (impact vs. volcanism). Table 1 summarizes some of the basic characteristics of the two geological features and lists similarities and differences.

The diameters of both structures, and the lake diameters, are about equal. In contrast, the apparent depth of the crater (ca. 0.5 km for Bosumtwi, but 1.2 km for Crater Lake) are quite different, and the current lake depths (ca. 80 m for Bosumtwi and ca. 620 m for Crater Lake) are quite different. The reason for this is primarily the contrasting ages of the structures (1.07 million years for Bosumtwi, versus ca. 6800 years for Crater Lake). The greater age of Bosumtwi would have led to a more pronounced infilling of the crater as a result of erosional processes (310 m of lake sediments on Bosumtwi versus 20 to 40 m of lake beds in Crater Lake). It is also possible that differences in the deformation and fracturing of the rocks when an explosion is induced from below by volcanic action, as opposed to an impact event, lead to differences in slope stability and infilling in the two types of structures.

The biological consequences of the two events have similarities and dissimilarities. In either case, the immediate effects are destruction within a radius of several tens of kilometers from ground zero. The effects are more severe and cover a wider area for impact because of the high-temperature fireball that expands radially around the crater. In the case of a volcanic eruption, the effects are much more directional because of topographic variations that allow directed explosions, pyroclastic flows and mud avalanches to travel in certain preferred directions.

In terms of recolonization, it is surprising that the sites of large volcanic eruptions, and the related destruction, were recolonized by flora and fauna after just a few decades. Thus, it can be inferred that the area around an impact crater of the size of Bosumtwi may have been recolonized after a similar span of time. What are not known are the length of time it took for the impact crater lake to become established, and the conditions in the newly formed lake. Impact-induced hydrothermal systems could have been active for hundreds to thousands of years, affecting life in the lake and re-colonization, and perhaps to the development of endemic species. The results from the 2004 drilling in Lake Bosumtwi should answer some of these questions.

Further comparisons between volcanic calderas and impact craters are clearly called for, and study of the regional destruction and the recolonization patterns associated with volcanic eruptions and small- to medium-sized impact craters, such as Bosumtwi, has important implications for the development of life on Earth and, possibly, also elsewhere in the solar system.

Acknowledgments

We thank David King and M. Böhme for helpful reviews. This work was partly supported by the Austrian Science Foundation (project P17194-N10), and is a contribution to the International Continental Scientific Drilling Program (ICDP) Bosumtwi drilling project. We are grateful to D. Jalufka (University of Vienna) for drafting the diagrams.

References

Artemieva N (2002) Tektite origin in oblique impacts: Numerical modeling of the initial stage. In: Plado J, Pesonen LJ (eds) Meteorite impacts in Precambrian Shields. Impact Studies, vol. 2, Springer, Heidelberg, pp 257-276

Bacon CR (1983) Eruptive history of Mount Mazama and Crater Lake caldera, Cascade Range, U.S.A. Journal of Volcanology and Geothermal Research 18: 57-115

Bacon CR, Lamphere MA (1990) The geological setting of Crater Lake, Oregon. In: Drake ET, Larson GL, Dymon, J, Collier R (eds) Crater Lake: An Ecosystem Study. Pacific Division, American Association for the Advancement of Science, San Francisco, California, pp 19-28

Boamah D, Koeberl C (1999) Shallow drilling around the Bosumtwi crater, Ghana: Preliminary results [abs.]. Meteoritics and Planetary Science 34: A13-A14

Boamah D, Koeberl C (2002) Geochemistry of soils from the Bosumtwi impact structure, Ghana, and relationship to radiometric airborne geophysical data. In: Plado J, Pesonen LJ (eds) Meteorite Impacts in Precambrian Shields. Impact Studies, vol. 2, Springer, Heidelberg, pp 211-255

Boamah D, Koeberl C (2003) Geology and geochemistry of shallow drill cores from the Bosumtwi impact structure, Ghana. Meteoritics and Planetary Science 38: 1137-1159

Bootsman CS, Reimold WU, Brandt D (1999) Evolution of the Molopo drainage and its possible disruption by the Morokweng impact event at the Jurassic-Cretaceous boundary. Journal of African Earth Sciences 29: 669-678

Chao ECT (1968) Pressure and temperature histories of impact metamorphosed rocks - Based on petrographic observations. Neues Jahrbuch Mineralogische Abhandlungen 108: 209-246

Chapman CR, Morrison D (1994) Impacts on the Earth by asteroids and comets: Assessing the hazards. Nature 367: 33-40

Cockell CS, Lee P (2002) The biology of impact craters—a review. Biological Reviews 77: 279-310

Del Moral R, Wood DM (1988) Dynamics of herbaceous vegetation recovery on Mount St. Helens, Washington, USA, after a volcanic eruption. Vegetatio 74: 11-27

Druitt TH, Bacon CR (1986) Lithic breccia and ignimbrite erupted during the collapse of Crater Lake caldera, Oregon. Journal of Volcanology and Geothermal Research 29: 1-32

El Goresy A (1966) Metallic spherules in Bosumtwi crater glasses. Earth and Planetary Science Letters 1: 23-24

El Goresy A, Fechtig H, Ottemann T (1968) The opaque minerals in impactite glasses. In: French BM, Short NM (eds) Shock Metamorphism of Natural Materials. Mono Book Corp., Baltimore, pp 531-554

Garvin JB, Schnetzler CC (1994) The Zhamanshin impact feature: A new class of complex crater? In: Dressler BO, Grieve RAF, Sharpton VL (eds) Large Meteorite Impacts and Planetary Evolution. Geological Society of America Special Paper 293: 249-257

Glass BP (1968) Glassy objects (microtektites?) from deep-sea sediments near the Ivory Coast. Science 161: 891-893

Glass BP (1969) Chemical composition of Ivory Coast microtektites. Geochimica et Cosmochimica Acta 33: 1135-1147

Gurov EP (1996) The Boltysh impact crater: Lake basin with a heated bottom (abs) International Workshop, Tunguska 96. Bologna, Italy, p 41

Hall JB, Swaine MD, Talbot MR (1978) An early Holocene leaf flora from Lake Bosumtwi, Ghana. Palaeogeography, Palaeoclimatology, Palaeoecology 24: 247-262

Jones WB (1985) Chemical analyses of Bosumtwi crater target rocks compared with Ivory Coast tektites. Geochimica et Cosmochimica Acta 49: 2569-2576

Jones WB, Bacon M, Hastings DA (1981) The Lake Bosumtwi impact crater, Ghana. Geological Society of America Bulletin 92: 342-349

Junner NR (1937) The geology of the Bosumtwi caldera and surrounding country. Gold Coast Geological Survey Bulletin 8: 1-38

Klimasauskas E, Bacon CR, Alexander J (2002) Mount Mazama and Crater Lake: Growth and Destruction of a Cascade Volcano. U.S. Geological Survey Fact Sheet 092-02, 4 pp

Karp T, Milkereit B, Janle P, Danuor SK, Pohl J, Berckhemer H, Scholz CA (2002) Seismic investigation of the Lake Bosumtwi impact crater: preliminary results. Planetary and Space Science 50: 735-743

Koeberl C (1994) African meteorite impact craters: Characteristics and geological importance. Journal of African Earth Science 18: 263-295

Koeberl C, Bottomley RJ, Glass BP, Storzer D (1997a) Geochemistry and age of Ivory Coast tektites and microtektites. Geochimica et Cosmochimica Acta 61: 1745-1772

Koeberl C, Reimold WU, Pesonen LJ, Brandt D (1997b) New studies of the Bosumtwi impact structure, Ghana: The 1997 field season [abs.]. Meteoritics and Planetary Science 32: A72-A73

Koeberl C, Reimold WU, Blum JD, Chamberlain CP (1998) Petrology and geochemistry of target rocks from the Bosumtwi impact structure, Ghana, and comparison with Ivory Coast tektites. Geochimica et Cosmochimica Acta 62: 2179-2196

Kolbe P, Pinson WH, Saul JM, Miller EW (1967) Rb-Sr study on country rocks of the Bosumtwi crater, Ghana. Geochimica et Cosmochimica Acta 31: 869-875

Leube A, Hirdes W, Mauer R, Kesse GO (1990) The early Proterozoic Birimian Supergroup of Ghana and some aspects of its associated gold mineralization. Precambrian Research 46: 139-165

Littler J, Fahey JJ, Dietz RS, Chao ECT (1961) Coesite from the Lake Bosumtwi crater, Ashanti, Ghana. Geological Society of America Special Paper 68: 218

Master S, Reimold WU (2000) The impact cratering record of Africa: An updated inventory of proven, probable, possible, and discredited impact structures on the African continent [abs.]. In: Catastrophic Events and Mass Extinctions: Impacts and Beyond. LPI Contribution No. 1053, Lunar and Planetary Institute, Houston, pp 133-134

Melosh HJ (1989) Impact Cratering: A Geologic Process. Oxford University Press, Oxford, 245 pp

Moon PA, Mason D (1967) The geology of ¼° field sheets 129 and 131, Bompata SW. and NW. Ghana Geological Survey Bulletin 31: 1-51

Nakada S (2000) Hazards from pyroclastic flows and surges. In: Sigurdsson H (ed) Encyclopedia of Volcanoes. Academic Press, San Diego, pp 945-955

Nelson CH, Carlson PR, Bacon CR (1988) The Mount Mazama climactic eruption (~6900 BP) and resulting convulsive sedimentation on the Crater Lake caldera floor, continent, and ocean basin. In: Clifton HE (ed) Sedimentologic Consequences of Convulsive Geologic Events, Geological Society of America Special Paper 229: 37-57

Nelson CH, Bacon CR, Robinson SW, Adam DP, Bradbury JP, Barber JH Jr, Schwartz D, Vagenas G (1994) The volcanic, sedimentologic, and paleolimnologic history of the Crater Lake caldera floor, Oregon: Evidence for small caldera evolution. Geological Society of America Bulletin 106: 684-704

Ojamo H, Pesonen LJ, Elo S, Hautaniemi H, Koeberl C, Reimold WU, Plado J (1997) The Bosumtwi Impact Structure, Ghana: International geophysical co-operation at the best [abs.]. In: Kaikkonen P, Komminaho K, Salmirinne H (eds) Sovelletun geofysiikan XI neuvottelupäivät. Oulun yliopisto, Oulu, pp 10-11

Osinski GR, Spray JG, Lee P (2001) Impact-induced hydrothermal activity within the Haughton impact structure, arctic Canada: generation of a transient, warm, wet oasis. Meteoritics and Planetary Science 36: 731-745.

Pesonen LJ, Koeberl C, Ojamo H, Hautaniemi H, Elo S, Plado J (1998) Aerogeophysical studies of the Bosumtwi impact structure, Ghana [abs.]. Geological Society of America, Abstracts with Programs 30(7): A190

Pesonen LJ, Koeberl C, Elo S (2000) The Bosumtwi meteorite impact structure, Ghana: A magnetic model. Meteoritics and Planetary Science 35: 723-732

Pesonen LJ, Koeberl C, Hautaniemi H (2003) Airborne geophysical survey of the Lake Bosumtwi meteorite-impact structure, southern Ghana - the geophysical maps with descriptions. Jahrbuch der Geologischen Bundesanstalt Wien (Yearbook of the Austrian Geological Survey, Vienna) 143: 581-604

Phillips KN, Van Denburgh AS (1968) Hydrology of Crater, East and Davis Lakes, Oregon. U.S. Geological Survey Water Supply Paper 1859 –E, 160 pp

Pyle DM (1995) Mass and energy budgets of explosive volcanic eruptions. Geophysical Research Letters 22: 563-566

Pyle DM (2000) Sizes of volcanic eruptions. In: Sigurdsson H (ed) Encyclopedia of Volcanoes, Academic Press, San Diego, pp 263-269

Reimold WU, Brandt D, Koeberl C (1998) Detailed structural analysis of the rim of a large, complex impact crater: Bosumtwi crater, Ghana. Geology 26: 543-546

Saul JM (1969) Field investigations at Lake Bosumtwi (Ghana) and in the Ivory Coast Tektite strewn field. National Geographic Research Reports, 1964 Projects: 201-212.

Scholz CA, Karp T, Brooks KM, Milkereit B, Amoako PYO, Arko JA (2002) Pronounced central uplift identified in the Bosumtwi impact structure, Ghana, using multichannel seismic reflection data. Geology 30: 939-942

Shiro T, Roger M (1995) Species attributes in early primary succession on volcanoes. Journal of Vegetation Science 6: 517-522

Simonds CH, Kieffer SW (1993) Impact and volcanism: A momentum scaling law for erosion. Journal of Geophysical Research 98: 14,321-14,337

Smith RCM (1991) Landscape response to a major ignimbrite eruption, Taupo volcanic center, New Zealand. Society of Economic Paleontologists and Mineralogists Special Publication 45: 12-137

Stauffer RE, Thompson JM (1978) Phosphorus in hydrothermal waters of Yellowstone National Park, Wyoming. U.S. Geological Survey Journal of Research 6: 755-763

Talbot MR, Delibrias G (1977) Holocene variations in the level of Lake Bosumtwi, Ghana. Nature 268: 722-724

Talbot MR, Johannessen T (1992) A high-resolution paleoclimatic record for the last 27,500 years in tropical West Africa from the carbon and nitrogen isotopic composition of lacustrine organic matter. Earth and Planetary Science Letters 110: 23-37

Thornton IWB (1996) Krakatau—The Destruction and Reassembly of an Island Ecosystem. Harvard University Press, Cambridge, Massachusetts, USA, 346 pp

Thornton IWB (2000) The ecology of volcanoes: Recovery and reassembly of living communities. In: Sigurdsson H (ed) Encyclopedia of Volcanoes. Academic Press, San Diego, pp 1027-1081

Turner BF, Gardner LR, Sharp WE (1996) The hydrology of Lake Bosumtwi, a climate-sensitive lake in Ghana, West Africa. Journal of Hydrology 183: 243-262

Wagner R, Reimold WU, Brandt D (2002) Bosumtwi impact crater, Ghana: A remote sensing investigation. In: Plado J, Pesonen LJ (eds) Meteorite Impacts in Precambrian Shields. Impact Studies vol. 2, Springer Verlag, Heidelberg, pp 189-210

Ward SA, Thornton IWB (2000) Chance and determinism in the development of isolated communities. Global Ecology and Biogeography 9: 7-18

Williams DL, Von Herzen RP (1983) On the terrestrial heat flow and physical limnology of Crater Lake, Oregon. Journal of Geophysical Research 88: 1094-1104

Williams H (1942) Geology of Crater Lake National Park, Oregon. Carnegie Institute of Washington Publication No. 540, 162 pp

Wohletz KH, McGetchin TR, Sanford MT II, Jones EM (1984) Hydrodynamic aspects of caldera-forming eruptions: Journal of Geophysical Research 89: 8269-8286

Wood CA, Baldridge S (1990) Volcano tectonics of the Western United States. In: Wood CA, Kienle J (eds) Volcanoes of North America. Cambridge University Press, Cambridge, pp 147-313

Woodfield PD (1966) The geology of the ¼° field sheet 91, Fumso N. W. Ghana Geological Survey Bulletin 30: 1-66

Wright JB, Hastings DA, Jones WB, Williams HR (1985) Geology and Mineral Resources of West Africa. Allen and Unwin, 187 pp

Paleobiologic Effects of the Late Cretaceous Wetumpka Marine Impact, a 7.6-km-Diameter Impact Structure, Gulf Coastal Plain, USA

David T. King, Jr.[1], Lucille W. Petruny[2], and Thornton L. Neathery[3]

[1]Department of Geology, Auburn University, Auburn, Alabama 36849-5305 USA (kingdat@auburn.edu)
[2]Astra-Terra Research, Auburn, Alabama 36831-3323, USA, and Department of Curriculum and Teaching, Auburn University, Auburn, Alabama 36849 USA (lpetruny@att.net)
[3]Neathery and Associates, 1212-H Veterans Parkway, Tuscaloosa, AL 35404 USA (tlneathery@prodigy.net)

Abstract. The 7.6-km-diameter Wetumpka impact structure in Alabama, USA, formed by a bolide impact within the shallow epicontinental Gulf of Mexico, during the Late Cretaceous (late Santonian to early Campanian). Water depths for the epicenter of this event are estimated to have been 30 to 100 m, and this feature probably formed within ~ 25 km of the local barrier-island shoreline. All indications are that this impact would have been a locally devastating event. For example, the infra-red flash-burn radius extended to the local shore area and substantial seismic waves followed by a strong atmospheric blast wave would have reached the shoreline tropical forest very shortly after impact. Further, the radius of discontinuous ejecta would have extended beyond the local shoreline and other effects such as local tsunami run-up extended into the local shore-area tropical forest as well. The lower crater-filling unit of Wetumpka contains an impact-entombed fossil record: there is considerably more lignite (fragments and finely divided material) in some of the impact breccias and sands than existed in any of the target strata. This implies that some comminuted wood from coeval, shore-area tropical forests was swept up within the returning marine-water washback-surgeback event -- or was air-borne in returning winds -- and thus became incorporated into the deep parts of the crater fill. Wetumpka's crater fill also includes mixed trace and body fossils from slumped target materials and blocks displaying relic sedimentary facies of target units. Since impact, Wetumpka's rim structure, mainly composed of crystalline basement rocks, appears to have been an enduring subaerial feature. Rim height estimates indicate that this crystalline feature (comprised today of a remnant ~ 270° arc of elevated schists and gneisses) was subaerially exposed after impact. Deep tropical soil development (saprolitization) characterizes most of the higher elevations of the rim, probably attesting to long-term rim exposure since Late Cretaceous. We theorize that a limited terrestrial ecosystem could have existed at higher elevations upon the crater floor area as well during the interval between end of fall-back sedimentation and subsequent late-stage rim collapse; however only an intra-crater paleosol unit atop the fall-back crater fill remains as evidence of this possible period of stasis. Late-stage rim collapse resulted in catastrophic sedimentation across the crater floor area.

1 Introduction

The Wetumpka impact structure is a 7.6-km diameter impact feature of the inner coastal plain of the state of Alabama, USA (N32° 31.3', W86° 10.4'; Figure 1). Suggested as an impact crater for ~ 30 years (Neathery et al. 1976), Wetumpka has been documented in recent reports (King et al. 2002; King et al. 2003) as a marine impact feature containing shocked materials. Impact at Wetumpka probably occurred in marine waters that were between ~ 30 and 100 meters depth (according to ostracode eye structure; Puckett 1991) and in a setting that was within ~ 25 kilometers of a local barrier-island shoreline (see rationale in King et al. 2002; Figure 2). The age of Wetumpka has been estimated through stratigraphic and paleontologic relationships as late Santonian to early Campanian (King 1997). Wetumpka is a well-preserved example of a type of terrestrial impact crater wherein the overlying marine water layer and water-saturated sediments were involved in the impact deformation and subsequent impact-related sedimentation (cf. Ormö and Lindström 2000; Ormö et al. 2002; Poag et al. 2004). Wetumpka is a good example of a modest-sized, near-shore impact event.

Wetumpka has a two-part crater-filling stratigraphy (Figure 3), which includes an upper catastrophic megablock and sand unit that was deposited by late-stage rim collapse and a lower impact breccia unit, which is composed of three basic facies: washback- and surgeback-deposited sands and breccias, fall-back breccias, and amalgamated slumped target-rock blocks (King et al. 2003; washback and surgeback terminology *sensu* Poag et al. 2004). The lower washback-surgeback breccias are interpreted to have formed in part by the rapid return of excluded sea water (i.e., the collapsing water crater), which probably washed over the early-formed crater rim and mixed with fall-back debris and other impact-related materials (King et al. 2003; cf. Poag et al. 2004). Such washback-surgeback breccias contain distinctive sedimentary structures such as contorted lamination, clay injections, complexly distorted contact geometries, concentrically layered ball-like structures, and crudely graded bedding (cf. Poag et al. 2004).

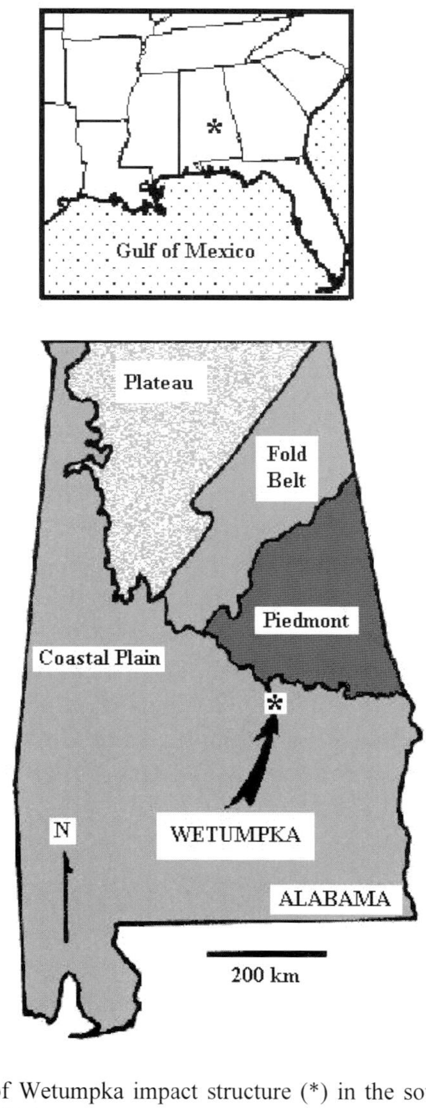

Fig. 1. Location of Wetumpka impact structure (*) in the southeastern United States and within the state of Alabama.

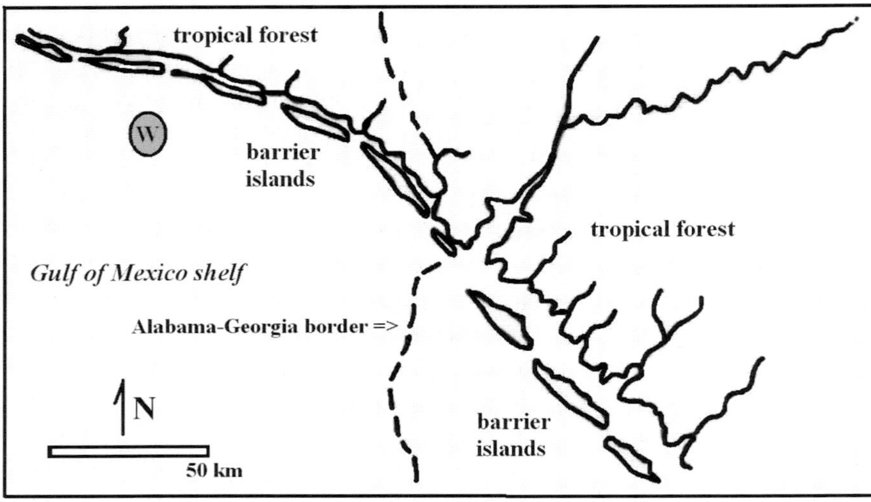

Fig. 2. Local Late Cretaceous paleogeography of the Alabama-Georgia state border region (border is dashed). Wetumpka impact site (W inside circle) is indicated. After a drawing by W. J. Frazier in Schwimmer (2002).

The Wetumpka impact event had a paleobiologic effect upon the fossil record: it served as a local reservoir for an impact-entombed fossil record in two main ways. First, coarse to fine fragments of terrestrial vegetation, probably derived from the adjacent tropical forest, were swept up and incorporated into Wetumpka washback- and surgeback-deposited breccias and sands. Second, intact blocks of target sedimentary units, which contain an internal fossil component of their own, are part of the slump and fallback debris that partially fills Wetumpka impact structure. Two Upper Cretaceous target sedimentary formations, which are recognizable as blocks in the drill cores and/or outcrop from near crater center, no longer crop out in the vicinity of the structure owing to extensive Tertiary erosion (i.e., Eutaw Formation and Mooreville Chalk; see King 1997). In fact, these slumped, target sedimentary rock blocks include some up-dip sedimentary facies of these formations that no longer exist in outcrops anywhere in the region. In yet another biological effect, the Wetumpka impact crater apparently functioned as a minor terrestrial (island) ecosystem embedded within the shelfal marine realm for an unknown, but potentially substantial, length of time. During this period, marine waters were largely excluded from inside the rim, and a potentially unique fresh-water ecosystem may have thrived at higher elevations (i.e., at or near the central peak) within Wetumpka's rim.

As one might expect from the relatively small size of this structure, there is no apparent regional or global biotic extinction event associated with this local catastrophe. The impact's inferred chronostratigraphic position (in the lowermost part of the Mooreville Chalk; see King 1997 for analysis) is interpreted to be near the local boundary between two planktonic foraminiferal biozones: *Dicarinella asymetirca* range zone and *Globotruncanita elevata* interval zone (see Mancini et al. 1998 for recent biostratigraphy of the area). However, we have no evidence – and do not suggest – that the impact was the cause of faunal change associated with this boundary. This biozone boundary has a geochronometric age of ~ 83.5 million years (see global synthesis of biostratigraphy in Haq et al. 1988). Based upon these relations, Wetumpka's inferred impact horizon would be, by most accounts, within lower Campanian (e.g., see the widely referenced synthesis by Haq et al. 1988). However, according to Puckett (1994) this horizon is instead within upper Santonian in this area. The inferred Wetumpka impact horizon is also stratigraphically near (1) the ostracode biozone boundary between *Veenia quadrialira* and *Pterygocythereis cheethami* interval zones and (2) the nannoplankton biozone boundary between *Lucianorhabdus cayeuxi* and *Calculites obscurus* interval zones. A thorough analysis of ostracode and planktonic foraminiferal biostratigraphic distributions in the impact area (Puckett 1994) showed that there were no coincident extinctions among species of either taxonomic group within the area's uppermost Santonian or Campanian strata.

Regarding megafauna and the Wetumpka impact, we note with caution that the inferred impact horizon is approximately at the same level as the last occurrence horizon of two large oysters, *Pyncnodonte aucella* and *Exogyra upatoiensis* (ranges from Sohl and Smith 1980). Whereas this co-disappearance of large species may seem more than coincidental, a similar large oyster in these strata, *Exogyra ponderosa*, was not so affected and it ranges upward to the uppermost Campanian. It is important to note that, at present, it is not possible to exactly show coincidence between the last occurrence of the two large oysters and the level of Wetumpka impact or any other demonstrable biostratigraphic effect in the local sedimentary record.

In this paper, we will examine three aspects of Wetumpka's paleobiologic effects: (1) catastrophic effects upon the surrounding marine and terrestrial system; (2) Wetumpka as an "impact reservoir for fossils;" and (3) Wetumpka as a unique post-impact terrestrial ecosystem.

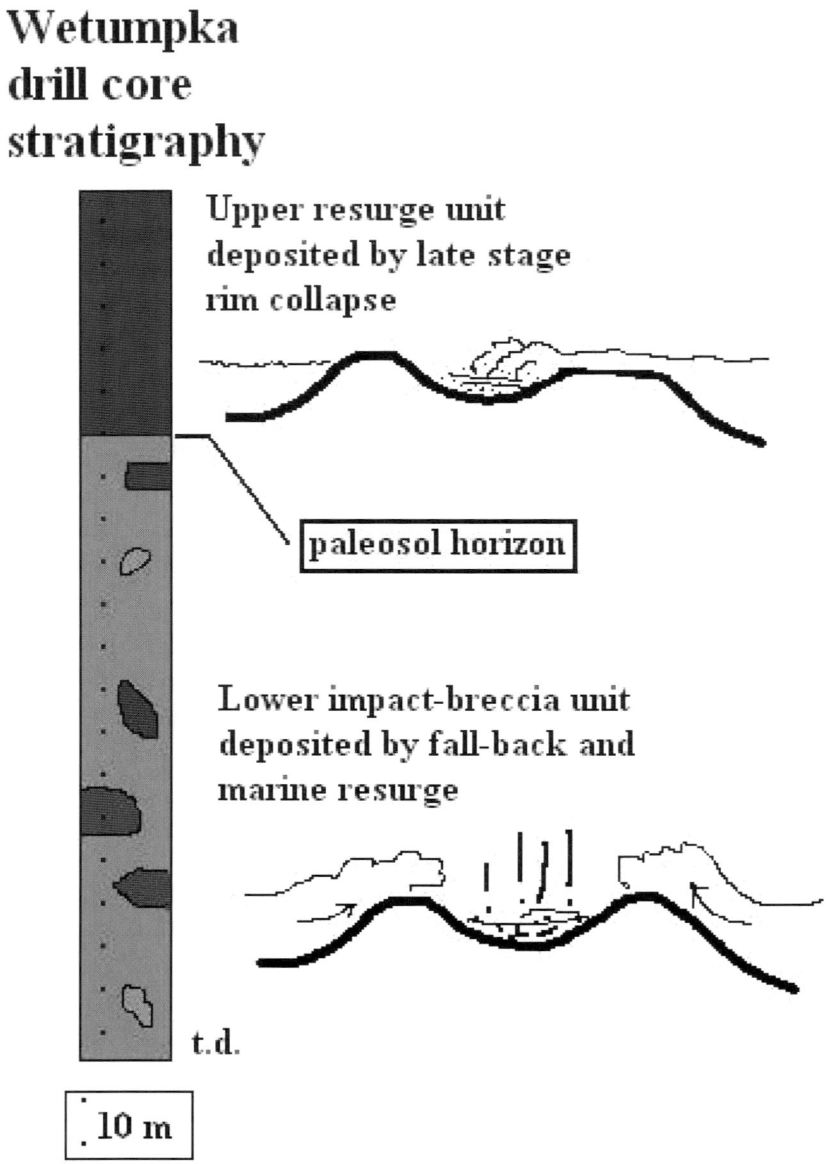

Fig. 3. Wetumpka drill-core stratigraphy (synthesis of two wells; King et al. 2002; 2003) showing two main units and interpreted origin of each as a stage in crater filling. t.d. = total depth of drilling.

2 Catastrophic effects of impact

Morrison et al. (1994) present a "summary of impact effects as a function of energy" showing that impact craters with diameters of ~ 6 to 8 km (i.e., comparable to Wetumpka) result from impact of ~ 350-m diameter cosmic objects (average mass = ~ 3.4 x 10^{14} kg). Cosmic objects are presumed to penetrate the atmosphere at average intra-solar system velocities of ~ 20 km/sec, and thus an impact on the scale of Wetumpka would yield ~ 10^2 to 10^3 MT equivalents of TNT (or yield = ~ 4.2 x 10^{24} to 4.2 x 10^{25} erg). Such sudden energy release is well below the "nominal threshold for global disaster" (estimated to be ~ 3 x 10^5 MT or ~ 8.4 x 10^{27} erg; Toon et al. 1994), but is within the range that Morrison et al. (1994) describe as "subglobal" disasters wherein "ocean tsunamis become significant (and) impacts destroy the area of a small state."

According to Morrison et al. (1994), the area of potential forest devastation around a terrestrial impact is given approximately by the equation $A = 10^4 Y^{0.666}$, where A is the devastated area in hectares and Y is the yield in MT equivalent of TNT. Devastated area A thus calculated for Wetumpka (where Y = 10^2 to 10^3) would have been ~ 2.15 to 9.95 x 10^5 hectares (radius of disaster circle = ~ 26 to 56 km, ranging according to Y). Wetumpka's proximity (within ~ 25 km) to a low, forested coastal plain shoreline (including low-lying barrier islands and a proximal low-lying alluvial coast; King 1997; King et al. 2002) make this observation rather pertinent. In this instance, forest devastation would have been related mainly to – in probable order of arrival upon land – infra-red flash burn, seismic energy, impact blast waves, fall-back of ejecta, and tsunami run-up, all of which had potentially damaging effects upon the adjacent land.

2.1 Infra-red flash burn

Infra-red wavelength emission accompanying hypervelocity impact causes a thermal radiation pulse whose fire-ignition potential is subject to a scaling-law function. According to Adushkin and Nemchinov (1994), the threshold of fire ignition based upon nuclear weapons testing is 10^9 erg/cm^2 and therefore the maximum burn area (A_f) and maximum burn radius (R_f) can be calculated assuming a clear day. Rough correction factors can be applied if the atmosphere was not clear at time of impact. We do not know how the energy budget of Wetumpka was partitioned among the various energy effects, but assuming a 25 percent thermal-radiation budget (as per King 1976) of Wetumpka's maximum kinetic-energy yield (i.e., 10^3 MT), E_r, the energy relegated to thermal radiation was probably ~ 250 MT. Adushkin and Nemchinov (1994) scale $A_f = 30E_r$ km^2 and $R_f = 3E^{0.5}$ km. For Wetumpka on a clear day, the maximum burn area, A_f, would have been ~ 7500 km^2 (or ~ 7.5 x 10^5 hectares), and the maximum burn radius, R_f, ~ 27 km.

Flash fires occur when direct, radiant thermal energy exceeds 10^9 erg/cm^2 upon combustible natural materials. These effects would have occurred within the ~ 7500 km^2 area computed above. The combustible natural materials Wetumpka were likely tropical forest cycads, conifers, angiosperms, and other lush vegetation thriving on Alabama's low coastal plain (flora described by Mancini 1981). Subsequent re-entry of particles sent into low orbits would have secondarily heated the atmosphere, and could have caused additional radiant thermal damage (Toon et al. 1994).

Adushkin and Nemchinov (1994) note that some impact events may have flash-burn radii larger than the size of the area affected by the atmospheric blast wave. We really do not know exactly how energy is budgeted in impacts of any significant size because impact events yield much more energy than any nuclear tests (our source of empirical data) ever observed. If more energy were budgeted to thermal radiation than was supposed above, the maximum burn area would have exceeded the area destroyed by the shock wave. If less an area, the atmospheric shock-wave pressure might have immediately extinguished the flash-burn area and thus actually limiting overall burn damage.

Wetumpka lignite fragments show no flash-burn effects, such as charcoalization. This sort of negative evidence leads us to speculate that flash-burn effects may have been somewhat limited in this instance or that perhaps atmospheric blast-wave effects may have negated fire effects.

2.2 Seismic energy effects

Seismic energy from Wetumpka's ground impact was sizable and can be viewed in two ways: (1) released seismic energy, E, and (2) surface-wave magnitude (M_s). There is a logarithmic relationship between E and M_s such that one step increase in M_s is a 30-fold increase in E (see Bolt 1993). Assuming a 75-percent conversion of impact energy into seismic energy, seismic energy (E) at Wetumpka would have ranged from ~ 3.2 x 10^{24} to 3.2 x 10^{25} erg. Thus, an earthquake surface-wave magnitude (M_s), a rough proxy for Richter magnitude, would have been ~ 8.4 to 9.0 (inferred from examples in Bolt 1993). On a low-lying coastal plain, such as that adjacent to Wetumpka marine impact, the effect of strong seismic waves probably had a potential to topple larger terrestrial vegetation (Bolt 1993).

Physical evidence of seismic effects from Wetumpka may be present in the local stratigraphic section. Frazier (1979) and other investigators have noted that there are thin, seismically disturbed zones and clastic dike swarms within the Upper Cretaceous section of Alabama and adjacent Georgia, but owing to a lack of precise correlation, we as yet have no good way of separating tectonic earthquake effects that disturbed the region throughout Late Cretaceous from any potential Wetumpka impact earthquake seismite. In a regional synthesis of Upper Cretaceous stratigraphy completed before current studies at Wetumpka resumed, King (1994) pointed out that there appears to be a regionally significant disturbance, the AGCD ("Alabama-Georgia clastic-dike injection") event, which was also noted by Reinhardt (1980), Frazier (1987), and others. The age of the AGCD event is estimated to have been at least 83 million years ago (King 1994) based upon cross-cutting relationships, and thus potentially may be explained by the Wetumpka impact event (note biostratigraphic age data above). Further study of this dike-impact relationship is indicated.

2.3 Impact blast waves

Impact blast waves possess "an abrupt pressure pulse ... followed immediately by a substantial wind" (Toon et al. 1994). Peak over-pressure, defined as the difference between ambient pressure and pressure of the shock front, characterizes atmospheric shock waves. There is also a quantifiable peak over-pressure-to-maximum wind-speed relationship, e.g., peak over-pressure of 14 kPascal accompanies a maximum wind speed of ~ 30 m/sec (for comparison, minimum wind speed in a force-5 hurricane force is 70 m/s). Key peak over-pressure of 28 kPa produces maximum wind speed of ~ 70 m/s (Toon et al. 1994), thus in a forest, near total devastation of the tree mass typically results. Key peak over-pressure, as above, for a ground impact occurs in an area with maximum radius r according to $r = 5.08$ km $(E^{0.333})$, where E is the adjusted impact energy in MT. E is adjusted according to the equation $E = qY$, where q is an empirically determined constant (0.5 for nuclear weapons) and Y is the kinetic energy yield in MT (Toon et al. 1994). At Wetumpka, a ground impact could have set up an atmospheric blast wave that delivered key peak over-pressure (28 kPa) at a maximum radius (r) of ~ 19 to 40 km.

From this discussion, we infer that tropical forests on the adjacent shore would have been profoundly affected by this blast wave, which would have struck the area with straight-line wind speeds somewhat like that of a force-5 hurricane. The result may have been splintering of tropical forest wood, which could help us account for the elongate lignite fragments observed in some intervals of the impact breccia within Wetumpka drill cores. Return air currents drawn back by intensive low pressure at the crater and also moving quite rapidly may have helped transport some splintered wood seaward toward the impact crater and ultimately into it.

2.4 Fall-back of ejecta

The expected limit of continuous ejecta from terrestrial impacts is likely ~ 2.35 crater radii (Melosh 1989) or, in this instance, radially out to ~ 9 km from Wetumpka center. Discontinuous ejecta potentially fell over much larger area, including the nearby shoreline. Being a marine impact into a target with ~ 200-m of wet-sediment on top, most of the ejected material was likely of sedimentary origin and mainly disintegrated into fine particles. This disintegration may also help account for the lack of a distinctive, extra-crater breccia unit from Wetumpka within the local Upper Cretaceous section (i.e., such a layer remains elusive). The roughly comparable distances of the limit of continuous ejecta and the estimated distance to nearest shoreline also may be a factor in the presumed low effect of falling ejecta upon the surrounding marine system and tropical forest areas. With the crater center located ~ 25 km offshore, we would expect only discontinuous ejecta, which was mainly fine sediment, to fall upon the adjacent land. For this reason, perhaps we can say that such sedimentary ejecta was not strongly involved in comminution of terrestrial vegetation.

2.5 Tsunami run-up

Hills et al. (1994) and Melosh (2003) discuss potential effects of small asteroid impacts in shallow water, mainly by drawing inferences from shallow sea-floor nuclear-weapons testing. Hills et al. (1994) cite Glasstone and Dolan (1977) who found that terminal wave height (h_w) at distance r, in km, from a nuclear explosion in shallow water is given by the general equation $h_w = 6.5$ m $[(1000 \text{ km/r}) (Y/1 \text{ GT})^{0.54}]$, wherein d is depth of water in m and Y is yield in gigatons (GT). Entering the equation above with r equal to 25 and at a minimum yield ($Y_{min} = 10^2$ MT or 0.1 GT) with same radius, r, wave height would have been ~ 75 m. According to Glasstone and Dolan (1977), a higher yield would produce a higher wave.

However, Hills et al. (1994) also noted that "asteroids with radii larger than (target) ocean depths produce tsunami-like waves with amplitudes comparable to the ocean depth at a short distance from the impact." Similarly, Melosh (2003), drawing on a recently released naval warfare research paper (van Dorn et al. 1968), notes that experimental nuclear-weapons detonation in the ocean has shown that the amplitude of impact-generated water waves "can never exceed the depth of the ocean" in which they occur.

Thus, we might infer that Wetumpka's impactor (having a radius of ~ 175 m and striking in 30 to 100 m of water; King et al. 2002) would produce a tsunami-like wave of, at most, ~ 30 to 100 m height at some small distance r from the impact. Citing observations of weapons tests, van Dorn et al. (1968) say that nuclear explosion wave height may be limited to 0.39 times oceanic depth. If these tests are analogous regarding Wetumpka, this means an initial tsunami wave height of between ~ 12 and 39 m at some small distance r from impact.

In the instance of open-ocean impacts, Melosh (2003) points out that impact-generated waves, which typically possess periods of 20 to 100 seconds, would likely break upon the continental shelf area. Wetumpka was, however, an impact upon the continental shelf not open ocean, thus the generated waves would more likely break near the shoreline, which would be the closest site of significant submarine slope decrease.

Amplitudes of tsunami waves decrease with distance (r) from target, however, in this instance, the progressively decreasing water depth toward shore would cause amplitude to increase in a shoreward direction. Low shoaling factors (perhaps < 2) associated with impact-generated wavelengths tend to cause early wave breaking and any Wetumpka tsunami wave breaking in the nearshore realm might be quite limited in its potential tsunami run-up distance (van Dorn 1968; Melosh 2003).

Considering a relatively flat coastal plain, Hills et al. (1994) estimate a scaled equation for run-up distance, X_{max}, for tsunami waves as: $X_{max} = 1000 \text{ m } [(h_o/10 \text{ m})^{1.333}]$, where h_o is run up height. A good estimate for h_o in the Wetumpka instance is about 9 m (Hills et al. 1994). So, for Wetumpka's tsunami, which was generated, we assume, 25 km offshore, tsunami run-up distance, X_{max}, would have been, at most, ~ 869 m, or slightly less than one km. This distance is hardly enough to cover the nearby barrier island and reach the mainland shore with any significant force or effect. From this we infer that tsunami effects were likely dissipated upon the barrier island and not so much on the mainland. If tsunami effects were involved in transporting woody material from the land to the crater area or to shelf depositional sites, that material probably came from the barrier-island realm.

3 Impact reservoir for fossils

The concept of a terrestrial impact structure, particularly a wet-target or marine impact, as a reservoir or "sink" for preserving fossils is not difficult to imagine. However, there are not many established examples. Terrestrial impact craters that are fossil reservoirs typically have this distinction because of a subsequent lacustrine ecosystem that developed within the crater's surficial depression (e.g., Ries crater; Pösges and Schieber 1997; and Boltysh crater; Gurov et al. 2003).

At Wetumpka, the previously mentioned lignitic component within some impact breccias and sands is *prima facie* evidence of the fossil reservoir effect. The impact crater fill apparently was not a very good environment of preservation for other coeval marine or terrestrial species – as none have been found – but an allochthonous component of terrestrial organic material (wood fragments and fine debris, now all lignite) found its way into the crater fill (Figure 4).

In addition, broken pieces of impacted formations, some of which are several meters across, retain constituent trace and body fossils, which occur together in a well-mixed, (i.e., non-stratigraphic) order (King et al. 2002). In fact, mixing of fossil types and biostratigraphic ages characterizes the Wetumpka intra-crater assemblage. Body fossils include various pelecypods (including *Exogyra*) and some trace fossils (including *Taenidium* (Savrda et al. 2000; Figure 5) within intact blocks of Tuscaloosa Group sediments) and marine trace fossils *Planolites* and *Thalassinoides* (within intact blocks of Eutaw Formation and Mooreville Chalk, both marine units). Notably, there are no trace fossils within structure-filling impact breccias or sands themselves. Absence of trace fossils, which are typically quite abundant in other Upper Cretaceous terrestrial and marine units in the area, is taken as strong negative evidence favoring the interpreted catastrophic nature of the crater-filling sediments. This negative evidence also favors the interpretation, noted above, that marine waters did not immediately occupy the crater depression. If marine waters had filled the crater depression after impact, one would expect marine burrowing organisms to have penetrated ~ 30 cm to ~ 1 m into any crater filling sediment at or near any hypothetical intra-crater, sediment-water interface. Further, flat-lying marine sediments should be lying directly upon impact breccias, and this is not the case at Wetumpka.

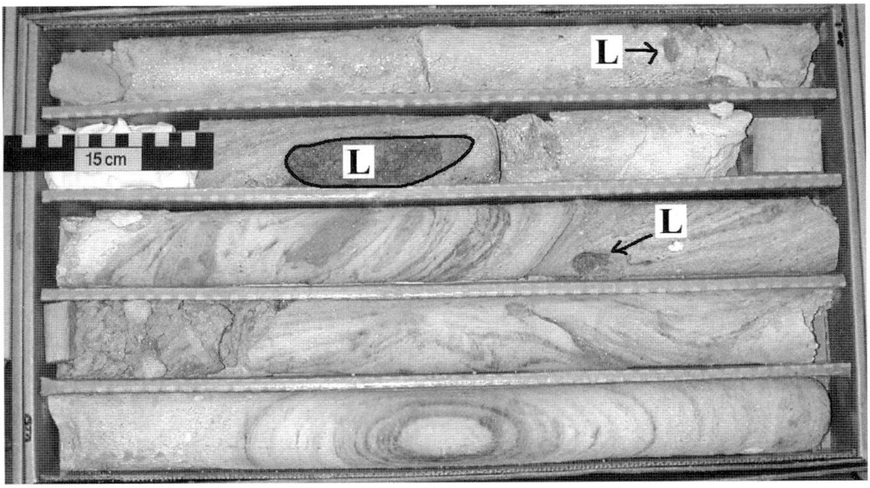

Fig. 4. Lignitic component in impact sands. Dark material (L) is lignite; dark component in sands (swirled and contorted laminations) is finely divided lignitic material as well. Cored interval: depth = -118.5 m (upper right) to -121.8 m (lower left). 15-cm scale bar.

Within the Wetumpka impact structure's sedimentary fill, there is a reported instance of vertebrate megafossil stratigraphic mixing or leakage. Specifically, a polycotylid pliosaur's (*Discosaurus*) vertebra, which must have originated within the Mooreville Chalk judging from some attached chalky matrix, was recovered from within a target-rock block of sandy Tuscaloosa Group sediment (Thurmond and Jones 1981). Tuscaloosa is a terrestrial fluvial deposit (Savrda et al. 2000), therefore, in this rather odd instance, a Campanian marine reptile's vertebra ended up within a Cenomanian fluvial sand deposit (age relations from King 1997) owing to the forces of this impact.

Fig. 5. Core from Schroeder well, at Wetumpka impact crater center, showing most of a ~ 6-m Tuscaloosa Group target rock block. Interval is -164.1 m to -169.7 m (upper left to lower right). Mottled interval in upper part of center three cores contain *Taenidium* in bioturbation. Light grey = sand; Dark grey = red clays. 15-cm scale bar.

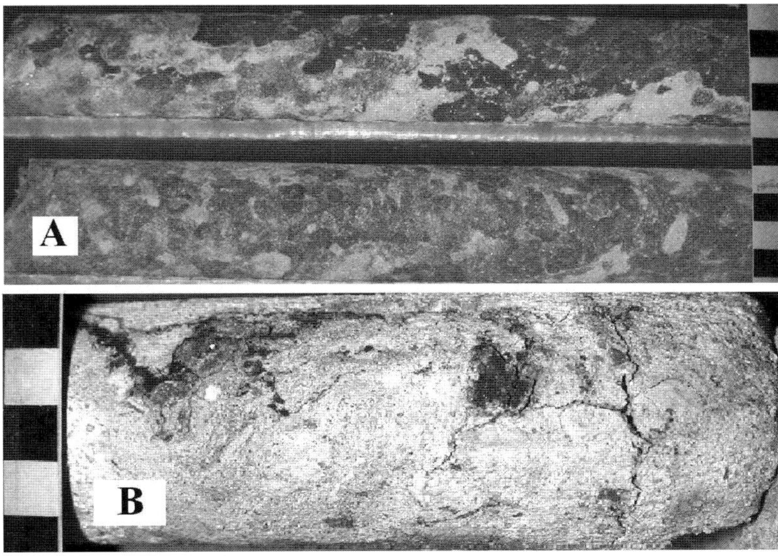

Fig. 6. Biotic components in Wetumpka core. (A) *Taenidium* in bioturbation from intact block of Cenomanian fluvial facies (target unit Tuscaloosa Group; depth = ~ -167.2 m); (B) dark material is lignitic component within Tuscaloosa Group sands (lignite does not occur in Tuscaloosa Group target material; lignite is embedded in periphery of intact block of target sediment; depth = -157.5 m). Both samples from the Schroeder well. Scales are = 1 cm per square block.

4 Wetumpka's unique ecosystem

After transient crater collapse, estimated rim height at Wetumpka, based upon terrestrial examples, was likely at most ~ 287 m (King, 1997; using formulae in Melosh 1989). As pointed out before, this height is more than sufficient to exclude local sea water, which was estimated to have been ~ 30 to 100 m in depth (King et al. 2002). Whereas it is possible that returning sea water from the collapsing water crater and impact-related tsunamis washed over all or part of this early-formed rim (King et al. 2003), the rim may very well have been stable for some time after impact and thus excluded sea water to the extent that at least some high ground near crater center was terrestrial. The best evidence to date for this hypothesis comes from one of the drill cores noted above (Schroeder well), which shows a distinctive two-part stratigraphy. The lower part is comprised of slumpback and marine washback-surgeback deposits (terms of Poag et al. 2004), but the upper part is composed entirely of catastrophically redeposited blocks of sedimentary target formations and resedimented sands of mixed target origin. This upper unit has been interpreted as the result of subsequent late modification-stage rim collapse and violent marine flooding (King et al. 2003). This catastrophic event may be partly responsible for the gap in Wetumpka's rim (currently spanning ~ 270° of arc around the crater; see map in King et al. 2002; 2003). Rather firm evidence for a period of marine-water exclusion is found in a peculiar 28-cm thick, mottled zone of red, clayey siltstone, which is interpreted as a lateritic paleosol horizon based upon comparison with lateritic paleosols in nearby terrestrial Upper Cretaceous deposits (Figure 7; cf. Reinhardt and Sigleo 1983, and Sigleo and Reinhardt 1988). Reinhardt and Sigleo (1983) and Sigleo and Rienhardt (1988) noted sesquioxides, mottles, pedotubules, blocky peds, slickensides, and plinthites among the criteria for paleosols developed on sedimentary materials in the U.S. Gulf Coastal Plain (see also Nettleton et al. 1989, for more on these criteria). Remarkably, we have observed all these features in the Wetumpka intra-crater paleosol. The interpreted, intra-crater paleosol unit has a gradual lower contact, and a very sharp upper contact, suggesting its in-situ origin and rapid burial during the rim-collapse catastrophe (Figure 7). Paleosol development on sedimentary material which contains extensive mottles suggests plant-root effects in soil development (Mack and James 1992). The development of Holocene paleosols with similar features in the Gulf coastal plain (Sigleo and Reinhardt 1988) suggests that they can form in 11,000 years, or less.

As noted above, Wetumpka's crystalline rim would have been a feature of significant relief above the surrounding sea level. Consistent with this interpretation, we note extensive saprolitization (i.e., formation of deep tropical soil C zone units in crystalline bedrock; Figure 8a; cf. Reinhardt and Sigleo 1983; Sigleo and Reinhardt 1988) within parts of the rim, which are indicative of the rim's long history of exposure and erosion (cf. Blank 1978). As saprolitization is aided by plant growth, this is taken as further evidence of tropical vegetation development upon the rim during this interval. On a part of the western rim, inside the rim flank, a coarse, sandy beach deposit lies upon a truncated bench on the crater rim (Figure 8b). This feature suggests vertical movement of the crater rim during late modification stage (or, alternatively, sea-level change) causing a slight wave-cut notch to develop on the rim.

Based upon intra-crater stratigraphy and the paleosol evidence above, we infer that Wetumpka's rim and higher parts of the interior probably existed for some time as a terrestrial island ecosystem adjacent to the mainland. It is notable that an intra-crater paleosol, not a lacustrine deposit, lignite-bearing swamp deposit, or some other specific depositional facies represents the sedimentologic record of early modification stage at the center of Wetumpka. All this may provide some evidence for a rather brief time interval of marine exclusion (a few decades or centuries, versus many millennia perhaps required for notable sedimentary facies development).

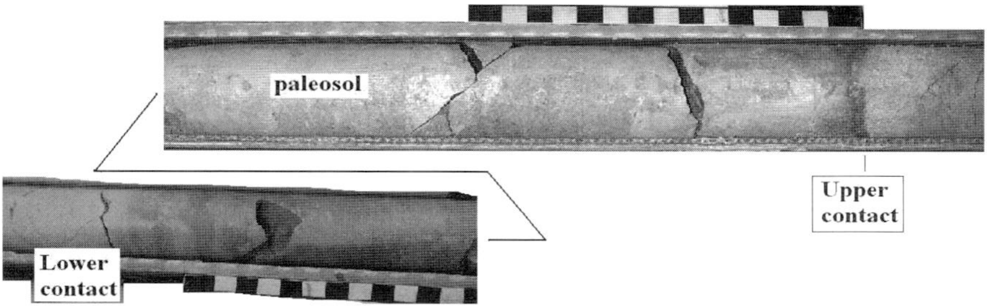

Fig. 7. Core segments showing most of the paleosol interval that begins at depth = -99.6 m in Schroeder well. The upper contact is sharp and iron stained whereas the lower contact is vague and mottled. Each square on the scale = 1 cm.

5 Conclusions

Wetumpka impact was a locally devastating event, which involved the local marine ecosystem and an adjacent terrestrial tropical-forest ecosystem. From the crater's proximity to shore (~ 25 km), we can infer than the local shoreline probably was initially devastated by infra-red flash burn and then by earthquake effects. Shortly thereafter, atmospheric blast wave, falling ejecta, and tsunami run-up also affected the area. We attribute splintered wood fragments and fine lignitic debris, found in relatively high concentrations in Wetumpka impactites as compared to nearby target strata, primarily to splintering and comminution by the atmospheric blast wave, which would have reached the shoreline prior to arrival of most ejecta particles and the tsunami run up. Returning waters from washback and surgeback events perhaps brought this comminuted woody material into the crater fill very early on in the impact process, or alternatively all or part of the wood component could have been moved by return air flow from the blast-wave passage. In addition to the wood from extant sources, body and trace fossils within and apart from their original target sedimentary units were deposited in the lower crater-filling unit. These fossils are *not* in original biostratigraphic order.

A

B

Fig. 8. Outcrops on northwestern part of Wetumpka's rim showing weathering features. (A) Upper Cretaceous saprolite over hard crystalline rim, height of outcrop = 10 m; (B) deeply weathered zone of crystalline rim is overlain by Upper Cretaceous saprolite, which is truncated by a coarse, white beach sand deposit, height of outcrop = 15 m.

There is evidence of extensive chemical weathering, including paleosol formation upon the crater fill and saprolitization of the crystalline rim, during the post-impact exposure interval. Paleosol and saprolite formation are strongly aided by vegetation cover, which suggests to us that there was a tropical island ecosystem upon the rim and within the crater while these areas were exposed above sea level.

Acknowledgements

We thank Vulcan Materials Company of Birmingham, Alabama, for providing the professional staff and truck-mounted equipment needed to obtain the drill cores at the Schroeder and Reeves wells. Further, we thank the respective landowners for their help in allowing drilling on their lands. This work was supported by donors to the Wetumpka Impact Crater Fund at Auburn University and a Dean's Research Initiative Grant from the College of Sciences and Mathematics at Auburn University. We thank reviewers Jens Ormö and Jared Morrow for their thoughtful comments. This paper is dedicated to the memory of David T. King, Sr.

References

Adushkin VV, Nemchinov IV (1994) Consequences of impacts of cosmic bodies on the surface of the Earth. In: Gehrels T (ed) Hazards due to Comets and Asteroids, University of Arizona Press, Tucson, pp 721-778

Bolt BA (1993) Earthquakes. W.H. Freeman and Company, New York, 331 pp

Blank HR (1978) Fossil laterite on bedrock in Brooklyn, New York. Geology 6: 21-24

Frazier WJ (1979) Sedimentology and paleoenvironmental analysis of the Upper Cretaceous Tuscaloosa and Eutaw Formations in western Georgia. In: Arden DD, Beck BF, Morrow E (eds) Second Symposium of the Geology of the Southeastern Coastal Plain, Georgia Geologic Survey, Information Series 53, pp 39-52

Frazier WJ (1987) A guide to facies stratigraphy of the Eutaw Formation in western Georgia and eastern Alabama. In: Frazier WJ, Hanley TB (eds) Geology of the Fall Line: a Field Guide to Structure and Petrology of the Uchee Belt and Facies Stratigraphy of the Eutaw Formation in southwestern Georgia and adjacent Alabama, Georgia Geological Society, Atlanta, pp B1-B25

Glasstone S, Dolan PJ (1977) The effects of nuclear weapons, 3d ed. United States Government Printing Office, Washington, 653 pp

Gurov EP, Kelley SP, Koeberl C (2003) Ejecta of the Boltysh impact crater in the Ukranian shield. In: Koeberl C, Martínez-Ruiz F (eds) Impact Markers in the Stratigraphic Record, Impact Studies volume 2, Springer Verlag, Berlin Heidelberg, pp 179-202

Haq BU, Hardenbol J, Vail PR (1988) Mesozoic and Cenozoic chronostratigraphy and eustatic cycles. In: Wilgus CK, Hastings BS, Kendall CGStCK, Posamentier HW, Ross CA, Van Wagoner JC (eds) Sea-level Changes: an Integrated Approach, Society of Economic Paleontologists and Mineralogists, Special Publication 42, pp 71-108

Hills JG, Nemchinov IV, Popov, SP, Teterev A (1994) Tsunami generated by small asteroid impacts. In: Gehrels T (ed) Hazards Due to Comets and Asteroids, University of Arizona Press, Tucson, pp 779-789

King Jr DT (1994) Upper Cretaceous depositional sequences in the Alabama Gulf Coastal Plain: their characteristics, origin, and constituent clastic aquifers. Journal of Sedimentary Research 64: 258-265

King Jr DT (1997) The Wetumpka impact crater and the Late Cretaceous impact record. In: Neathery TL, King DT Jr, Wolf LW (eds) The Wetumpka impact structure and related features, Alabama Geological Society Guidebook 34c, pp 25-56

King Jr DT, Neathery TL, Petruny LW, Koeberl C, Hames WE (2002) Shallow marine-impact origin for the Wetumpka structure (Alabama, USA). Earth and Planetary Science Letters 202: 41-549

King Jr DT, Neathery TL, Petruny LW (2003) Crater-filling sediments of the Wetumpka marine-target impact crater (Alabama, USA). In: Dypvik H, Burchell MJ, Claeys P (eds) Cratering in Marine Environments and on Ice. Impact Studies, Springer Verlag, Berlin Heidelberg, pp 97-113

King Jr EA (1976) Space Geology, an Introduction. Wiley and Sons, New York, 349 pp

Mack GH, James WC (1992) Paleosols for Sedimentologists. Geological Society of America, Short Course Notes, 127 pp

Mancini EA (1981) Paleobotanical studies in Alabama: 1845-1980. Geological Survey of Alabama Circular 106, 21 pp

Mancini EA, Puckett TM, Parcell WC, Crow CJ, Smith CC (1998) Sequence stratigraphy and biostratigraphy of Upper Cretaceous strata of the Alabama Coastal Plain. In: Mancini EA, Puckett TM (eds) Sequence Stratigraphy and Biostratigraphy of Upper Cretaceous Strata of the Alabama Coastal Plain, Alabama Geological Society Guidebook 35, pp 1-12

Melosh HJ (1989) Impact Cratering, a Geologic Process. Oxford University Press, New York, 245 pp

Melosh HJ (2003) Impact-generated tsunamis: an over-rated hazard [abs.]. Lunar and Planetary Science 34: abs. #2013 (CD-ROM)

Morrison D, Chapman CR, Slovic P (1994) The impact hazard. In: Gehrels T (ed) Hazards Due to Comets and Asteroids, University of Arizona Press, Tucson, pp 59-92

Neathery TL, Bentley RD, Lines GC (1976) Cryptoexplosive structure near Wetumpka, Alabama. Geological Society of America Bulletin 87: 567-573

Nettleton WD, Gamble EE, Allen BL, Borst G, Peterson FF (1989) Relict soils of subtropical regions of the United States. In: Bronger A, Catt JA (eds) Paleopedology, Nature and Application of aaleosols. Catena – a cooperating journal of the International Society of Soil Science, Supplement 16, pp 59-94

Ormö J, Lindström M (2000) When a cosmic impact strikes the sea bed. Geological Magazine 137: 67-80

Ormö J, Shuvalov VV, Lindström M (2002) Numerical modeling for target water depth estimation of marine-target craters. Journal of Geophysical Research 107 (E12): 5120-5128

Poag CW, Koeberl C, Reimold WU (2004) The Chesapeake Bay Crater, Geology and Geophysics of a late Eocene Submarine Impact Structure. Impact Studies, Springer Verlag, Berlin Heidelberg, 522 pp

Pösges G, Schieber M (1997) The Ries Crater – Museum Nördlingen. Bavarian Academy for Teacher Training, Academy Bulletin 253, 80 pp

Puckett TM (1991) Absolute paleobathymetry of Upper Cretaceous chalks based on ostracodes: Evidence from the Demopolis chalk (Campanian-Maastrichtian) of the northern Gulf Coastal Plain. Geology 19: 449-452

Puckett TM (1994) Planktonic foraminiferal and ostracode biostratigraphy of upper Santonian through lower Maastrichtian strata in central Alabama. Gulf Coast Association of Geological Societies Transactions 44: 585-595

Reinhardt J (1980) Upper Cretaceous stratigraphy and depositional environments. In: Frey RW (ed) American Geological Institute, Excursions in Southeastern Geology II, pp 386-392

Reinhardt J, Sigleo WR (1983) Mesozoic paleosols: examples from the Chattahoochee River valley. In: Carrington TJ (ed) Current studies of Cretaceous formations in eastern Alabama and western Georgia, Alabama Geological Society, Guidebook 20: 3-10

Savrda CE, Blanton-Hooks AD, Collier JW, Drake RA, Graves RL, Hall AG, Nelson AI, Slone JC, Williams DD, Wood HAR (2000) *Taenidium* and associated ichnofossils in fluvial deposits, Tuscaloosa Formation, eastern Alabama, southeastern USA. Ichnos 7: 227-242

Schwimmer DR (2002) King of the Crocodylians, the Paleobiology of *Deinoschuchus*, Indiana University Press, Bloomington, 220 pp

Sigleo WR, Reinhardt J (1988) Paleosols from some Cretaceous environments in the southeastern United States. In: Reinhardt J, Sigleo WR (eds) Paleosols and Weathering through Geologic Time: Principles and Applications, Geological Society of America, Special Paper 216, pp 123-142

Sohl NF, Smith CC (1980) Notes on Cretaceous biostratigraphy. In: Frey RW (ed) American Geological Institute, Excursions in Southeastern Geology II, pp 392-402

Thurmond JT, Jones DE (1981) Fossil Vertebrates of Alabama. University of Alabama Press, Tuscaloosa, 242 pp

Toon OB, Zahnle K, Turco RP, Covey C (1994) Environmental perturbations caused by impacts. In: Gehrels T (ed) Hazards Due to Comets and Asteroids, University of Arizona Press, Tucson, pp 791-827 van Dorn WG, LeMéhauté B, Hwang L-S (1968) Handbook of Explosion-generated Water Waves, volume 1 – State of the art, Tetra Tech, Pasadena. Report No TC-130, 174 pp

The Sweet Aftermath: Environmental Changes and Biotic Restoration Following the Marine Mjølnir Impact (Volgian-Ryazanian Boundary, Barents Shelf)

Morten Smelror[1] and Henning Dypvik[2]

[1]Geological Survey of Norway, N-7491 Trondheim, Norway (morten.smelror@ngu.no)
[2]Department of Geosciences, University of Oslo, P.O. Box 1047, Blindern, N-0316 Oslo, Norway

Abstract. During the Late Jurassic and earliest Cretaceous the Barents Shelf was dominated by fine-grained clay sedimentation, with mostly anoxic to hypoxic depositional conditions. The stratified water-masses contained typically relatively rich, but low diversity, nectonic faunas and marine microfloras above the pycnocline. In contrast the benthic faunas contained only a few bivalve species and low diversity communities of foraminifera. At the time of the Volgian-Ryazanian boundary (142.2 ±2.6 Ma) a 1.5-2 km-diameter bolide hit the paleo-Barents Sea and created the 40 km-diameter Mjølnir Crater. The central peak of the crater formed an island, and the high standing crater rims and annular ridges further led to significant changes in the sea-bed topography. The impact and crater formation led to significant disturbance and environmental changes, both at the crater site and over large distances of the paleo-Barents Shelf. Tsunamis were formed and travelled back and forth across the seas for a day or two after the impact. Continuing collapse of unstable, unconsolidated highs and rims formed avalanches, slumps and slides that developed into gravity flows in the crater surroundings. Computer simulations of ejecta formation and distribution indicate that major ejecta transportation occurred along the trajectory of the incoming bolide, i.e., toward the northeast. No evidence exists of any major biotic extinction or changes in diversity related to the impact event, but the overall compositions of the microfossil assemblages show a significant change within the impact-influenced strata. In the lowermost post-impact deposits in the Mjølnir Crater, and in association with the ejecta-bearing strata on the adjacent shelf, a conspicuous acme of the marine prasinophyte

Leiosphaeridia combined with an influx of abundant juvenile freshwater algae of the genus *Botryococcus* occur. The prolific blooms of *Leiosphaeridia* suggest that these algae had a behavioral pattern typical for so-called disaster species. The recovery of the algal bloom in deposits off Troms, 500 km to the south of the Mjølnir Crater, and on Svalbard, 450 km to the north, suggest that a regional eutrophication event was induced in the impact-ocean. The duration of the environmental change and the biotic turnover is currently difficult to estimate, but was most likely relatively short. Depositional conditions comparable to those found on the shelf prior to the impact (i.e., stratified water-masses, with anoxic – hypoxic bottom conditions and low diversity marine benthic faunas) were restored during the earliest Ryazanian (i.e., prior to the time corresponding to the *Heteroceras kochi* ammonite zone).

1 Introduction

The effects of bolide impacts on the biosphere may range from global catastrophes with mass extinctions to more local responses, depending on the size of the impacting bolide, the nature of the target area, and the subsequent environmental changes developed in response the impact. Effects from the spread of huge amounts of impact dust in the atmosphere may include shutdown of photosynthesis and enhanced greenhouse effects. Release of water vapor derived from oceanic impacts or CO_2 from impact into carbonate rocks may also contribute to stronger greenhouse conditions. Acid rain derived from large amounts of sulphuric acid, aerosols and production of nitric oxides may further lead to hostile, and even lethal, conditions on local or regional scales. Wildfires caused by an impact on land could threaten life near the impact site, and not least by causing large amounts of smoke to combine with the impact dust in the atmosphere to form an intense photochemical smog.

In addition to the above scenarios, a number of lethal biotic disruptions of the biosphere have been proposed to result from large bolide impacts. These disruptions include months of global darkness and reduced surface temperatures from impact-derived atmospheric dust, nitric acidification of surface waters, poisoning by metals derived from the volatilised impactor, and a combination of strongly enhanced global darkness, cooling and surface-water acidification from volatilisation of sulfate from an evaporite target (Alvarez et al. 1980; Lewis et al. 1982; Pollack et al. 1983; Wolbach et al. 1985; Erickson and Dickson 1987; O'Keefe and Ahrens 1989; Sigurdsson et al. 1992; Pope et al. 1993; D'Hondt et al. 1994). Large impacts into oceans could also cause significant, rapid heating (up to several 1000 C° for an impactor of 10 km in diameter) and generate large amounts of vapour and steam, which could rise explosively upward into the atmosphere from the target area (Melosh 1982). Tsunamis resulting from impact into an ocean could cause a total mixing of the water column and could inundate and erode low-lying continental areas at considerable distances from the impact site.

An extensive record exists of the possible effects of the giant impact that created the Chicxulub crater at the Cretaceous-Tertiary (K/T) boundary (Rampino and Haggerty 1996, and references therein). However, the effects on the impact from medium and smaller bolides are not well documented. The same is the case for marine impacts, which are underrepresented in the geological record (Grieve 1998). The objectives of the present study have been to document the environmental changes caused by the Mjølnir impact (Dypvik et al. 1996), which struck the present central Barents Shelf at the time of the Volgian-Ryazanian boundary (Smelror et al. 2001a). The good preservation of the Mjølnir marine impact crater and the good recovery of impact ejecta from wells and boreholes adjacent to the target area make this an excellent area to study the chronological sequence of biological changes. The biotic restoration following the marine impact (Volgian-Ryazanian boundary, Barents Shelf) can also be investigated.

The present study is based on data from the Mjølnir Crater (core 7329/03-U-01), corehole 7430/10-U-01 drilled 30 km northeast from the crater, and corehole 7018/05-U-01 from near Troms (the Troms III area, 500 km south of the crater) (Fig. 1). In addition, observations from the Volgian-Ryazanin boundary strata on Svalbard, and from borehole 6814/04-U-02 in the Nordland VII area (800 km south of the crater), are included.

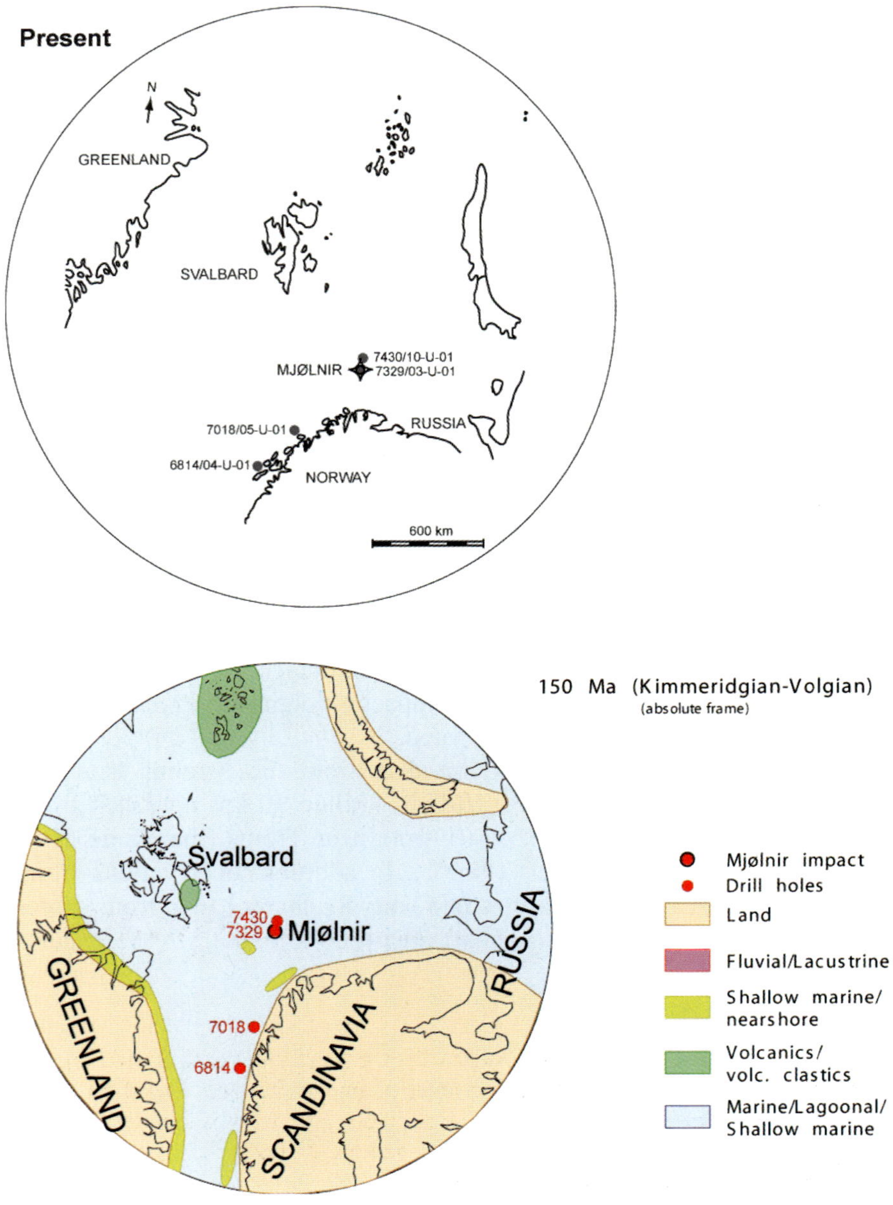

Fig. 1. A: Location map of studies cores and outcrops in the Barents Sea region (left), and Late Jurassic paleo-geographical reconstruction of the Arctic (based on plate tectonic reconstructions of Larry Lawver, personal communication).

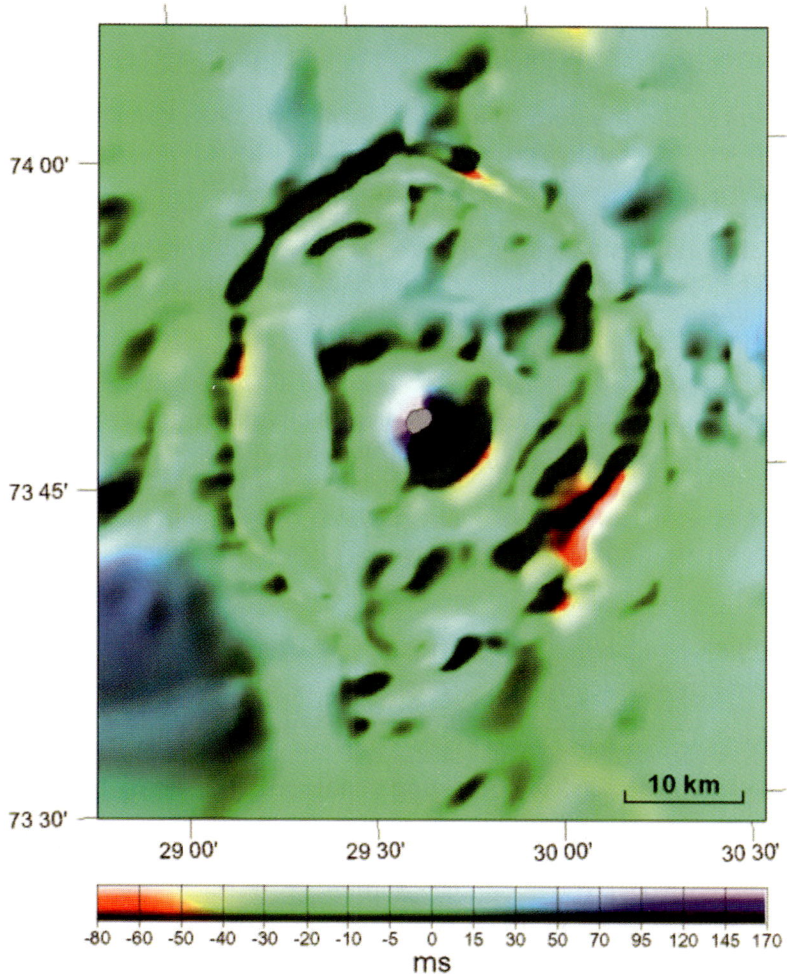

Fig. 2. Illuminated perspective image of the present structural morphology of the Mjølnir crater approximately at the level of impact horizon, based on the entire (2127 km) seismic reflection database available. The view is directly from above; light sources are at azimuths 30°, 290°, and 340°. The grey area on top of the central high shows truncation by erosion. Vertical exaggeration ~20x (from Tsikalas et al. 1998b).

2 Morphology of the Mjølnir Crater and lithostratigraphy of the studied core sections

The Mjølnir Crater was formed at the Volgian-Ryazanian boundary (142.2 ± 2.6 Myr ago) when an asteroid in the range of 1.5 - 2 km in diameter hit the 300-400 m-deep northern part of the "Kimmeridgian Clay Sea" (i.e. the paleo-Barents Sea) (Gudlaugsson 1993; Dypvik et al. 1996; Smelror et al. 2001a). The crater is 40 km in diameter (Fig. 2) and roughly circular in shape, exhibiting a clear radial zonation consisting of: 1) a central high with a diameter of 8 km, 2) a 4 km-wide annular basin, and 3) a 12 km-wide structurally disturbed outer zone, limited outward by distinct boundary faults (Dypvik et al. 1996; Tsikalas et al. 1998 a, b, c). The steep boundary faults form a ca. 150 m-high, nearly circular rim wall around the internal structural features. Reconstruction of the original crater morphology reveals extensive post-impact deformation. This includes near-field erosion, loading of an extensive overburden and structural reactivation and differential subsidence (Tsikalas et al. 1998b; Tsikalas and Faleide 2003).

Core 7329/03-U-01

This core from the Mjølnir Crater can be divided into three main lithostratigaphic units: the Ragnarok Formation representing the re-deposited crater infill, the Hekkingen Fomation representing the oldest post-impact deposits, and the overlying condensed carbonates assigned to the Klippfisk Formation (Fig. 3).

The Ragnarok Formation constitutes the interval from the base of the core at 171.08 m below sea-bottom up to 74.05 m (Dypvik et al. 2004). This interval is divided into two depositional units: unit I (171.08-88.35 m), a mixture of Middle and Upper Triassic to Lower Jurassic target rocks impacted by the asteroid and re-deposited as fallout into the crater; unit II (88.35-74.05 m), a thin sequence of gravity flow deposits. The latter unit has three subunits: IIA (88.35-87.43 m) is a conglomeratic debris flow of sand and small pebbles, IIB (87.45-75.73 m) represents a mudflow deposit, and IIC (75.73-74.05 m) consists of at least three separate gravity flows that are mixtures of sand, silt and clay. The sediments composing unit II most likely originated from the uplifted central high of the crater.

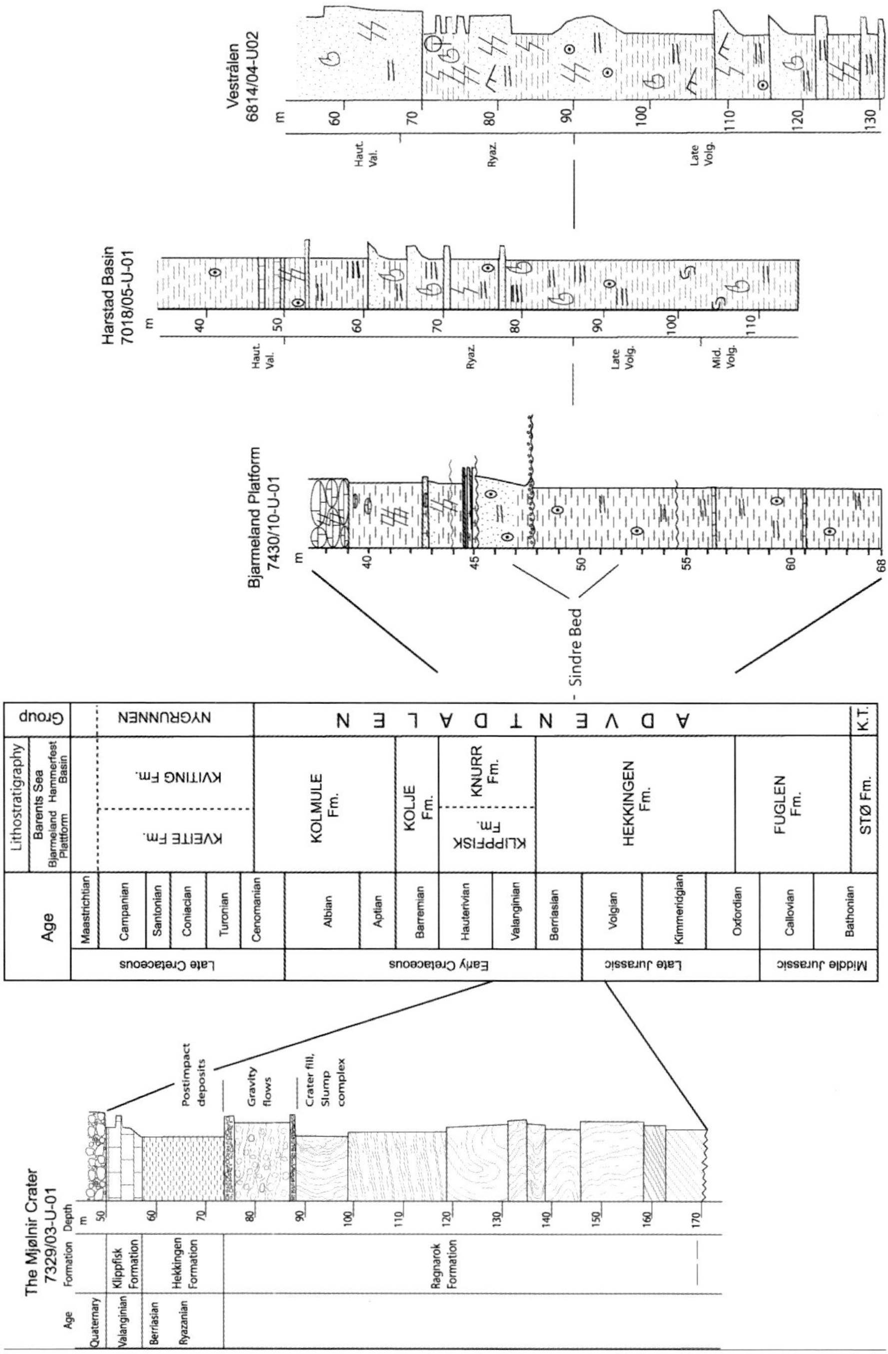
Vesterålen
6814/04-U02
Harstad Basin
7018/05-U-01
Bjarmeland Platform
7430/10-U-01
Sindre Bed
The Mjølnir Crater
7329/03-U-01
Postimpact deposits
Gravity flows
Crater fill, Slump complex
Ragnarok Formation
Hekkingen Formation
Klippfisk Formation
Age
Lithostratigraphy
Barents Sea
Bjarmeland Platform
Hammerfest Basin
Group
NYGRUNNEN
ADVENTDALEN
K.T.
KVITING Fm.
KVEITE Fm.
KOLMULE Fm.
KOLJE Fm.
KNURR Fm.
KLIPPFISK Fm.
HEKKINGEN Fm.
FUGLEN Fm.
STØ Fm.
Late Cretaceous
Early Cretaceous
Late Jurassic
Middle Jurassic
Maastrichtian
Campanian
Santonian
Coniacian
Turonian
Cenomanian
Albian
Aptian
Barremian
Hauterivian
Valanginian
Berriasian
Volgian
Kimmeridgian
Oxfordian
Callovian
Bathonian
Haut. Val.
Ryaz.
Late Volg.
Mid. Volg.
Quaternary
Ryazanian

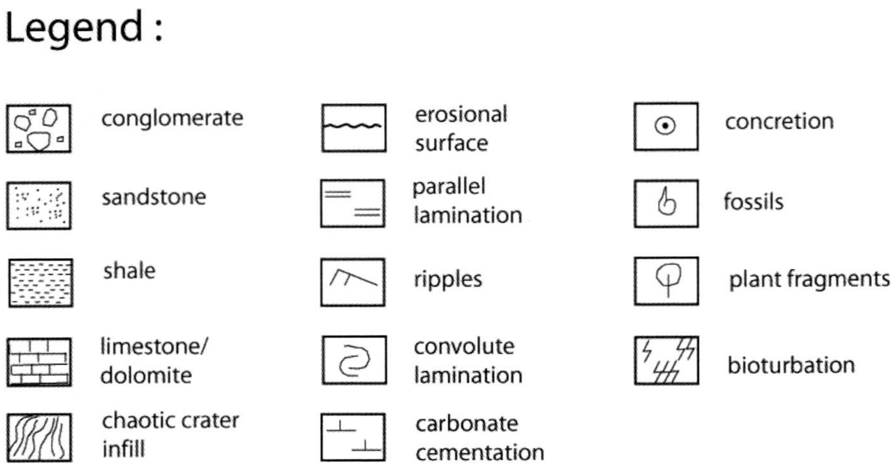

Fig. 3. Core logs and lithostratigraphic correlations of cores 7329/03-U-01, 7430/10-U-01, 7018/05-U-01 and 6814/04-U-02.

The Ragnarok Formation is capped by Lower Ryazanian dark brown to black organic-rich shale of the Hekkingen Formation. This lithological unit extends from 74.05-57.20 m in the core. The Hekkingen Formation has a wide distribution on the western Barents Shelf and is the most prolific hydrocarbon source rock in the area (Bugge et al. 2002; Dallmann 1999; Leith et al. 1993; Nøttvedt et al. 1993; Smelror et al. 2001c).

The Hekkingen Formation is capped by Valanginian condensed carbonates and marls of the Klippfisk Formation extending from 57.20 m to 50.00 m. The upper part of the drilled succession comprises Quaternary overburden.

Core 7430/10-U-01

This core is located 30 km NE of the Mjølnir Crater rim (Fig. 1). Sedimentological and lithostratigraphic descriptions, together with biostratigraphic information of this core have been published in Århus (1991), and by Dypvik et al. (1996), Smelror et al. (1998), and Smelror et al. (2001a, b).

The impact-influenced and ejecta-bearing strata within the Hekkingen Formation of this core are found between 52 m and 46.5 m (Fig. 3). This unit is now formally described as the Sindre Bed (Dypvik et al. 2004), and consists of dark grey, smectitic claystones and shales with dispersed mudflake conglomerates. The Sindre Bed directly succeeds the organic rich, finely laminated dark grey claystones of the Hekkingen Formation

(Dypvik et al. 1996; Dypvik and Ferrell 1998). Within the Sindre Bed between 47.6 and 47.4 m, three upward-fining conglomeratic units are found. They are separated by mm- to cm-scale, thin claystones/shales. The 19 cm-thick unit is succeeded by 120 cm of sandy and silty shales (47.4 - 46.5 m) that include a well-defined iridium peak at 46.85 m (Dypvik et al. 2004). Above this, dark claystone of the Hekkingen Formation continues up to 42.9 m, where it is succeeded by condensed carbonates and marls of the Klippfisk Formation (Smelror et al. 1998).

Core 7018/05-U-01

This core is located 500 km southwest of the Mjølnir Crater (Fig. 1). A brief lithostratigraphic description of core 7018/05-U-01 has previously been published in Smelror et al. (2001c). The Upper Volgian-Lower Ryazanian deposits of the Hekkingen Formation (the Krill Member) consist of dark to very dark grey claystones (Fig. 3). The claystones are finely laminated with abundant carbonate beds. Bioturbation is generally absent, except for a few observations close to the Volgian-Ryazanian boundary at around 88 m. Ammonites and bivalves are recovered at some levels, and a few coalified fragments are also found.

The laminated beds, the lack of bioturbation and the sparse benthic fauna, combined with the high organic content, indicate that sedimentation took place in a distal marine shelf environment, with fluctuating anoxic and hypoxic conditions at the sea bottom.

Core 6814/04-U-02

This core was drilled close to the margin of the paleo-Barents Sea, about 800 km southwest of the Mjølnir crater, in the coastal region of mainland Norway (Fig. 1). The studied samples represent Upper Volgian to Ryazanian beds of the Hekkingen Formation (Smelror et al. 2001c) (Fig. 3). The studied succession consists dominantly of shales and silt, with sandier sediments than in the other wells studied, reflecting its Late Jurassic, nearshore paleo-position. Several upward-coarsening sand units are observed in the core. A ca. 10 m-thick sand body is present at the Volgian/Ryazanian boundary. Trace fossils like *Planolites* and *Helminthopsis* are found at several levels, and *Zoophycos* are found in the uppermost part of the formation (Smelror et al. 2001c). A sparse benthic fauna with bivalves, together with ammonites, is occasionally present.

The Volgian-Ryazanian boundary beds at Janusfjellet, Svalbard

The Upper Jurassic-lowermost Cretaceous succession at Janusfjellet consists dominantly of shales and claystones, with some minor carbonate concretions. The section has been described in detail by Dypvik et al. (1991 a, b). The present study only deals with the Volgian-Ryazanian part of the succession, i.e., the uppermost part of the Agardhfjellet Formation. This formation consists of dark grey claystones and shales with some intervals of sandy shales. Dispersed carbonate lenses ranging from a few cm up to a meter in length are found at some levels. The carbonate lenses in this part of the Janusfjellet section consist mainly of calcite and siderite, with less than 5vol% dolomite present. Only very modest degrees of bioturbation are observed. The macrofaunas recovered from this interval comprise a few bivalves, ammonites and belemnites. At Janusfjellet the Agardhfjellet Formation is followed by soft, yellowish to greyish green claystones of the Mykelgardfjellet Bed.

3
Fossil material and methods

The cores and the fossil material treated in the present study have been collected through several shallow drilling programs in the Barents Sea conducted by Sintef Petroleum Research (formerly IKU Petroleum Research). Macrofossils and samples for palynological analyses from the Janusfjellet section on Spitsbergen were collected by Nils Århus and Otto Salvigsen.

The macrofossils from the split cores have been picked at site and have subsequently been determined by various experts. Samples for micropaleontological and palynological analyses have been taken randomly or at fixed intervals. The sample resolution is normally 0.5-4 m, but the oldest post-impact stratum in core 7329/03-U-01 from the Mjølnir Crater and the ejecta-bearing Sindre Bed in core 7430/10-U-01 (from 30 km NE of the Mjølnir Crater) have been sampled at shorter intervals (down to cm-scale sample spacing). The methods used for preparation of microfossils and palynomorphs are described in Smelror et al. (1998), Bremer et al. (2004), and Smelror and Dypvik (2005).

4 Environmental and biotic consequences of the Mjølnir impact

The Mjølnir Crater has been interpreted to have resulted from an oblique bolide impact from the south/southwest with an incidence angle of 45°-50° from horizontal. The physical impact resulted in an extensive disturbance both in the sedimentary target and the water column.These disturbances are manifested by:

(1) a 850-1400 km^3 seismically-disrupted volume at the impact-site (Tsikalas et al. 1998a) that is directly related to the impact-driven processes of brecciation, excavation, structural uplift, gravitational collapse and infilling;

(2) a 180-230 km^3 excavated/ejected material volume displaced from the impact site and deposited in the vicinity (Tsikalas et al. 1998a; Shuvalov et al. 2002); and

(3) collapse of the impact-induced water cavity giving rise to a large-amplitude tsunami. Dissipation of the energy released during the Mjølnir impact partly determined the amount of material ejected and the extent of tsunami generation.

4.1 Sedimentary environments and biofacies of the target area

The published and present sedimentological, geochemical and paleontological data suggest that the Mjølnir bolide impacted in the middle of a wide epicontinental shelf, with 300-500 m water depth, and with stratified water masses characterized by fluctuating anoxic to hypoxic bottom conditions. The diversity of macrofossils found in the sediments is relatively low (Tables 1-4). The observations made from the Upper Volgian-Lower Ryazanian beds, containing dominantly nekton, possible nektobenthic species and rare benthos (mostly bivalves), show that the deposits fall within biofacies 1-4 in the classification scheme of Wignall and Hallam (1991). Relatively similar and uniform depositional conditions had persisted since the initial regional transgression in the Late Oxfordian (Smelror et al. 2001c) that marked the onset of the dark clay deposition over the paleo-Barents Shelf. The Mjølnir impact created a sudden disruption of these depositional conditions. The biotic consequences for the inhabiting marine phytoplankton and faunas (micro and macro) are discussed below.

4.2 Dinoflagellates and marine prasinophytes

Environmental disruptions resulting from crater impact in marine settings will have both short- and long-term effects on the phytoplankton communities, depending on the size of the impacting bolide, the water depth and topography, and the nature of the bedrock beneath the seafloor at the target area. Since phytoplankton are the first link in the marine food chain, the phytoplankton responses to oceanic impacts will to a large degree control the fate of the zooplankton and other consumers in the marine food webs. Milne and McKay (1982) found, from two independent estimates, that temperate marine zooplankton would starve to death relatively quickly if phytoplankton photosynthesis were to cease. According to their calculations, a sudden atmospheric darkening, blocking 99% of sunlight, as might accompany the impact of a large asteroid, would be sufficient to stop global photosynthesis. Using present North Sea plankton data in their equation, they found that only 9 to 41 days of atmospheric darkening was needed to initiate a zooplankton starvation crisis during the spring season.

Except, perhaps, for the K/T boundary, very little published data exist on the distribution of phytoplankton in stratigraphical sections across documented impact boundaries, or in sections of impact ejecta. The well-studied K/T boundary impact is evident from the 180 km-diameter Chicxulub impact structure on the Yucatan Peninsula, coinciding with the biostratigraphic K/T boundary. Ejecta material from this giant impact is found globally and is confirmed by geochemical and mineralogical evidence. A number of studies have argued that this impact was the main triggering mechanism for the mass extinctions and severe biotic disruptions seen at the K/T boundary. Plankton extinctions in the oceans seem to have been abrupt and catastrophic (Smit 1990; Pospichal 1994; D'Hondt et al. 1994; Rampino and Haggerty 1996; Molina et al. 1998; Kaiho and Lamolda 1999; Arenillas et al. 2002). These extinctions were especialy common for carbonate-secreting species (D'Hondt et al. 1994), and most species of Late Cretaceous planktonic foraminifera (perhaps up to 90%) and several species of calcareous nannoplankton disappeared at the boundary. In contrast, Brinkhuis and Zachariasse (1988) found that only 11% of the organic-walled dinoflagellate cysts disappeared at the K/T boundary at El Haria, Tunisia.

In their study of marine biotic signals across an Upper Eocene impact layer at Massignano in Italy, Coccioni et al. (2000) found a marked increase of the dinoflagellate cyst *Thalassiphora pelagia* in the first 50 cm above the impact layer. This change may represent a cooling pulse of surface waters and/or an increase in productivity that followed the impact event, and persisted for about 60,000 years.

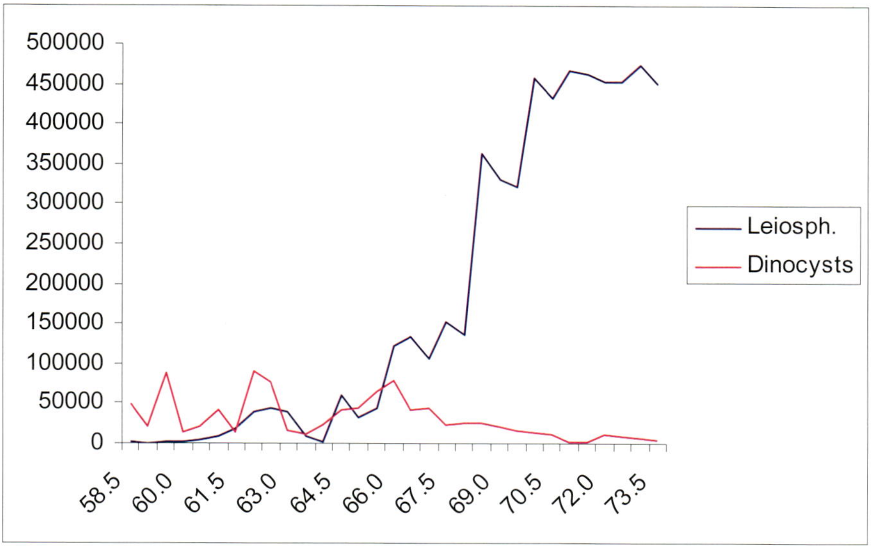

Fig. 4. Diagram showing the quantitative distribution (Y-axsis: cysts/gram sediments) of dinoflagellate cysts and leiosphaeres (leiosph.) in the Lower Ryazanian post-impact strata of the Mjølnir Crater (X-axis: Dept in core 7329/03-U-01)

Dinoflagellate diversity, originations and extinctions have fluctuated significantly during the Mesozoic (MacRae et al. 1996). The extinction of genera and species at the Jurassic-Cretaceous boundary is, however, only minor to moderate. In contrast, the genesis of species at the boundary is more profound, with around 20% new species in the earliest Cretaceous compared to latest Jurassic. In our data from the Volgian-Ryazanian boundary strata (142.2 ± 2.6 Ma) we have not found any evidence for abrupt extinction of dinoflagellate cyst species. The diversity is generally low to moderate below and above the boundary (Tables 5-8). All available data show that the Mjølnir impact apparently had no effect of species extinctions or genesis. Micropaleontological studies of the Montagnais impact structure, located on the shelf off Nova Scotia, showed similarly that the impact had neither regional nor global effects on biological diversity (Jansa et al. 1990; Jansa 1993). The Montagnais crater is 45 km in diameter and was formed at 50.8 Ma by the impact of a 3.4-km-diameter cometary nucleus into a shallow sea (< 600 m). This observation led Jansa et al. (1990) to suggest that impacting bolides into comparable oceanic target areas must be larger than 3 km in diameter to cause extinctions.

One important aspect in the study of the Mjølnir impact crater has been to see how the different groups of phytoplankton responded to the environmental changes caused by the impact. In a study of ejecta-bearing Volgian-Ryazanian boundary strata at core-site 7430/10-U-01, 30 km northeast of the Mjølnir Crater (Fig. 1), Smelror et al. (2001b) found high abundances of prasinophycean algae assigned to the genus *Leiosphaeridia*. A contemporaneous and similar acme of prasinophytes is also found in the Volgian-Ryazanian boundary strata at Janusfjellet on Svalbard (Table 8; Dypvik et al. 2000). These algal blooms are associated with layers of high iridium anomalies at both locations. In addition, grains of shocked quartz have been found within the prasinophyte beds of corehole 7430/10-U-01 (Dypvik et al. 1996). Both biostratigrapy (Smelror et al. 2001a) and seismic control give direct evidence for the ejecta-crater correlation between the ejecta-bearing strata in 7430/10-U-01 and the Mjølnir Crater.

Comparable to what has been found in core 7430/10-U-01 and in the Janusfjellet section on Svalbard, a distinct bloom of leiospheres has also now been found 500 km SE of the Mjølnir Crater in the Volgian-Ryazanian boundary strata in corehole 7018/05-01 offshore Troms. In this core the algal peak is recorded from level 89.3 m to 87.1 m (Table 7). A similar algal abundance peak has not been found in corehole 6814/04-U-02 located off Northern Nordland about 800 km SE of site of the Mjølnir impact.

In the oldest post-impact strata of core 7329/03-U-01 we have found a similar, and even more prolific abundance peak of *Leiosphaeridia* (Fig. 4). Here the acme of *Leiosphaeridia* reach 513,000 specimens/gram sediment (post-compacted) in the lowermost sample at level 74 m (below top of the core), and remains at abundances around 450,000 specimens/gram sediment up to about the 71 m level. From 69.5 to 68.5 m the abundance varies between 320,000-360,000 specimens/gram sediment, and from 68 to 66 m the abundance drop to between 107,000-152,000 specimens/gram sediment. From 65.5 m and to the uppermost studied sample at 58.2 m the abundance drops further and is reduced to between 50,000 specimens/gram (at 64.5 m) to around 500 specimens/gram sediment (at 59 m) (Fig. 4). The prolific abundance peaks documented here from the first post-impact deposits are comparable to what has been reported from "algal blooms" in modern and Holocene sediments.

The regional distribution of this bloom event, reaching from the Mjølnir Crater and up to Svalbard some 450 km to the north, points towards an extensive ocean eutrophication. Smelror et al. (2001a) suggested that the algal bloom was induced by the large amounts of nutrients released into the water column by the impact. Presumably the period of eutrophic conditions was relatively short. This means we have to assume a very high sedimentation rate to account for the more than 5.5 m of dark shales containing the very high algal abundances. This depositional rate is extreme for such fine-grained sediments, but not impossible given the fact that about 233 km^3 of ocean bottom sediments and underlying bedrock were thrown up and spread in the water column and air in a few seconds (Shuvalov et al. 2002). From the subsequent fall-back and re-surge much of the sediments were brought back to the crater, including the organic-rich sediments from the target area and more distal areas agitated by the tsunamis.

In contrast to the leiospheres, the dinoflagellate cysts recovered from the post-impact deposits in the Mjølnir Crater do not show any elevated abundances (Fig. 4). On the contrary, the deposits containing the highest abundances of *Leiosphaeridia* contain the lowest abundance of dinoflagellate cysts. From level 74 m up to 70.5 m the content of dinoflagellate cysts varies between 3,500-11,200 cysts/gram sediment (post-compacted), the lowest number being found in the lowermost examined sample at 74 m.

The prolific bloom of *Leiosphaeridia* and the very low content of dinoflagellate cysts in the sediments deposited just after the impact apparently show that these two groups of marine microplankton responded differently to the abrupt environmental change caused by the impact. It appears that the leiospheres were able to adapt rapidly to the new situation and were able to take advantage of the large amounts of suspended nutrients in the water column. In contrast the dinoflagellates were stressed by the sudden change, and the numbers of cysts produced became significantly reduced. There could be several reasons for this difference in response including different tolerance of salinities, different tolerance of seawater pH and different length of time used for reproduction and growth (i.e., difference in duration of their life cycles).

From level 70 m to 66 m the dinoflagellate cyst abundance increases from 14,300 to 79,100 cysts/gram sediment. Above 66 m in core 7329/03-U-01 a foraminiferal succession typical for the uppermost Hekkingen Formation elsewhere on the Barents Shelf and in the upper Agardhfjellet Formation on Svalbard is re-established (Bremer et al. 2001; Bremer et al. 2004). At approximately the same stratigraphic level the prasinophyte bloom also disappears. This suggests restoration of marine conditions similar to those existing on the shelf prior to the impact. The highest counts of 90,700 and 89,100 cysts/gram sediment are found at 62 m and 59.5 m, respectively (Fig. 4). The counts of 10,000 and 90,000 are comparable to the content of dinoflagellate cysts found in modern day shelf sediments.

Species specifically adapted to stressed environmental conditions associated with mass extinction intervals, or which have specific behavioral patterns that become prevalent during these stressed intervals, are often refereed to as *disaster species* (Harries et al. 1996). These species develop relatively large populations (i.e., blooms) during the early survival interval. Such disaster species generally have a short-lived population increase, and they are most often quickly replaced by opportunist and other survivors early in the repopulation (Harris et al. 1996). The prolific blooms of the prasinophyte algae *Leiosphaeridia* from the Mjølnir post-impact deposits suggest that these algae had such a behavioral pattern, and consequently can be categorized as disaster species.

4.3 Nannoplankton

The Late Jurassic-Early Cretaceous interval represents a particularly significant stage in the development of calcareous nannofossils in terms of both diversity and abundance, and several zonal schemes have been developed both for Boreal and Tethyan areas. In spite of this, we have not so far been able to find any nannofossils in the ejecta-bearing Sindre Bed. Consequently, the eventual response of this kind of phytoplankton to the impact event is unknown. The samples below are also barren of nannoplankton, and to our knowledge there exists no record of such phytoplankton in the pre-impact Upper Jurassic sediments of the western Barents Shelf.

One important observation from core 7430/10-U-01 is, however, the first influx of calcareous nannoplankton at 41.50 m, i.e., 5 m above the top of the Sindre Bed (see range chart in Smelror et al. 1998). The influx of calcareous nannoplankton here indicates a change to oceanographic conditions favourable for growing, and depositional conditions suitable for preserving, calcareous nannofossils in the Early Ryazanian. The consistent occurrence of *Buchia* cf. *okensis* from level 42.4-40.2 m suggests that this change took place just prior to the time corresponding to the *Heteroceras kochi* ammonite zone (Smelror et al. 2001a; Bremer et al. 2001; Bremer et al. 2004).

4.4 Freshwater algae

A pronounced and peculiar bloom of juvenile freshwater algae of the genus *Botryococcus* is present in the oldest post-impact sediments of the Mjølnir Crater core (73.7-68.5 m) (Table 5; Bremer et al. 2004). A similar acme has previously been recorded from the ejecta-bearing strata (Sindre Bed) in borehole 7430/10-U-01 (Table 6; Smelror et al. 2001a), and it is also observed in the Agardhfjellet Formation at Janusfjellet on Central Spitsbergen (Table 8; Dypvik et al. 2000), about 450 km north of the Mjølnir Crater. This bloom appeared short-lived, an assumption based on the fact that only juvenile specimens seem to be present.

Botryococcus are not able to reproduce in marine environments, but can be shed in relatively large quantities off major river plumes and transported for long distances on the shelf. The incoming of these algae in the oldest post-impact strata of the Mjølnir Crater and in ejecta-bearing strata on the western Barents Shelf is not easily explained. One theory is that the algae were brought to the shelf by the resurging tsunamis, returning from the shores after impact (Smelror et al. 2001b). Another theory involves the formation of a crater with a central peak and crater rims rising above sea level, forming an island on the paleo-Barents Shelf (Shuvalov et al. 2002). Such an island, or a trough with barrier rims, could have held accumulations of freshwater from the succeeding rainfalls. The first theory (tsunami resurge) assumes relatively short distances to the bordering shores and high influx of freshwater plumes forming a stable upper water layer over large areas. The latter theory (island, lagoon) is valid and suitable to the Mjølnir Crater area, but fails to explain the recovery of a similar and contemporaneous *Botryococcus* bloom as far from the crater as the Janusfjellet section on Svalbard (Dypvik et al. 2000).

4.5 Zooplankton

Extinction patterns and biotic turnovers of planktonic foraminifera are very well documented at the K/T boundary (Kiessling and Claeys 2001). Detailed and high resolution data sets exist from all major regions of the Earth, and from the available data it seems that planktonic foraminifera suffered the most pronounced species extinctions among the marine organisms across the K-T boundary. The magnitude of the extinction is, however, still a matter of debate, with published extinction rates varying between ca. 40 and 90% (Kiessling and Claeys 2001).

In contrast to the well-documented planktonic microfaunas from the many worldwide K-T boundary sections, planktonic foraminifera appear to be missing in the Volgian-Ryazanian boundary beds of the Western Barents Shelf. So far no planktonic foraminifera have been found in the Mjølnir Crater core 7329/03-U-01, in core 7430/10-U-01 or in core 7018/05-U-02. Whether this phenomenon reflects an absence of calcareous planktonic foraminifera in the upper water column on the paleo-Barents Shelf, or is due to post-depositional dissolution of the tests, is presently not known.

Radiolaria are found in the Upper Volgian-Lower Ryazanian of both cores 7430/10-U-01 and 7018/05-U-02, as well as in the Lower Ryazanian post-impact sediments of the Mjølnir Crater core. The recovery is rather poor, and detailed analyses have only been carried out in core 7430/10-U-01 (Table 9). The sample from the 51.5 m-level in the lowermost ejecta-bearing strata (i.e., the Sindre Bed defined between 52-46.5 m) contains an assemblage consisting of abundant but poorly preserved *Nassellaria* and *Spumellaria,* comprising 100% pyritised specimens from 14 different identified species. The sample at 47.5 m within the ejecta-bearing unit, contains only a few pyritised specimens of the genera *Spumellaria, Nassellaria*, *Praeconcaryomma* and *Parvicingula.* The sample at 46.8 m, just above the iridium peak at 46.85 m (Dypvik et al. 1996) contains a totally different assemblage comprising only a few calcified radiolaria of the *Tricolocapsa*. This marked change of the radiolarian assemblages clearly reflects the dramatic environmental changes linked to the impact.

4.6 Nekton (nekto-benthos)

Ammonites are occasionally found through the Upper Volgian and Lower Ryazanian strata of the Barents Shelf region. Belemnites and fish remains are also occasionally found in these deposits, but are generally subordinate compared to the ammonites.

Ammonites became totally extinct at the K/T boundary, and together with the dinosaurs, are the most prolific "victims" of the K/T impact. Several studies have suggested that the ammonites actually became extinct before the K/T boundary (see references in Kiessling and Claeys 2001). Recently an ammonite containing iridium and a few crystals of Ni-rich spinel was found just five centimetres below the K/T boundary clay layer in the Bidart section (French Basque Country). This finding supports the idea that the ammonites existed up to the very top of the Cretaceous and disappeared suddenly right at the K-T boundary (Rocchia et al. 2001). This occurrence is further supported by the recovery of ammonites some 20-15 cm below the boundary layer in the Baie de Loya section near Hendaye (France) (Rocchia et al. 2001).

The limited recovery of ammonites in the Upper Volgian-Lower Ryazanian of the Western Barents Shelf inhibits the possibility of evaluating the consequence of the Mjølnir impact by these nektonic organisms. No ammonites are found in the post-impact strata of the Mjølnir Crater core. In corehole 7430/10-U-01 the first post-impact ammonite is from 44.10 m (Table 2), 2.4 m above the top of the Sindre Bed (i.e., the ejecta-bearing strata). Possibly the environmental changes occurring immediately after the impact were unfavourable for the cephalopods. In corehole 7018/05-U-02 an ammonite of the genus *Borealites* sp. is found at 87.95 m (Table 3), within the beds with the algal bloom that serves as a marker unit between 89.28 m and 87.11 m. This finding supports the previous assignment (Smelror et al. 2001a) of an age close to the Volgian-Ryazanian boundary for the impact event. Farther upwards in the core ammonites of the genus *Surites* are found at 85.86 m and 85.80 m, providing additional evidence for an earliest Ryazanian age for the oldest post-impact deposits.

In a recent study of the effects of the K/T boundary events on bony fishes, Cavin (2001) found that the general trophic-dependent pattern of extinctions agrees better with the expected consequences of a short-term catastrophic event than with a long-term environmental change. Most of the victims of the boundary event were fast swimming and piscivorous predators. This supports the previous observations from terrestrial vertebrates that the important extinction involved organisms belonging to the food chains resting on primary production (Cavin 2001).

Fish remains are so far only reported from corehole 7430/10-U-01 (Table 10). Here an acme of fish ossciles is found at 46.45 m, only 5 cm above the top of the ejecta unit (Sindre Bed). The enrichment of fish ossicles just above the ejecta deposits may indicate a link between the impact and the abundance of fish remains, but presently this hypothesis is speculative and has to be further investigated. Possible mechanisms for causing the sudden death of the fishes may include direct impact from shock waves, release of poisonous H_2S from the anoxic bottom layers, abrupt changes in the sea water chemistry, salinity and/or oxygen content.

In the aftermath of the impact, the documented phytoplankton bloom apparently caused eutrophic conditions that could have been lethal to some of the fish stocks. Sudden and extensive ocean eutrophication can seriously affect several trophic levels in the marine food chains, and algal blooms may involve toxic species, lethal to shellfish and fishes (Anderson 1997; Johnsen et al. 1997; Fogg 2002). The effects of the algal blooms and ocean eutrophication induced by the Mjølnir impact on the fishes and other nektons still remains unanswered, but the magnitude and spatial extent of this bloom suggests that the effect could have been devastating.

4.7 Benthic foraminifera

Benthic foraminifera assemblages from the Mjølnir Crater core (7329/03-U-01) have been documented in detail by Bremer et al. (2004). The oldest post-impact sediments of the Hekkingen Formation occur from 74 m to around 67 m, and do not contain any foraminiferids except in a single sample at 73.0 m where a few specimens of *Trochammina* aff. *septentrionalis* are recorded. This monospecific and low abundance assemblage reflects the stressed conditions found in this part of the core (Bremer et al. 2004). As described above this unit contains a very prolific algal bloom of leiospheres and juvenile freshwater algae (*Botryococcus)*, suggesting eutrophic, and partly brackish, conditions in the upper water column.

Above 67.0 m both the foraminiferal diversity and abundance increase while the dominance decreases. In spite of this faunal expansion, the diversity is still relatively low. The assemblages are dominated by agglutinating taxa but there are also a few calcareous forms present in the samples at 66.0 m and 60.0 m. The faunas are dominated by *Evolutinella schleiferi, Evolutinella vallata*, *Gaudryina rostellata*, *Recurvoides obskiensis*, *Trochammina* cf. *annae* and *Trochammina* aff. *Quinquelocularis* (Bremer et al. 2004). These assemblages closely resemble those found in the Agardhfjellet Formation on Svalbard. The present observations of the benthic foraminifera faunas thus suggest that "normal" (pre-impact) oceanographic and depositional conditions were restored above 67 m.

In core 7430/10-U-01 a low diversity assemblage with *Haplophragmoides* spp. is found below, within (47.9 m and 47.5), and above the Sindre Bed (Table 10). The oldest examined post-impact sample at 43.8 m in the Hekkingen Formation contains only a single specimen of *Recurvoides* sp., while more diverse and abundant assemblages are found from 43.9 m and upwards in the overlying marls and carbonates of the Klippfisk Hekkingen Formation.

In core 7018/05-U-01 a relatively rich, but moderately diverse, assemblage of foraminifera is recorded at 93 m, below the beds with the algal peak (Table 11). No foraminifera were recovered from the interval containing the algae acme. From 88.5 m and up to 84.1 m only a low diversity assemblage with *Haplophragmoides* spp. and *Trochammina* spp. is found. This mirrors the situation in the oldest post-impact sediments in cores 7329/03-U-01 (Mjølnir Crater) and 7430/10-U-01 where foraminifera are nearly absent from the algal bloom deposits.

4.8 Bivalves

A low diversity fauna of bivalves is commonly found in the Upper Volgian-Lower Ryazanian strata of the Barents Sea region (Tables 1-4; see also: Århus 1991; Århus et al. 1990), and is also well documented from Andøya in northern Norway (Birkelund et al. 1978; Zakharov et al. 1981) and from Peary Land in North Greenland (Håkansson et al. 1981). These faunas are generally dominated by species of the genus *Buchia*. Buchiid bivalves are characterized by a rapid evolutionary change, showing a wide geographic distribution within the Late Jurassic and Early Cretaceous of the Boreal Realm. They occur in a wide variety of facies and depositional environments, from beach conglomerates and shallow marine sandstones, to offshore and deep-water mudstones and submarine fans.

In the Mjølnir Crater core (7329/03-U-01) *Buchia* first appears at 73.97 m, just 8 cm above the uppermost debris flow deposits of the Ragnarok Formation. Additional specimens are recorded at 73.65 m, and are further found at several levels from 72.95 m to 58.65 m (Table 1).

In core 7430/10-U-01 buchiid bivalves (*Buchia terebratuloides, B. unschensis*) occur throughout the ejecta-bearing beds from 52 m to 46.5 m (Table 2). The recovery of these buchiids allows a relatively precise age determination of the impact event and the post-impact strata (Smelror et al. 2001a). In core 7018/05-U-01 no bivalves are found immediately above the beds with the algal bloom (89.28-87.11 m), and the youngest post-impact buchiids are found here at 85.67 m.

The wide distribution of bivalves in the Boral Realm and their tolerance of different environmental conditions suggest that *Buchia* had broad adaptive ranges. An interesting behaviour from the present living species *Mytilus edulis* is the ability of the larvae to regulate their rate of metamorphosis and thereby increase their chances for survival in stressed situations. The bivalve employs this larval strategy, and is therefore able to drift for weeks in a physiologically suspended state under difficult environmental conditions before resuming metamorphosis when favourable planktic conditions are reached (Harries et al. 1996). According to Harries et al. (1996) bivalves of the family Mytilidae, in general, have high survival rates at the species and lineage level during Mesozoic and Cenozoic mass extinction intervals.

The present findings suggests that *Buchia* faunas were re-established relatively soon after the impact. The distribution shows that the buchiid bivalves were not notably affected by the impact and succeeding environmental changes. In this respect *Buchia* possibly had comparable survival strategies as the crisis progenitor bivalve *Mytiloides* (Family Inoceramidae) as described from the Cenomanian-Turonian boundary by Kauffman and Harries (1996).

5
Restoration of the "Kimmeridge Clay Sea" oceanographic conditions

In the Mjølnir Crater core 7329/3-U-01 the return to "normal black clay" depositional conditions, and the restoration of marine microfloras typical for these kinds of depositional environments, took place just above the stratigraphic level with the last occurrence of *Buchia unschensis* (67.14 m) and the first appearance of *Buchia okensis* (66.80 m). In corehole 7018/05-U-02 the recovery of the ammonite *Borealites* sp. 87.95 m (i.e. within the Sindre Bed with the algal bloom) supports the previous assignment (Smelror et al. 2001a) of an age close to the Volgian-Ryazanian boundary for the impact event. The present biostratigrahic data thus suggest that normal "Kimmeridgian Clay Sea" oceanographic conditions were restored in the Early Ryazanian prior to the time corresponding to the *Heteroceras kochi* ammonite zone (Smelror et al. 2001a; Bremer et al. 2004).

The regional distribution of this bloom event reaching from the Mjølnir Crater and up to Svalbard some 450 km to the north, points towards an extensive ocean eutrophication. The algal bloom possibly was induced by the large amounts of nutrients released into the water column by the impact (Smelror et al. 2001b).

The duration of the aftermath, covering the time of the prasinophycean blooms and ocean eutrophication, can not be determined precisely. The fact that only juvenile freshwater algae have been found in the oldest post-impact deposits in cores 7329/03-U-01 and 7430/10-U-01 may indicate that this aftermath lasted a relatively short time. The thickness of the algal bloom beds in the Mjølnir Crater is, however, 7.7 m, and the thickness of the Sindre Bed (impact ejecta) in core 7430/10-U-01 is 5.5 m. In borehole

7018/05-U-01 the algal bloom covers more than 2 m of the Volgian-Ryazanian boundary strata. This means we have to assume some vertical spread due to bioturbation or a very high sedimentation rate. This depositional rate is extreme for such fine-grained sediments, but not impossible given the fact that about 233 km^3 of ocean bottom sediments and underlying bedrock were thrown up and spread in the water-column and air in a few seconds (Shuvalov et al. 2002). From the subsequent fall-back and re-surge much sediment was brought back to the crater, including the organic-rich sediments from the target area and more distal areas agitated by the tsunamis.

The thickness of the succeeding Ryazanian black shale deposits overlying the beds with impact ejecta and the algal blooms are about 2 m in borehole 7329/03-U-01 and 3.6 m in borehole 7430/10-U-01. In contrast, more than 37.4 m of dark shales accumulated above the impact-influenced beds in borehole 7018/05-U-01 and about 25 m above the ejecta-bearing strata at Janusfjellet on Svalbard. These profound differences in thickness of the sequences are probably due to a combination of the different bathymetry and physiographic conditions at the sites during deposition. Subsequent erosion of the sediments on the structural high prior to deposition of the condensed carbonate of the Klippfisk Formation in the Mjølnir Crater and at sites 7330/10-U-01 and 7018/05-U-01 on the shelf, and the plastic clay of Myklegardfjellet Bed and overlying shale, siltstone and sand of the Ullaberget Member on Svalbard, also played a role.

6 Conclusions

- The Late Jurassic and earliest Cretaceous on the Barents Shelf was dominated by fine-grained clay sedimentation, with mostly anoxic to hypoxic depositional conditions. The stratified water masses contained typically relatively rich, but low diversity, nektonic faunas and marine microfloras above the pycnocline. In contrast, the benthic faunas contained only a few bivalve species and low diversity communities of foraminifera.

- The Mjølnir impact at the Volgian-Ryazanian boundary (142.2 ±2.6 Ma) created the 40 km in diameter Mjølnir Crater, with a central peak forming an island in the paleo-Barents Sea. The high standing crater rims and annular ridges further led to significant changes in the sea-bed topography.

- The impact and crater formation led to significant disturbances and environmental changes, both at the crater site and over large distances of the paleo-Barents Shelf. Tsunami were formed and travelled back and forth across the seas for a day or two after impact. Continuing collapse of unstable, unconsolidated highs and rims, formed avalanches, slumps and slides that contributed to tsunami formation in the surrounding seas as well as the development of gravity flows into the crater.

- There is no evidence of any major biotic extinction or changes in diversity related to the impact event, but the overall compositions of the microfossil assemblages show a significant turnover within the impact-influenced strata. In the lowermost post-impact deposits in the Mjølnir Crater, and in association with the ejecta-bearing strata on the adjacent shelf, a conspicuous acme of the marine prasinophyte *Leiosphaeridia* combined with an influx of abundant juvenile freshwater algae of the genus *Botryococcus* occur. The prolific blooms of *Leiosphaeridia* suggest that these algae had a behavioral pattern typical for so-called disaster species.

- The recovery of the algal bloom in deposits off Troms, 500 km to the south of the Mjølnir Crater, and on Svalbard, 450 km to the north, suggests that a regional eutrophication event was induced in the impact-ocean.

- In the post-impact "algal bloom" interval only a monospecific assemblage, with few foraminifera, is found. In contrast, bivalves of the genus *Buchia* appear to be relatively common in these algal-rich deposits.

- The duration of environmental change and the biotic turnover is currently not possible to estimate, but was most likely relatively short. Depositional conditions comparable to those found on the shelf prior to the impact (i.e., stratified water-masses, with anoxic – hypoxic bottom conditions and with low diversity marine benthic faunas) were restored during the earliest Ryazanian (i.e., prior to the time corresponding to the *Heteroceras kochi* ammonite zone).

Acknowledgements

Material from cores 7430/10-U-01, 7018/05-U-01 and 6814/04-02 have been made available from Sintef Petroleum Research Ltd. The data on

Radiolarians from core 7430/10-U-01 have been provided by Harris Kavouras, while data on macro- and microfaunas from core 7018/05-U-01 have been provided by Natascha Shulgina and Valery Basov, respectively. They are all thanked for their valuable contributions. We would also like to thank Eric Buffetaut (CNRS Paris) and Tonu Meidla (University of Tartu) for their critical review of the manuscript and for useful comments and suggestions.

References

Alvarez LW, Alvarez W, Asaro F, Michel HV (1980) Extraterrestrial cause for the Cretaceous-Tertiary extinction. Science 208: 1095-1108

Anderson D (1997) Turning back the harmful red tide. Nature 388: 513-514

Arenillas I, Arz JA, Molina E (2002) Quantifying the evolutionary turnover across the K-T boundary catastrophic planktic foraminiferal extinction event at El Kef, Tunisia. GFF 124: 121-126

Birkelund T, Thusu B, Vigran J (1978) Jurassic-Cretaceous biostratigraphy of Norway, with some comments on the *Rasenia cymadoce* Zone. Palaeontology 21: 31-63

Bremer GMA, Dypvik H, Smelror M, Nagy J (2001) Biotic responses to the marine Mjølnir meteorite impact (Volgian-Ryazaian boundary, Barents Sea). 7th Workshop of the ESF Impact Programme, Sbumarine Craters and Eject-Crater Correlations, and Icy Impacts and Icy Targets. NGF Abstracts and Proceedings 1 (2001): 11-12

Bremer GMA, Smelror M, Nagy J, Vigran JO (2004) Biotic responses to the Mjølnir meteorite impact, Barents Sea: Evidence from a core drilled within the crater. In: Dypvik H, Burchell M, Claeys P (eds) Cratering in Marine Environments and on Ice, Impact Studies vol. 5, Springer Verlag, Heidelberg, pp 21-38

Brinkhuis H, Zachariasse WJ (1988) Dinoflagellate cysts, sea level changes and planktonic foraminifers across the K/T boundary at El Haria, Northwest Tunisia. Marine Micropaleontology 13: 153-191

Bugge T, Elvebakk G, Fanavoll S, Mangerud G, Smelror M, Weiss HM, Gjelberg J, Kristensen SE, Nilsen K (2002) Shallow stratigraphic drilling applied in hydrocarbon exploration of the Nordkapp Basin, Barents Sea. Marine and Petroleum Geology 19: 13-37

Cavin L (2001) Effects of the Cretaceous-Tertiary boundary event on bony fishes. In: Buffetaut E, Koeberl C (eds) Geological and Biological Effects of Impact Events, Impact Studies vol. 1, Springer-Verlag, Heidelberg, pp141-158

Coccioni R, Basso D, Brinkhuis H, Galeotti S, Gardin S, Monechi S, Spezzaferri S (2000) Marine biotic signals across a late Eocene impact layer at Massignano,

Italy: evidence for long-term environmental perturbations? Terra Nova 12: 258-263

Dallmann W (ed) (1999) Lithostratigraphic Lexicon of Svalbard. Norsk Polarinstitutt, Tromsø, 214 pp

D'Hondt S, Pilson MEQ, Sigurdsson H, Hanson Jr AK, Carey S (1994) Surface-water acidification and extinction at the Cretaceous-Tertiary boundary. Geology 22: 983-986

Dypvik H, Ferrell RE Jr (1998) Clay mineral alteration associated with a meteroite impact in the marine environment (Barents Sea). Clay Minerals 33: 51-64

Dypvik H, Nagy J, Eikeland T-A, Backer-Owe K, Johansen H (1991a) Depositional conditions of the Bathonian to Hauterivian Janusfjellet Subgroup, Spitsbergen. Sedimentary Geology 72: 55-78

Dypvik H, Nagy J, Eikeland T-A, Backer-Owe K, Andresen A, Haremo P, Bjærke T, Johansen H (1991b) The Janusfjellet Subgroup (Bathonian to Hauterivian) on central Spitsbergen; a revised lithostratigraphy. Polar Research 9: 21-43

Dypvik H, Gudlaugsson ST, Tsikalas F, Attrep M Jr, Ferrell RE Jr, Kringsley DH, Mørk A, Faleide JI, Nagy J (1996) Mjølnir structure: An impact crater in the Barents Sea. Geology 24: 779-782

Dypvik H, Kyte FT, Smelror M (2000) Iridium peaks and algal blooms – The Mjølnir impact [abs.]. Lunar and Planetary Science 31, Abstract #1538, CD-ROM

Dypvik H, Mørk A, Smelror M, Sandbakken PT, Tsikalas F, Vigran JO, Bremer GMA, Nagy J, Gabrielsen RH, Faleide JI, Bahiru GM, Weiss HM (2004) Impact breccia and ejecta from the Mjølnir Crater in the Barents Sea – The Ragnarok Formation and Sindre Bed. Norwegian Journal of Geology 84: 143-167

Erickson DJ, Dickson SM (1987) Global trace-element biogeochemistry at the K/T boundary, oceanic and biotic response to a hypothetical meteorite impact. Geology 15: 1014-1017

Fogg GE (2002) Harmful algae - a perspective. Harmful Algae 1: 1-4

Grieve RAF (1998) Extraterrestrial impacts on earth: the evidence and consequences. In: Grady MM, Hutchison R, McCall GJH, Rothery DA (eds) Meteorites: Flux with Time and Impact Effects. Geological Society, London, Special Publication 140, 105-131

Gudlaugsson ST (1993) Large impact crater in the Barents Sea. Geology 21: 291-294

Harries PJ, Kauffman EG, Hansen TA (1996) Models for biotic survival following mass extinction. In: Hart MB (ed) Biotic Recovery from Mass Extinction Events. Geological Society, London, Special Publication 102: 41-60

Håkansson E, Birkelund T, Piasecki S, Zakharov V (1981) Jurassic-Cretaceous boundary strata of the extreme Arctic (Peary Land, North Greenland). Bulletin of the Geological Society of Denmark 30: 11-42

Jansa LF (1993) Cometary impacts into ocean: their recognition and the threshold constraint for biological extinctions. Palaeogeography, Palaeoclimatology, Palaeoecology 104:271-286

Jansa LF, Aubry P-M, Gradstein FM (1990) Comets and extinctions: Cause and effects? In: Sharpton VL, Ward PD (eds) Global Catastrophes in Earth History. Geological Society of America Special Paper 247: 223-232

Johnsen G, Volent Z, Tangen K, Sakshaug E (1997) Time Series of Harmful and Benign Phytoplankton Blooms in Northwest European Waters Using the Seawatch Buoy System. In: Kahru M and Brown CW (eds) Monitoring Algal Blooms: New Techniques for Dectecting Large-Scale Environmental Change, Landes Bioscience, pp 113-141

Kaiho K, Lamolda M (1999) Catastrophic extinction of planktonic foraminifera at the Cretaceous-Tertiary boundary evidenced by stable isotopes and foraminiferal abundance at Caravaca, Spain. Geology 27: 355-358

Kauffman EG, Harries PJ (1996) The importance of crisis progenitors in recovery from mass extinctions. In: Hart, MB (ed) Biotic Recovery from Mass Extinction Events, Geological Society of London Special Publication 102: 15-39

Kiessling W, Claeys P (2001) A Geographic Database Approach to the KT boundary. In: Buffetaut E, Koeberl C (eds) Geological and Biological Effects of Impact events, Impact Studies vol. 1, Springer-Verlag, Heidelberg, pp 141 - 158

Leith TL, Weiss HM, Mørk A., Århus N, Elvebakk G, Embry AF, Brooks PW, Stewart KR, Pchlina TM, Bro EG, Verba ML, Danyushevskaya A, Borisov AV (1993) Mesozoic hydrocarbon source-rocks of the Arctic region. In: Vorren TO, Bergsaker E, Dahl-Stamnes ØA, Holter, E, Johansen B, Lie E, Lund TB (eds) Arctic Geology and Petroleum Potential. Norwegian Petroleum Society Special Publication 2, Elsevier, Amsterdam, pp 1-25

Lewis JS, Hampton Watkins G, Hartman H, Prinn RG (1982) Chemical consequences of major impact events on Earth. In: Silver LT, Schultz, PH (eds) Geological implications of impacts of large asteroids and comets on the Earth. Geological Society of America Special Paper 190: 215-221

MacRae RA, Fensome RA, Williams GL (1996) Fossil dinoflagellate diversity, originations, and extinctions and their significance. Canadian Journal of Botany 74: 1687-1694

Melosh HJ (1982) The mechanics of large meteoroid impacts in the Earth's oceans. In: Silver LT, Schultz PH (eds) Geological implications of impacts of large asteroids and comets on the Earth. Geological Society of America Special Paper 190: 121-127

Milne DH, McKay CP (1982) Response of marine plankton communities to a global atmospheric darkening. In: Silver LT, Schultz PH (eds) Geological implications of impacts of large asteroids and comets on the Earth. Geological Society of America Special Paper 190: 297-303

Molina E, Arenillas I, Arz JA (1998) Mass extinction in planktic foraminifera at the Cretaceous/Tertiary boundary in subtropical and temperate latitudes. Bulletin de la Société géologique de France 169: 351-363

Nøttvedt A, Cecchi M, Gjeldberg JG, Kristensen SE, Lønøy A, Rasmussen A, Skott PH, van Veen PM (1993) Svalbard-Barents Sea correlation: a short review. In: Vorren TO, Bergsaker E, Dahl-Stamnes ØA, Holter, E, Johansen

B, Lie E, Lund TB (eds) Arctic Geology and Petroleum Potential. Norwegian Petroleum Society Special Publication 2. Elsevier, Amsterdam, pp 363-375

O'Keefe JD, Ahrens TJ (1989) Impact production of CO2 by the Cretaceous/Tertiary extinction bolide and resulting heating of the Earth. Nature 338: 247-249

Pollack JB, Toon OB, Ackerman TP, McKay CP, Turco RP (1993) Environmental effects of an impact-generated dust cloud: Implications for the Cretaceous-Tertiary extinctions. Science 219: 287-289

Pospichal JJ (1994) Calcareous nannofossils and the K/T boundary: An update: New developments regarding the KT event and other catastrophes in Earth history [abs.]. Lunar and Planetary Institute Contribution 825, p 90

Rampino MR, Haggerty BM (1996) Impact crises and mass extinctions: A working hypothesis. In: Ryder G, Fastovsky D, Gartner S (eds) The Cretaceous-Tertiary Event and Other Catastrophes in Earth History. Geological Society of America, Special Paper 307, pp 11-30

Rocchia R, Robin E, Smit J, Pierrard O, Lefevre I (2001) K/T impact remains in an ammonite from the uppermost Maastrichtian of Bidart section (French Basque Country). In: Buffetaut E, Koeberl C (eds) Geological and Biological Effects of Impact Events, Impact Studies vol. 1, Springer Verlag, Berlin Heidelberg, pp 141-158

Sandbakken P, Dypvik H (2001) The Mjølnir Crater – A core description. 7th Workshop of the ESF Impact Program. NGF Abstracts and Proceedings of the Norwegian Geological Society 1: 69-70

Shuvalov V, Dypvik H, (2004) Ejecta formation and crater development of the Mjølnir impact. Meteoritics and Planetary Sciences 39: 467-479

Shuvalov V, Dypvik H, Tsikalas F (2002) Numerical simulations of the Mjølnir impact crater. Journal of Geophysical Research 107: doi: 10.1029/2001JE001698, 1-1-1-12

Sigurdsson H, D'Hondt S, Carey S (1992) The impact of the Cretaceous/Tertiary bolide on evaporite terrane and generation of major sulfuric aerosol. Earth and Planetary Science Letters 109: 543-559

Smelror M, Dypvik H (2005) Dinoflagellate cyst and prasinophyte biostratigraphy of the Volgian-Ryazanian boundary strata, western Barents Shelf. Norges geologiske undersøkelse Bulletin 443:61-69

Smelror M, Mørk A, Monteil E, Rutledge D, Leereveld H (1998) The Klippfisk Formation - a lithostratigraphic unit of Lower Cretaceous platform carbonates on the Western Barents Shelf. Polar Research 17: 181-202

Smelror M, Kelly SRA, Dypvik H, Mørk A, Nagy J, Tsikalas F (2001a) Mjølnir (Barents Sea) meteorite impact offers a Volgian-Ryazanian boundary marker. Newsletter on Stratigraphy 38: 129-140

Smelror M, Dypvik H, Mørk A (2001b) Phytoplankton Blooms in the Jurassic-Cretaceous Boundary Beds of the Barents Sea Possibly Induced by the Mjølnir Impact. In: Buffetaut E, Koeberl C (eds) Geological and Biological Effects of Impact Events, Impact Studies vol. 1, Springer Verlag, Heidelberg, pp 69-81

Smelror M, Mørk MBE, Mørk A, Løseth H, Weiss HM (2001c) Middle Jurassic-Lower Cretaceous transgressive-regressive sequences and facies distribution off Troms, northern Norway. In: Martinsen OJ, Dreyer T (eds) Sedimentary Environments Offshore Norway -Palaeozoic to Recent. Norwegian Petroleum Society Special Publication 10, Elsevier, Amsterdam, pp 211-232

Smit J (1990) Meteorite impact, extinctions and the Cretaceous-Tertiary boundary. Geologie en Mijnbouw 69: 187-204

Tsikalas F, Faleide JI (2003) Near-field Erosional Features at the Mjølnir Impact Crater: the Role of Marine Sedimentary Target. In: Dypvik H, Burchell M, Claeys P (eds) Cratering in Marine Environments and on Ice, Impact Studies vol. 5, Springer Verlag, Heidelberg, pp 39-55

Tsikalas F, Gudlaugsson ST, Faleide JI (1998a) The anatomy of a buried complex impact structure: the Mjølnir structure, Barents Sea. Journal of Geophysical Research 103: 30 469-30 484

Tsikalas F, Gudlaugsson ST, Faleide JI (1998b) Collapse, infilling, and post-impact deformation at the Mjølnir impact structure, Barents Sea. Geological Society of America Bulletin 110: 537-552

Tsikalas F, Gudlaugsson ST, Eldholm O, Faleide JI (1998c) Integrated geophysical analysis supporting the impact origin of the Mjølnir Structure, Barents Sea. Tectonophysics 289: 257-280

Wignall PB, Hallam A (1991) Biofacies, stratigraphic distribution and depositional models of British onshore Jurassic black shales. In: Tyson RV, Pearson TH (eds) Modern and Ancient Continental Shelf Anoxia. Geological Society of London Special Publication 58, pp 291-309

Wollbach WS, Lewis RS, Anders E (1985) Cretaceous extinctions: Evidence for wildfires and search for meteoritic material. Science 230: 167-170

Zakharov V, Surlyk F, Dalland A (1981) Upper Jurassic-Lower Cretaceous *Buchia* from Andøya, northern Norway. Norsk Geologisk Tidsskrift 61: 261-269

Århus N (1991) The transition from deposition of condensed carbonates to dark claystones in the Lower Cretaceous succession of the southwestern Barents Sea. Norsk Geologisk Tidsskrift 71: 259-263

Århus N, Kelly SRA, Collins JSH, Sandy MR (1990) Systematic palaeontology and biostratigraphy of two Early Cretaceous condensed sections from the Barents Sea. Polar Research 8: 165-194

Tables

Table 1. Distribution of macrofossils in the Volgian-Ryazanian strata of core 7329/03-U-01 (Data from Smelror et al. 2001a).

7329/03-U-01	Early Ryazanian- Late Volgian													
Sample depth (m):	58.56	60.12	60.13	60.37	66.80	67.14	67.20	68.05	69.08	69.21	69.22	71.05	71.90	72.08
Macrofossils:														
Buchia spp.	•						•				•		•	
Borealites sp.		•		•										
Buchia okensis			•		•									
Buchia unschensis						•		•	•	•		•		•

Table 2. Distribution of macrofossils in the Volgian-Ryazanian strata of core 7430/10-U-01 (Data from Smelror et al. 2001a).

7430/10-U-01	Early Ryazanian- Late Volgian													
Sample depth (m):	44.10	44.35	46.45	46.90	47.16	47.20	47.25	47.75	47.82	49.95	50.06	50.23	50.40	51.88
Macrofossils:														
Borealites sp.	•													
Buchia ex. Gr. volgensis		•			•									
Buchia cf. volgensis			•	•										
Buchia ex. Gr. terebratuloides							•							
Buchia unschensis			•	•		•	•	•	•	•				
Buchia cf. terebratuloides										•	•		•	•
Buchia cf. unschensis											•			
Buchia terebratuloides												•		

Table 3. Distribution of macrofossils in the Volgian-Ryazanian strata of core 7018/05-U-01 (Data from Smelror et al. 2001c and from Natascha Shulgina, unpublished).

7018/05-U-01	Early Ryazanian- Late Volgian																			M. Volg.	
Sample depth (m):	77.18	82.16	82.55	82.86	83.44	83.55	83.60	83.74	83.84	84.03	84.16	84.37	84.77	85.07	85.67	85.76	85.80	85.86	87.95	102.11	107.80
Macrofossils:																					
Buchia spp.	•	•		•	•																
Surites spp.			•	•		•	•	•			•						•	•			
Buchia fischeriana			•						•												
Buchia cf. fischeriana						•	•														
Buchia okensis							•														
Borealites sp. indet							•														
Buchia aff. okensis								•													
Buchia cf. okensis									•			•			•						
Borealites sp. (ex. gr. suprasubditus)										•						•					
Ammonites gen. Indet.													•	•							
Borealites cf. suprasubditus																			•		
Laugeites sp.																				•	
Laugeites cf. groenlandicus																					•

Table 4. Distribution of macrofossils in the Volgian-Ryazanian strata at Janusfjellet, Svalbard (Data from Simon R. A. Kelly, pers. comm.).

Janusfjellet	**E. Ryazanian- L. Volgian**			
Sample depth (m):	200.5	190.0	184.0	182.0
Macrofossils:				
Buchia volgensis	●			
Paleonucula isfjordica	●		●	
Dorsoplanites spp.		●	●	
Discinisca sp.			●	
Buchia taimyrensis				●

Table 5. Distribution of dinoflagellate cysts and other aquatic palynomorphs in the earliest Ryazanian (post-impact) strata of core 7329/03-U-01.

7329/03-U-01	**Early Ryazanian**												
Sample depth (m):	58.5	59.0	59.5	60.0	60.5	61.5	62.0	63.0	64.9	67.1	70.5	72.8	73.9
Dinoflagellate cysts:													
Atopodinium haromense	•	•	•	•	•		•	•		•	•	•	•
Cassiculosphaeridia magna	•	•	•	•	•	•	•	•		•	•	•	•
Chlamydophorella sp. A	•	•						•	•	•	•	•	•
Cribroperidinium spp.	•	•	•	•	•	•	•	•			•	•	•
Escharisphaeridia pocockii	•	•		•	•	•			•			•	•
Paragonyaulacysta borealis	•	•	•	•	•	•	•	•			•	•	•
Sentusidinium sp. A	•	•	•	•	•								
Sentusidinium spp.	•	•		•	•	•		•			•	•	•
Sirmiodinium grossii	•	•	•	•	•	•	•	•	•		•	•	•
Chytroeisphaeridia grossa		R											
Escharisphaeridia psilata		•	•				•	•		•		•	•
G. jurassica var. longicornis		R											
Pareodinia halosa		•		•	•						•		•
Rhynchodiniopsis martonense		•						•					
Scriniodinium crystallinum		R											
Senoniasphaera jurassica		•	•	•	•					•		•	•
Tubotuerella apatela		•	•	•	•	•		•	•	•	•	•	•
Valensiella ovula		•			•			•			•	•	•
Isthmocystis distincta			•					•					
Leptodinium sp. A			•		•						•		
Dingodinium tuberosum				•							•		•
Endoscrinium anceps				•									
Kalyptea stegasta				•		•							
Pareodinia ceratophora				•				•					
Ctenidodinium sp. A				•									
Cribroperidinium globatum					•			•			•		•
Escharisphaeridia rudis							•	•					•
Gochteodinia villosa							•						
Prolixisphaer. parvispinum							•				•	•	
Wallodinium krutzschii								•					
Scriniodinium inritibilum											•	•	•
Marine algae:													
Leiosphaeridia spp.	•	•	•	•	•	•	•	•	•	●	●	●	●
Pterospermopsis spp.					•						•		
Veryhachium sp.					•								
Freshwater algae:													
Botryococcus spp.										•	•	•	•

Table 6. Distribution of dinoflagellate cysts and other aquatic palynomorphs in the Volgian-Ryazanian strata of core 7430/10-U-01 (Data from Smelror et al. 1998).

7430/10-U-01	**Early Ryazanian- Late Volgian**									
Sample depth (m):	43.30	43.85	44.50	45.15	47.50	47.87	49.0	51.0	53.0	55.0
Dinoflagellate cysts:										
Adnatosphaeridium sp.	•									
Cassiculosphaeridia reticulata	•									
Tubotuberella apatela	•	•	•	•	•	•	•	•	•	•
Sirmiodinium grossii	•	•			•	•	•	•	•	•
Paragonyaulacysta borealis	•		•	•	•	•			•	•
Prolixosphaeridium spissum	•									
Trichodinium sp.	•									
Gonyaulacysta spp	•	•	•				•		•	•
Leptodinium milloudii	•	•								
Stiphrosphaeridium aff. S. arbustum		•						•		
Cribroperidinium spp.		•	•	•			•	•	•	•
Sentusidinium spp.			•					•		•
Cleistosphaeridium sp.				•						
Chlamydophorella sp.				•	•					
Apteodinium sp.							•			
Rhynchodiniopsis sp.									•	
Tubotuberella dangeardii									•	
Valensiella ovula									•	
Atopodinium prostatum									•	•
Senoniasphaera jurassica									•	•
Aldorfia dictyota									•	•
Pareodinia sp.									•	•
Paragonyaulacysta capillosa									•	•
Cometodinium spp.										•
Marine algae:										
Leiosphaeridia spp.	•	•	•	•	•	•	•	•	•	•
Freshwater algae:										
Botryococcus spp.				•	•	•	•	•		

Table 7. Distribution of dinoflagellate cysts and other aquatic palynomorphs in the Volgian-Ryazanian strata of core 7018/05-U-01 (Data from Smelror et al. 2001c and unpublished).

7018/05-U-01	**Early Ryazanian- Late Volgian**											
Sample depth (m):	83.1	85.5	86.3	87.1	88.5	89.3	90.1	90.9	91.1	92.3	93.6	94.6
Dinoflagellate cysts:												
Aldorfia dictyota	•	•										
Apteodinium daveyi	•	•	•	•	•	•						
Atopodinium haromense	•	•			•	•	•	•	•	•		
Cassiculosphaeridia magna	•					•		•	•	•		
Chlamydophorella sp.	•	•	•		•				•			
Circulodinium spp.	•	•	•	•	•	•	•	•	•	•	•	•
Cribroperidinium globatum	•	•				•	•		•			•
Cribroperidinium spp.	•	•	•		•				•		•	•
Dichadogonyaulax pannea	•	•										
Escharisphaeridia pocockii	•		•				•					
Paragonyaulacysta borealis	•	•	•	•	•	•	•	•	•	•	•	
Sentusidinium spp.	•	•						•		•		•
Sirmiodinium grossii	•	•	•	•	•	•	•	•	•		•	•
Escharisphaeridia psilata	•	•		•		•		•	•			•
Pareodinia halosa	•	•	•		•		•	•	•	•	•	•
Tubotuerella apatela	•		•	•	•	•	•			•	•	•
Tubotuberella dangeardii	•	•	•		•							
Scriniodinium spp.		•	•	•	•	•	•	•	•		•	•
Apteodinium sp. indet			•		•	•						
Gonyaulacysta cf. helicoidea					•		•		•	•	•	•
Valensiella ovula						•	•	•	•	•	•	•
Heslertonia? pellucida							•					
Cometodinium habibii								•				
Pareodinia ceratophora									•			
Cribroperidinium gigas									•			
Prolixosphaeridium anasillum									•			
Lanterna saturnalis										•	•	
Scriniodinium cf. attadalense										•	•	
Pareodinia asperata												•
Wallodinium krutzschii												•
Prasinophytes / Acritarchs:												
Leiosphaeridia spp.	•	•	•	•	•	•	•	•	•	•	•	•

Table 8. Distribution of dinoflagellate cysts and other aquatic palynomorphs in the Volgian-Ryazanian strata at Janusfjellet, Svalbard (Data from Nils Århus, personal communication and Smelror, unpublished data).

Janusfjellet	Early Ryazanian- Late Volgian												
Sample depth (m):	217.7	216.9	214.8	212.8	211.0	204.7	200.5	198.0	194.5	190.0	184.0	182.0	171.5
Dinoflagellate cysts:													
Sentusidinium spp.	●	●		●	●	●			●	●		●	
Sirmiodinium grossii	●	●	●	●	●	●	●	●	●	●	●	●	●
Paragonyualacysta borealis	●	●	●	●	●	●	●	●	●	●	●	●	●
Valensiella ovula	●			●	●					●			
Gonuaulacysta spp.	●					●	●		●			●	
Comasphaeridium sp.	●									●		●	
Cribroperidinium spp.	●		●	●				●			●	●	●
Horologinella spinosigibberosa	●												
Tanyosphaeridium variecalamus		●						●					
Wallodinium krutzchii		●		●					●	●			
Tubotuberella apatela		●	●		●	●	●		●	●	●	●	●
Marine algae:													
Leiosphaeridia spp.								●	●	●			
Tasmanites spp.								●	●	●			
Pterospermella sp.									●				
Freshwater algae:													
Botryococcus spp.									●	●			

Table 9. Distribution chart of Radiolarians in the Upper Volgian - Lower Ryazanian strata of core 7430/10-U-01. (Data from Harris Kavouras, pers. comm.).

7410/10-U-01	Early Ryazanian- Late Volgian							
Sample depth (m):	43.7	45.0	46.8	47.5	49.0	51.5	54.6	56.8
Radiolaria:								
Nassellaria gen et sp. idet	●			●		●		
Spumellaria gen et sp. indet	●			●		●		
Tricolocapsa sp. A			●					
Tricolocapsa sp. D			●					
Parvicingula sp. 3				●				
Praeconocaryomma sp.				●		●		
Parvicingula spp.				●		●		
Stichocapsa sp. aff. decorata						●		
Cryptamporela (?) sp						●		
Orbuculiforma spp.						●		
Acaeniotyle (?) sp.						●		
Stichocapsa sp.						●		
Spongodiscus sp. B.						●		
% calcified Radiolaria			100%					
% pyritised Radiolaria	100%			100%		100%		

Table 10. Distribution of foraminifera in the Volgian-Ryazanian strata of core 7430/10-U-01 (Data from Valery Basov, pers. comm.).

7430/10-U-01	Early Ryazanian- Late Volgian											
Sample depth (m):	43.8	45.1	46.0	46.5	47.2	47.5	47.9	50.0	51.0	51.9	54.2	56.1
Foraminifera:												
Recurvoides sp.	•											
Haplophragmoides spp.		•	•				•		•		•	•
Lenticulina spp.		•							•			•
Haplophragmoides infracalloviensis						•	•					
Haplophrogmoides goodenoughensis							•					
Others:												
Fish ossicles				•	•			•		•		
Sponge spicules							•					

Table 11. Distribution of foraminifera in the Volgian-Ryazanian strata of core 7018/05-U-01 (Data from Valery Basov, pers. comm.).

7018/05-U-01	Early Ryazanian- Late Volgian										
Sample depth (m):	84.1	85.0	86.0	88.5	90.0	91.0	92.2	93.0	94.0	96.3	98.5
Foraminifera:											
Haplophragmoides spp.	•	•				•			•	•	•
Trochammina spp.	•	•	•				•		•		
Trochammina ex gr. gyroidiniformis								•			
Ammobaculites ex gr. fontinensis								•			
Ammodiscus zasplelovae								•			
Haplophragmoides cf. mutabilis								•	•		
Hyperammina sp.								•	•		
Recurvoides obskiensis								•			
Lituotuba spp.								•		•	
Others:											
Radiolaria indet.								•			

Guembelitria irregularis Bloom at the K-T Boundary: Morphological Abnormalities Induced by Impact-related Extreme Environmental Stress?

Rodolfo Coccioni[1] and Valeria Luciani[2]

[1]Istituto di Geologia e Centro di Geobiologia dell'Università "Carlo Bo", Campus Scientifico, Località Crocicchia 61029 Urbino, Italy (cron@info-net.it)
[2]Dipartimento di Scienze della Terra dell'Università, Corso Ercole I d'Este, 32 44100 Ferrara, Italy (lcv.@unife.it)

Abstract. The planktonic foraminiferal species *Guembelitria irregularis* displays an aberrant test due to the irregular disposition and growth of the chambers which suggest a morphological malformation. Available data across the Cretaceous-Tertiary (K-T) transition from three Tunisian sections (El Kef II, Elles II, Ain Settara) and Kazakhstan (Koshak) and new data from Spain (Caravaca), and Italy (Erto), show a dramatic and remarkable increase in abundance of *G. irregularis* (up to 93% at El Kef II) in the small 38-63 μm fraction of the assemblages from the lower Danian planktonic foraminiferal Zones P0-P1a. Positive peaks in the abundance of this form are also recorded in the latest Maastrichtian, even though with minor percentages (up to 16% at El Kef II). Fossil and Recent foraminiferal tests showing morphological abnormalities have long been reported from stressed environments. Accordingly, we speculate that the morphological abnormalities shown by the *G. irregularis* test across the K-T boundary are the result of the extremely stressful environmental conditions related to the complex interplay of different events (rapid and extreme climate fluctuactions, sea-level changes, intense volcanism, and impact events, which characterize the last hundreds of thousand of years of the Cretaceous and the beginning of the Danian. In particular, the post-K-T morphological abnormality of *G. irregularis* may be related to high stress conditions induced by the K-T impact.

1 Introduction

The planktonic foraminiferal genus *Guembelitria* is characterized by a triserial chamber arrangement, small size, thin-walls and simple morphology of the test, with little or no surface ornamentation. In shallow marginal marine environments, as well as in open marine and restricted marginal settings, the guembelitrids thrived across the Cretaceous-Tertiary boundary (KTB) (e.g., Smit 1982; Liu and Olsson 1992; Keller et al. 2002b), during times of extremely stressful environmental conditions for any living organisms. It is proper to highlight that within the twenty three-year debate surrounding the KTB mass extinction, the impact of a large extraterrestrial body (Alvarez et al. 1980) has played a major role.

Remarkably, *Guembelitria cretacea* Cushman, 1933 and *Guembelitria irregularis* Morozova, 1961 dominated in the small 38-63 μm fraction of the lowermost Danian planktonic foraminiferal Zones P0 and P1a and Subzone P1a(1) *sensu* Keller (1993) and Keller et al (1995) (see Smit 1982; Keller et al. 1995; Pardo and Keller 1999; Pardo et al. 1999; Keller et al. 2002b; Luciani 2002). High abundances of *G. cretacea* and *Guembelitria dammula* (up to 90%) have however been found across the KTB from Bulgaria (Rögl et al. 1996; Adatte et al. 2002) and Madagascar (Abramovich et al. 2002). The consistent occurrence of the guembelitrids in lowermost Danian sediments indicates that they were survivors, as well as ecological opportunist, able to tolerate unstable environments and significant fluctuactions in temperature, salinity, oxygen, and nutrients. In particular, the species *G. irregularis* displays an aberrant test due to the irregular disposition and growth of the chambers that suggest a morphological malformation. Morphological abnormalities of fossil and Recent foraminiferal tests have long been reported from stressed environments. Accordingly, *G. irregularis* can be interpreted as useful indicators for high environmental stress.

In order to better document the development of pathologic planktonic foraminiferal tests at the KTB, we present and discuss here available data from Tunisia (El Kef II, Elles II, and Ain Settara sections) (Keller et al. 1995 and 2002b; Luciani 2002) and Kazakhstan (Koshak section) (Pardo and Keller 1999) together with new data from Spain (Caravaca section) and Italy (Erto section) (Fig. 1).

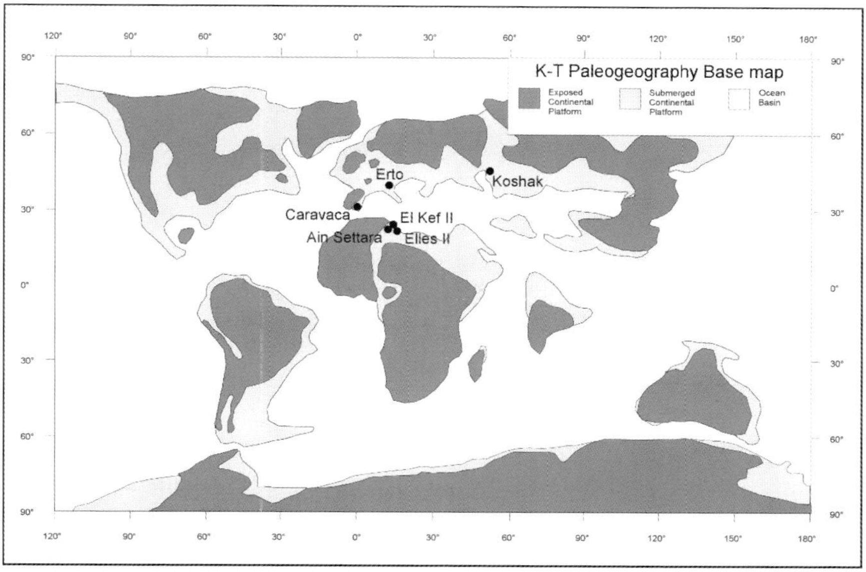

Fig. 1. Location of the K-T boundary sections discussed and studied. The paleogeographic reconstruction at the KTB time is based on Denham and Scotese (1987).

2
Materials and methods

The planktonic foraminiferal analyses recently performed across the KTB are largely based on the > 63 µm size fraction. However, some studies (Keller 1993; Keller et al. 1995) have shown that some species can be present only in the finest size fraction. Moreover, the relative species abundances may differ widely between the > 63 µm size fraction, usually analyzed, and the > 38 µm size fraction. In particular, these differences involve primarily the *Guembelitria* group. In fact, the *G. irregularis* tests are consistently found within the smallest fraction, and its absence or scarcity in some of the most complete KTB sections appear related to the adopted type of analysis. For this reason, we selected the available high-resolution most complete KTB sections, for which the > 38 µm size fraction was quantitatively analyzed. Furthermore, new quantitative studies for the Erto (Italy) and Caravaca (Spain) sections were carried out by counting in the 38 – 63 µm fraction the relative abundance of *G. cretacea, G. irregularis, Guembelitria danica* (Hofker 1978) and *Guembelitria trifolia* (Morozova, 1961) with respect to the total planktonic foraminifera in a statistical population of at least 300 specimens for each sample. Morphotypes resembling *G. dammula* were not observed in the sections studied. A total of 69 samples collected at a very high resolution (cm-scale) were analyzed (Caravaca: 40 samples; Erto: 29 samples). The very close rate of sampling is comparable with that of the Tunisia and Kazakhstan sections. The biostratigraphic scheme here adopted follows that of Keller (1993), Keller et al. (1995), and Pardo et al. (1996).

2.1 The Tunisian sections

The El Kef II, Elles II, and Ain Settara sections are located in the Atlas mountains of north-central Tunisia (Kalaat-Senan area) (Keller et al. 1995 and 2002b; Luciani 2002) (Fig. 1). The palaeogeographic setting of these three sections is similar. All are located on the continental shelf, but Ain Settara and Elles II were deposited in a slightly more proximal position to the Kasserine Island and at a shallower depth (middle to outer neritic) with respect to El Kef II (outer neritic to upper bathyal) (Keller et al., 2002b). The sampling spacing is very small for all the sections (5-10 cm or less) allowing a very high-resolution analysis.

The El Kef II section crops out about 600 m north of the El Kef stratotype section, located about 7 km west of the town of El Kef. The data here shown are from Keller et al. (1995) and are referred to a 1-m thick segment across the KTB. The stratigraphic interval spans the *Plummerita hantkeninoides* (CF1) and P0 zones (Fig. 2).

The Elles II section is located about 75 km southeast of El Kef near the hamlet of Elles, in a valley cut by the Karma river. Data here reported are from Keller et al. (2002b) and are referred to a 1.8-m thick interval across the K-T transition. The stratigraphic interval spans the *P. hantkeninoides* (CF1), P0 and P1a (1) Zones (Fig. 3).

The Ain Settara section crops out about 50 km southeast of the El Kef section and west of Elles. The segment analyzed across the KTB is 2.2-m thick, and spans the CF1, P0, and P1a Zones (Luciani 2002) (Fig. 4).

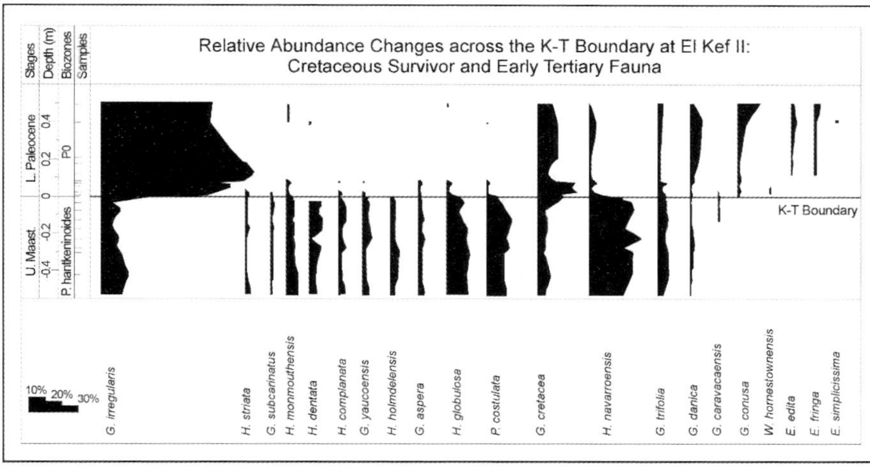

Fig. 2. Relative species abundances of Cretaceous survivor and early Tertiary planktonic foraminifera in the > 38 µm from the outer neritic-upper bathyal El Kef II section, Tunisia (after Keller et al. 1995). Note the remarkable increase in abundance of *Guembelitria irregularis* in the P0 Zone.

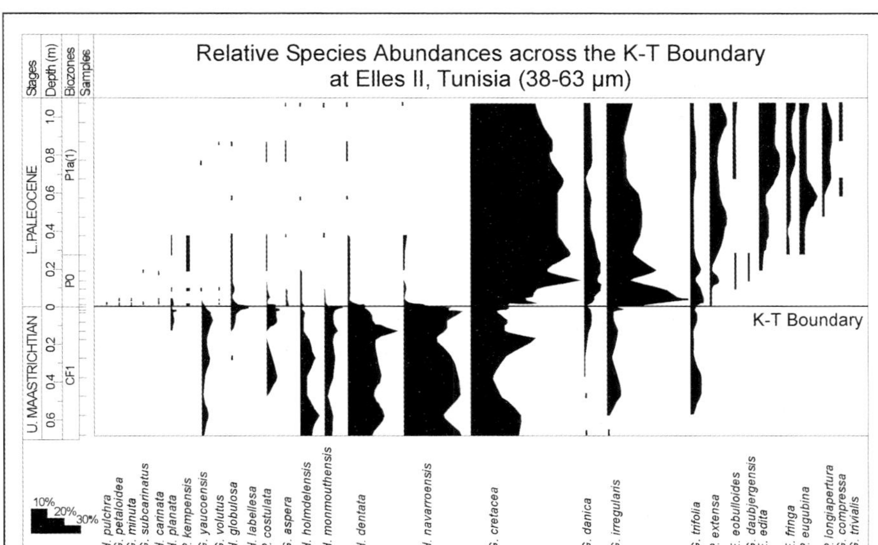

Fig. 3. Relative species abundances of Cretaceous survivor and evolving early Tertiary planktonic foraminifera planktonic foraminifera in the 38-63 µm from the middle-outer neritic Elles II section, Tunisia (after Keller et al. 2002b). Note the high percentage of *Guembelitria irregularis* in the P0 Zone.

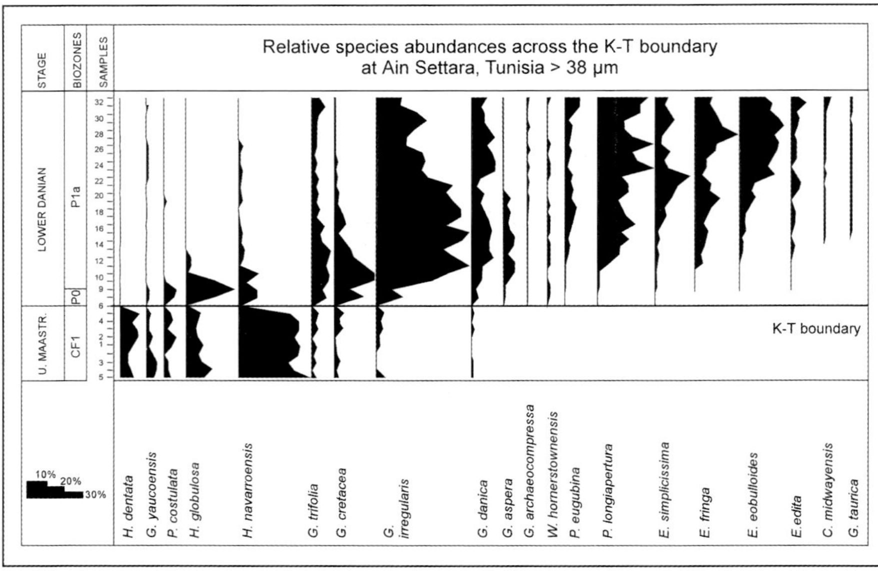

Fig. 4. Relative abundance changes of planktonic foraminifera species in the > 38 µm size fraction from the middle neritic Ain Settara section, Tunisia (after Luciani 2002). Note the marked increase in percentage of *Guembelitria irregularis* in the P0 and P1a Zones.

2.2 The Koshak section (Kazakhstan)

The inner-middle neritic Koshak section is located on the northern slope of the Aktau Mountains of the Mangyshlack Peninsula, Kazakhstan (Sarkar et al. 1992) (Fig. 1). The data here reported are from Pardo and Keller (1999) and Pardo et al. (1999) and are referred to a 5 m-thick segment across the K-T transition (Fig. 5). The stratigraphic interval spans the unzoned uppermost Maastrichtian and the P0, P1a, P1b, and P1c Zones.

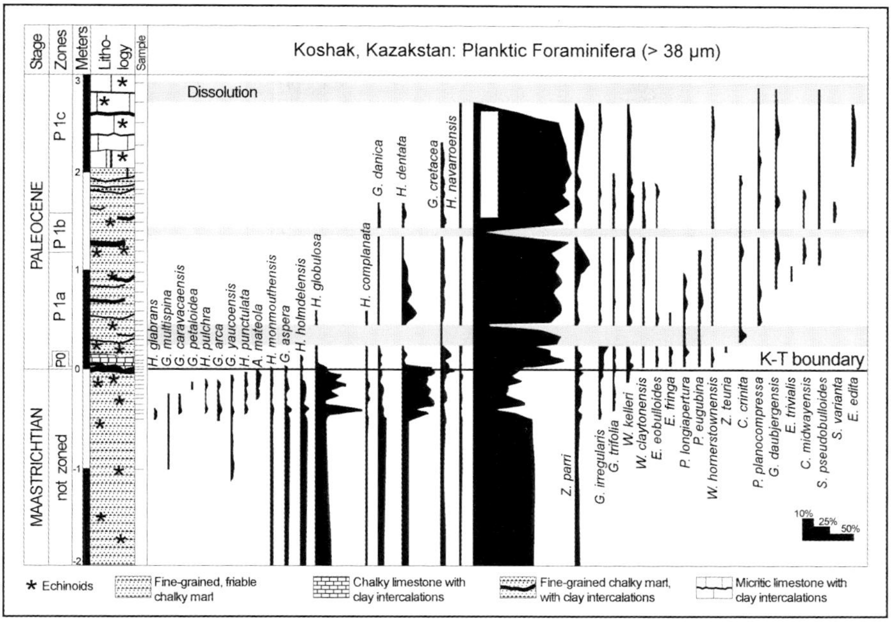

Fig. 5. Relative species abundances of planktonic foraminifera in the > 38 μm from the inner-middle neritic Koshak section, Kazakstan (after Pardo and Keller 1999). Note the increase in abundance of *Guembelitria irregularis* just above the KTB.

2.3 The Caravaca section (Spain)

The middle bathyal Caravaca section crops out in southeast Spain (Fig. 1). Planktonic foraminifera were previously analyzed by Canudo et al. (1991), but the fraction > 38 μm is for the first time here investigated from the same sample set studied by Coccioni and Galeotti (1994) (Fig. 6). The section studied consists of a stratigraphical interval 2-m-thick across the KTB and spans the *P. hantkeninoides, G. cretacea, Parvularugoglobigerina eugubina*, and *Parasubbotina pseudobulloides* Zones that is the CF1, P0, P1a, and P1b Zones (Fig. 6).

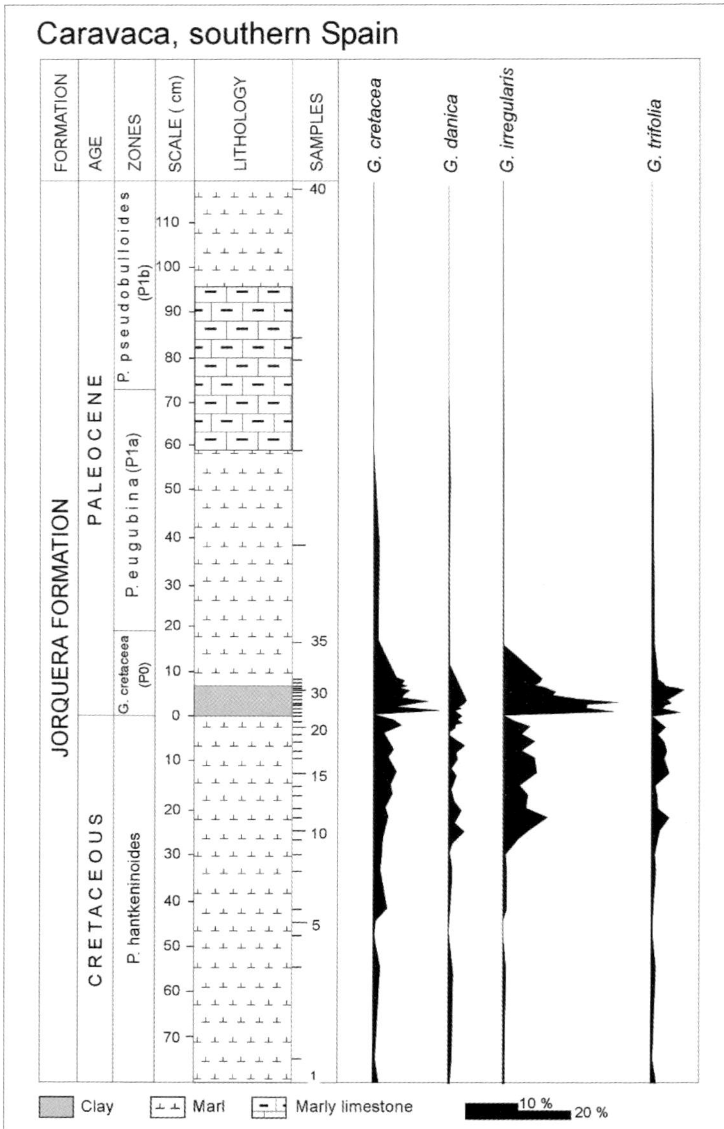

Fig. 6. Relative guembelitrids abundance across the KTB in the > 38 μm size fraction from the middle bathyal Caravaca section (southern Spain). Litho- and biostratigraphy modified after Molina et al. (1998) and Arz et al. (2000). Samples are from Coccioni and Galeotti (1994). Note the high percentage of *Guembelitria irregularis* in the *G. cretacea* Zone that is P0 Zone.

2.4 The Erto section (Italy)

The middle-lower bathyal Erto section crops out in a gorge near the village of Erto in the Vajont Valley (Fig. 1) not far from the site of a disastrous landslide which killed 2000 people in October 1963. The Vajont Valley is located in the eastern southern Alps, which are interpreted as a south-verging thrust belt of Neogene age (e.g., Castellarin 1979; Doglioni and Bosellini 1987). The section examined belongs to a major thrust unit and is overturned. The planktonic foraminifera from the K-T transition was previously studied by Luciani (1997). However, the fraction > 38 μm was here analysed from twenty-nine samples coming from a 3 m-thick segment across the K-T transition which spans the *P. hantkeninoides* (CF1), P0 and P1a Zones (Fig. 7).

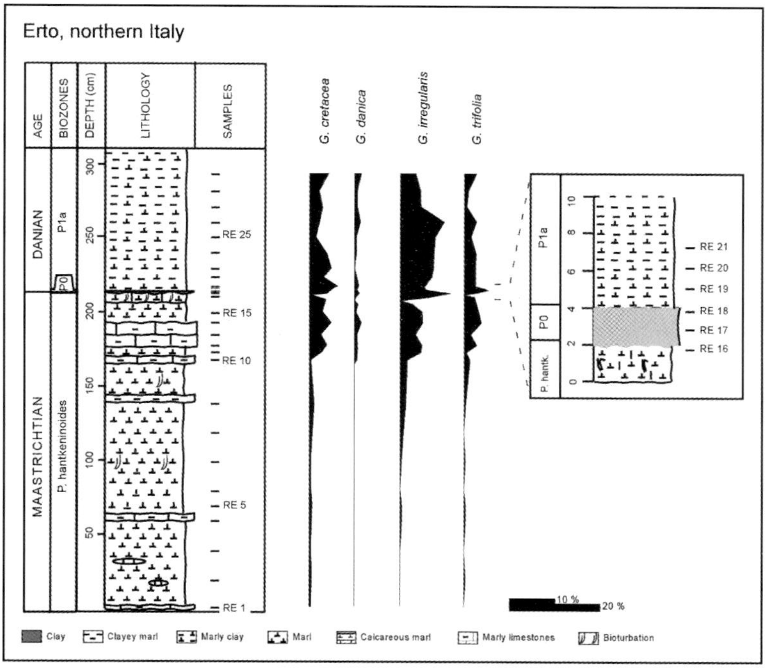

Fig. 7. Stratigraphic column of the middle-lower bathyal Erto section (Vajont Valley, Belluno province, northern Italy) plotted against relative abundance changes of the *Guembelitria* species with respect to the total planktonic foraminifera in a statistical population of at least 300 specimens for each sample. across the KTB in the fraction > 38 μm. Note the positive peak of *Guembelitria irregularis* abundance just above the KTB.

3 Results and discussion

3.1 The opportunistic genus Guembelitria across the K-T boundary

The KTB mass extinction in planktonic foraminifera is one of the most severe biotic effects generally attributed to the impact of a large extraterrestrial body (Alvarez et al. 1980) on the northeastern part of the Yucatan Peninsula (Mexico) resulting in the discovery of the Chicxulub crater (Hildebrand et al. 1991; Pope et al. 1991) dated at 65 Ma (Swisher et al. 1992). However, according to Keller (2003) and Keller et al. (2002a, 2004) impact ejecta (microtektites) from this crater have been discovered interlayered in late Maastrichtian marls of Zone CF1 in northeastern Mexico, indicating that the Chicxulub crater would represents a second impact that predated the KTB by about 300 kyr.

The K-T transition was a time of extremely stressful environmental conditions for any living organisms due to rapid and extreme climate fluctuactions from the tropics to the high latitudes, sea-level changes, intense volcanism, and impact events (see Sharpton and Ward 1990; Ryder et al. 1996; Koeberl and MacLeod 2002). At these times of critical, high stress environmental and biotic conditions, ecological generalists dominated and a few opportunistic taxa thrived including the disaster, opportunistic genus *Guembelitria,* which was able to tolerate unstable environments (Olsson 1970; Smit 1982; Liu and Olsson 1992; Abramovich and Keller 2002; Keller 2002; Keller 2003).

The palaeoecology of the genus *Guembelitria* is still poorly understood, and the environmental conditions in which this genus thrived are still speculative. Accordingly, further studies of its occurrences across the K-T transition are needed to conclusively determine the real ecological meaning of this genus. However, Kroon and Nederbragt (1990) suggested that *G. cretacea* is an upwelling species. According to Koutsoukos (1996), *G. cretacea* and *G. irregularis* were relatively more abundant in highly eutrophic to eutrophic waters. *Guembelitria* species appear to have been surface dwellers, as is suggested by their consistently common occurrence in shallow water settings (e.g., New Jersey, Texas, Alabama, Denmark, Tunisia; Olsson 1970; Keller 1989; Olsson and Liu 1993; Smit 1982; Keller et al. 1993, 1998, 2002b), small size and very light carbon isotopic values (see Keller 2002).

In the studied sections, the late Cretaceous-earliest Danian representatives of this genus are *Guembelitria cretacea, G. danica, G. irregularis*, and *G. trifolia.* In particular, the species *G. irregularis* is a microperforate elongate (height exceeding two/three times the maximum diameter) form with an irregular disposition of the chambers (Fig. 8). The ratio height/diameter of the test is not constant and the number of chambers in each whorl are not uniform so that the chambers do not form regular rows as in *G. cretacea* and *G. danica*. These lines of evidences allow us to regard *G. irregularis* as a morphological abnormalities-bearing guembelitrid.

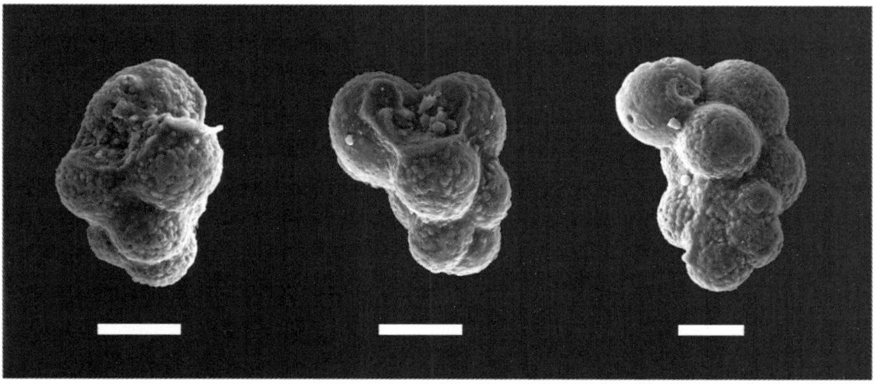

Fig. 8. *Guembelitria irregularis* from the lowermost part of Zone P1a of the Ain Settara section, Tunisia. This species displays an abnormal test due to the irregular disposition and growth of the chambers. The ratio height/diameter of the test is not constant and the number of chambers in each whorl are not uniform so that the chambers do not form regular rows as in *G. cretacea* and *G. danica*. These lines of evidences allow us to regard *G. irregularis* as a morphological abnormalities-bearing guembelitrid. Scale bars represent 30 μm.

Guembelitria irregularis was first reported from the Russian Danian-Montian (Morozova 1961) and then also found in upper Cretaceous successions, with the known earliest record in the upper Maastrichtian CF3 Zone (Keller 2002). However, due to the very small size of this form, it is possible that its earlier occurrences do not appear evident unless the size fraction smaller than 63 μm of foraminiferal assemblages is analyzed.

Relative abundance of *G. irregularis* is variable across the KTB. The highest proportions occur both in the uppermost Maastrichtian and in the lower Danian at the outer neritic to upper bathyal setting of the El Kef II section (up to 16% and up to 93% respectively) (Figs. 2 to 7). For a few tens of thousand of years after the KTB bolide impact, guembelitrids dominate (up to 80-90%) in the small 38-63 μm size fraction of the planktonic foraminiferal assemblages from lower Danian Zones P0 and P1a, with *G. irregularis* as prevailing species within this group (Figs. 2 to 7). This is at present recognizable from low to middle latitudes and at different water–depth settings, from the inner-middle neritic environment (Koshak section) to the middle-lower bathyal (Erto section) one.

The late Maastrichtian *Guembelitria* blooms are associated with highly eutrophic but low productivity environments (Keller 2002), rapid climatic warm-cool transition and Deccan volcanism (Keller 2001; Abramovich and Keller 2002; Keller 2003). The post-KTB *Guembelitria* blooms around the world occurred both in shallow and deep water environments, near shores and in open ocean, at high and low latitude even if higher abundances are found in relatively shallow water at low latitudes. *Guembelitria* blooms are therefore not specific to temperature, water depths, or salinity, but seem to occur during times of eutrophic waters and disruption of normal water mass stratification conditions.

3.2 Morphological abnormalities of fossil and recent foraminiferal tests

The foraminifera have short reproductive cycles and rapid growth and their tests are readily preserved. Moreover, because the cytoskeleton controls both the shape of the cell and the transport of organelles or vesicular structures, injury to the skeletons may alter the shape and arrangement of chambers and the transport of proteins and calcium to the test wall. Accordingly, the development of abnormal test morphology (mainly abnormal shapes, sizes, or dispositions of one or several chambers) can record evidence through time of environmental stress of ecological influences perturbing the test construction, although severe malformations may also follow other processes including reproduction or regeneration after mechanical damage.

Morphological abnormalities in number, shape, size and disposition of test chambers were detected and described in fossil and modern planktonic and benthic foramininifera (see reviews in Alve 1991, 1995; Yanko et al. 1994, 1998, 2000; Geslin et al. 1998, 2000, 2002; Stouff et al. 1999a,b; Coccioni 2000; Samir 2000). Notably, aberrant forms of *P. eugubina,*

Eoglobigerina eobulloides and *Eoglobigerina edita* occur in high abundance in the basal Danian section of DSDP Sites 577 and 47.2 (Shatsky Rise, North Pacific), and 524 (South Atlantic) (Gerstel et al. 1986). These forms, characterized by the development of secondary apertures, bullae and abnormal final chambers, were considered by Gerstel et al (1986) to be ecophenotypic variants reflecting ecologic stress or instability in the earliest Cenozoic marine environment.

Following the recent literature, morphological abnormalities in benthic foraminifera are considered to result mainly from various anthropogenic activities even if it is complicated to determine exactly the possible causes and origins of abnormal test formation. Taking into account that the normal rate of abnormal tests in a non-stressed, modern population is about 1% (Alve 1991), the relative abundance of malformed tests is a useful proxy for the reconstruction of ecological changes and may be proportional to the strength of the environmental stress.

4 Conclusions: A speculative hypothesis

Within the opportunistic genus *Guembelitria* known to tolerate unstable environments, the species *G. irregularis* displays an aberrant test due to the irregular disposition and growth of the chambers which suggest a morphological malformation. Available data from three Tunisia sections (El Kef II, Elles II, Ain Settara) and Kazakstan (Koshak) and new data from Spain (Caravaca), Italy (Erto), show a dramatic increase of *G. irregularis* (up to 80-90%) in the small 38-63 μm fraction of the early Danian assemblages (planktonic foraminiferal Zones P0-P1a). Positive peaks in abundance of this form are also recorded in the latest Maastrichtian, even though with minor percentages (up to 16% at El Kef II).

Morphological abnormalities of fossil and Recent foraminiferal tests have long been reported from stressed environments. Accordingly, we conjecture that the morphological abnormalities displayed by the *G. irregularis* test across the KTB are the result of the extremely stressful environmental conditions related to the complex interplay of different events. We speculate also that the relative abundances of *G. irregularis* may be proportional to the strength of the environmental stress. Following this hypothesis, the data here reported would suggest that the post-KTB environmental stress in surface waters reached its climax around the transition between the planktonic foraminiferal Zones P0 and P1a, that is just a few thousands of years after the bolide impact. The occurrence of large population of aberrant *P. eugubina* and *Eoglobigerina* in the earliest Danian at some DSDP Sites (Gerstel et al. 1986) strongly support our results. Also taking into account a multi-event mass extinction scenario [e.g., extraterrestrial body impact(s), volcanism, climate changes, sea-level changes] at the KTB (see Sharpton and Ward 1990; Ryder et al. 1996; Keller 2001, 2003; Koeberl and MacLeod 2002; Keller et al. 2004), the combination of all the physical effects (e.g., blast damage, extensive fires, global cooling, earthquakes, acid rain, sulfate release, destroyment of the ozone shield) induced by the largest impact in the past 65 million years would surely represent a devastating stress on the global biosphere, lasting for some hundreds of thousands of years (see Toon et al. 1997).

During the latest Maastrichtian rapid and extreme climate fluctuations, sea-level changes, impact events, rapid climatic warm-cool transition and Deccan volcanism (Keller 2001, Keller 2002; Abramovich and Keller 2002; Ravizza and Peucker-Ehrenbrink 2003) may have induced unstable perturbated environmental conditions inducing increased abundance of the aberrant *G. irregularis,* though with minor amounts with respect to the post-KTB.

Undoubtedly further studies are necessary to shed light on our hypothesis even if the complexity of the events that took place across the KTB is so great that considerable uncertainty will probably remain after even the most exhaustive analyis.

Acknowledgments

We thank M.A. Kaminski, G. Keller, and J. Smit for their helpful reviews. C. Koeberl made very helpful additional comments on the manuscript. This research was supported by the MIUR (Ministero dell'Istruzione, Università e Ricerca) COFIN (Cofinanziamento Progetti di Ricerca di interesse nazionale) 2001 to RC and MIUR ex 60% to RC and VL.

References

Abramovich S, Keller G (2002) High stress late Maastrichtian paleoenvironment: inference from planktonic foraminifera in Tunisia. Palaeogeography, Palaeoclimatology, Palaeoecology 178: 145-164

Adatte T, Keller G, Burns S, Stoykova KH, Ivanov MI, Vangelov D, Kramer U, Stueben D (2002) Paleoenvironment across the Cretaceous-Tertiary transition in Eastern Bulgaria. In: Koeberl C, MacLeod KG (eds) Catastrophic Events and Mass Extinctions: Impacts and Beyond. Boulder, Colorado, Geological Society of America Special Paper 356: 231-251

Alvarez LW, Alvarez W, Asaro F, Michel HV (1980) Extraterrestrial cause for the Cretaceous-Tertiary extinction. Science 208: 1095-1108

Alve E (1991) Benthic foraminifera in sediment cores reflecting heavy metal pollution in Sørfjord, Western Norway. Journal of Foraminiferal Research 21:1-19

Alve E (1995) Benthic foraminifera response to estuarine pollution: a review. Journal of Foraminiferal Research 25: 190-203

Arz JA, Arenillas I, Molina E, Sepúlveda R (2000) La estabilidad evolutiva de los foraminiferos planctónicos en el Maastrichtiense Superior y su extinctión en el limite Cretácico/Terciario de Caravaca, España. Revista Geológica de Chile 27: 27-47

Canudo JI, Keller G, Molina E (1991) Cretaceous/Tertiary extinction pattern and faunal turnover at Agost and Caravaca: S.E. Spain. Marine Micropaleontology 17: 319-341

Castellarin A (1979) Il problema dei raccorciamenti crostali del Sudalpino. Rendiconti della Società Geologica Italiana 1: 21-33

Coccioni R (2000) Benthic foraminifera as bioindicators of heavy metal pollution. In: Martin RE (ed) Environmental Micropaleontology: The Application of Microfossils to Environmental Geology, Kluwer Academic/Plenum Publishers, New York, pp 71-103

Coccioni R, Galeotti S (1994) K-T boundary extinction: geologically instantaneous or gradual event? Evidence from deep-sea benthic foraminifera. Geology 22: 779-782

Denham CR, Scotese CR (1987) Terra Mobilis: A plate tectonic program for the Macintosh, version 1.1. Geoimages, Earth in Motion Technology, Austin, TX: 26 pp

Doglioni C, Bosellini A (1987) Eoalpine and mesoalpine tectonis in the Southern Alps. Geologische Rundschau 77: 734-754

Gerstel J, Thunell RC, Zachos JC, Arthur MA (1986) The Cretaceous/Tertiary boundary event in the North Pacific: planktonic foraminiferal results from Deep Sea Drilling Project Site 577, Shatsky Rise. Paleoceanography 1: 97-117

Geslin E, Debenay JP, Lesourd M (1998) Abnormal wall textures and test deformation in *Ammonia* (hyaline foraminifer). Journal of Foraminiferal Research 28: 148-156

Geslin E, Stouff V, Debenay JP, Lesourd M. (2000) Environmental variation and foraminiferal test abnormalities. In: Martin RE (ed) Environmental Micropaleontology: The Application of Microfossils to Environmental Geology, Kluwer Academic/Plenum Publishers, New York, pp 191-215

Geslin E, Debenay J-P, Dulela W, Bonetti C (2002) Morphological abnormalities of foraminiferal tests in Brazilian environments: comparison between polluted and non-polluted areas. Marine Micropaleontology 45: 151-168

Hildebrand AR, Penfield GT, Kring D, Pilkington M, Camargo A, Jacobsen SB, Boynton W (1991) Chicxulub crater: A possible Cretaceous-Tertiary boundary impact crater on the Yucatan Peninsula, Mexico. Geology 19: 867-871

Keller G (1989) Extended Cretaceous/Tertiary boundary extinctions and delayed population change in planktonic foraminifera from Brazos River, Texas. Paleoceanography 4: 287-332

Keller G (1993) The Cretaceous/Tertiary boundary transition in the Antartic Ocean and its global implications. Marine Micropaleontology 12: 1-45

Keller G (2001) The end-Cretaceous mass extinction in the marine realm: year 2000 assessment. Planetary and Space Science 49: 817-830

Keller G (2002) *Guembelitria*-dominated late Maastrichtian planktic foraminiferal assemblages mimic early Danian in central Egypt. Marine Micropaleontology 47: 71-99

Keller G (2003) Biotic effects of impacts and volcanism. Earth and Planetary Science Letters 215: 249-264

Keller G, Barrera E, Schmitz B, Matsson E (1993) Gradual mass extinction, species survivorship, and long term environmental changes across the Cretaceous-Tertiary boundary in high latitudes. Geological Society of America Bulletin 105: 979-997

Keller G, Li L, MacLeod N (1995) The Cretaceous/Tertiary boundary stratotype section at El Kef, Tunisia: how catastrophic was the mass extinction?. Palaeogeography, Palaeoclimatology, Palaeoecology 119: 221-254

Keller G, Adatte T, Stinnesbeck W, Stüben D, Kramar U, Berner Z, Li L, von Salis Perch-Nielsen K (1998) The Cretaceous-Tertiary transition on the shallow Saharan platform of southern Tunisia. Geobios 30: 951-975

Keller G, Adatte T, Stinnesbeck W, Affolter M, Schilli L, Lopez-Oliva JG (2002a) Multiple spherule layers in the late Maastrichtian of northeastern Mexico. In: Koeberl C, MacLeod KG (eds), Catastrophic Events and Mass Extinctions: Impacts and Beyond, Boulder, Colorado, Geological Society of America Special Paper 356: 145-161

Keller G, Adatte T, Stinnesbeck W, Luciani V, Karoui-Yaakoub N, Zaghbib-Turki D (2002b) Paleoecology of the Cretaceous-Tertiary mass extinction in planktonic foraminifera. Palaeogeography, Palaeoclimatology, Palaeoecology 178: 257-297

Keller G, Adatte T, Stinnesbeck W, Rebolledo-Vieyra M, Urrutia Fucugauchi J, Kramar U, Stüben D (2004) Chicxulub impact predates the K-T boundary mass extinction. Proceedings of the National Academy of Sciences of the United States of America 101: 3753-3758.

Koeberl C, MacLeod KG (eds) (2002) Catastrophic Events and Mass Extinctions: Impacts and Beyond. Boulder, Colorado, Geological Society of America Special Paper 356, 746 pp

Koutsoukos EAM (1996) The Cretaceous-Tertiary in NE Brazil; high resolution event stratigraphy. Anais da Academia Brasileira de Ciencias 68(2): 265-266

Kroon D, Nederbragt AJ (1990) Ecology and Paleoecology of triserial planktic foraminifera. Marine Micropaleontology 16: 25-38

Liu C, Olsson RK (1992) Evolutionary radiation of microperforate planktonic foraminifera following the K/T mass extinction event. Journal of Foraminiferal Research 22: 328-346

Luciani V (l997) Planktonic foraminiferal turnover across the Cretaceous-Tertiary boundary in the Vajont valley (Southern Alps, northern Italy). Cretaceous Research 18: 799-821

Luciani V (2002) High resolution planktonic foraminiferal analysis from the Cretaceous/Tertiary boundary at Ain Settara (Tunisia): Evidence of an extended mass extinction. Palaeogeography, Palaeoclimatology, Palaeoecology 178: 299-319

Morozova V.G. (1961) Datsko-Monskie Planktonye Foraminifery juga SSSR (Planktonic foraminifera from the Danian-Montian of the Southern Soviet Union). Akademia Nauk SSSR, Paleontologicheskiy Zhurnal 2: 8-19 (in russian)

Olsson RK (1970) Paleocene planktonic foraminiferal biostratigraphy and Paleozoogeographic of New Jersey. Journal of Paleontology 44: 589-597

Olsson RK, Liu C (1993) Controversies on the placement of the Cretaceous-Paleogene boundary and the K/T mass extinction of planktonic foraminifera. Palaios 8: 127-139

Pardo A, Keller G (1999) Aspectos paeoceanográficos y paleoecológicos del límite Cretácico/Terciario en la peninsula de Mangyshlak (Kazakstán): inferencias a partir de foraminíferos planctónicos. Revista Española de Micropaleontologia 31: 265-278

Pardo A, Ortiz N, Keller G (1996) Latest Maastrichtian and K/T boundary foraminiferal turnover and environmental changes at Agost, Spain. In: MacLeod N, Keller G (eds), Biotic and Environmental Events across the Cretaceous/Tertiary Boundary, Norton, New York, pp 139-171

Pardo A, Adatte T, Keller G, Oberhänsli H (1999) Paleoenvironmental changes across the Cretaceous-Tertiary boundary at Koshak, Kazakhstan, based on planktic foraminifera and clay mineralogy. Palaeogeography, Palaeoclimatology, Palaeoecology 154: 247-273

Pope KO, Ocampo A, Duller D (1991) Mexican site for the K/T crater? Nature 351: 105

Ravizza G, Peucker-Ehrenbrink B (2003) Chemostratigraphic evidence of Deccan volcanism from the marine osmium isotope record. Science 302: 1392-1395

Rögl F, Salis Kv, Preisinger A, Aslanian S, Summesberger H (1996) Stratigraphy across the Cretaceous/Paleogene boundary near Bjala, Bulgaria. Bulletin des Centres de Recherches Exploration-Production Elf Aquitaine, Mémoire 16: 673-683

Ryder G, Fastovsky D, Gartner S (eds) (1996) The Cretaceous-Tertiary event and other catastrophes in earth history. Boulder, Colorado, Geological Society of America Special Paper 307, 569 pp

Samir AM (2000) The response of benthic foraminifera and ostracods to various pollution sources: a study from two lagoons in Egypt. Journal of Foraminiferal Research 30: 83-98

Sarkar A, Bhattacharya SK, Shukla PN, Bhandaru N, Naidin D (1992) High-resolution profile of stable isotopes and iridium across a K/T boundary section from Kyzylsai Hill, Mangyslak, Kazakhstan. Terra Nova 5: 585-590

Sharpton V, Ward P (eds) (1990) Global catastrophes in earth history: An interdisciplinary conference on impacts, volcanism, and mass mortality. Boulder, Colorado, Geological Society of America Special Paper 247, 631 pp

Smit J (1982) Extinction and evolution of planktonic foraminifera after a major impact at the Cretaceous/Tertiary boundary. Boulder, Colorado, Geological Society of America Special Paper 190: 329-352

Stouff V, Debenay JP, Lesourd M. (1999a) Origin of double and multiple tests in benthic foraminifera: observations in laboratory cultures. Marine Micropaleontology 36: 189-204

Stouff V, Geslin E, Debenay JP, Lesourd M (1999b) Origin of morphological abnormalities in *Ammonia* (foraminifera): studies in laboratory and natural environments. Journal of Foraminiferal Research 29: 152-170

Swisher CC III, Grajales-Nishimura JM, Montanari A, Margolis SV, Claeys P, Alvarez W, Renne P, Cedilo-Pardo E, Maurrasse FJ-MR, Curtis GH, Smit J, McWilliams MO (1992) Coeval $^{40}Ar/^{39}Ar$ ages of 65.0 million years ago from Chicxulub crater melt rock and Cretaceous-Tertiary boundary tektites. Science 257: 954-958

Toon OB, Zahnle K, Morrison D, Turco RP, Covey C (1997) Environmental perturbations caused by the impacts of asteroids and comets. Reviews of Geophysics 35: 41-78

Yanko V, Kronfeld J, Flexer A (1994) Response of benthic foraminifera to various pollution sources: implications for pollution monitoring. Journal of Foraminiferal Research 24: 1-17

Yanko V, Ahmad M, Kaminski MA (1998) Morphological deformities of benthic foraminiferal tests in response to pollution by heavy metals: implications for pollution monitoring. Journal of Foraminiferal Research 28: 177-200

Yanko V, Arnold A, Parker W (2000) Effects of marine pollution on benthic foraminifera. In: Sen Gupta BK (ed) Modern Foraminifera, Kluwer Academic Publishers, Dordrecht, pp 217-235

Unravelling the Cretaceous-Paleogene (KT) Turnover, Evidence from Flora, Fauna and Geology

Adriana Ocampo[1], Vivi Vajda[2], Eric Buffetaut[3]

[1]California Institute of Technology Jet Propulsion Laboratory, 4800 Oak Grove Dr. Pasadena California 91109 USA (adriana.c.ocampo@jpl.nasa.gov)
[2]GeoBiosphere Science Centre, University of Lund, Sölvegatan 12, 223 62 Lund, Sweden
[3] CNRS, 16 cour du Liégat, 75013 Paris, France

Abstract. The global devastation of ecosystems as a consequence of a meteorite impact 65 million years ago is clearly detectable in palaeontological and geological records all over the globe. Here we compare and contrast the consequences of the impact expressed in the vegetation, vertebrate fossil record and geological signatures left by the devastation, including information from new proximal KT boundary exposures and new palynological data. The geological evidence of the Chicxulub impact crater shows that the target rock was composed by higher percentages of anhydrite (sulfur source) than carbonates. Atmospheric radiative transfer models suggest that the vaporized target rock rapidly converted into sulfuric acid H_2SO_4 aerosols where it was injected in the stratosphere by the force of the impact and globally distributed. It took at least 10 years for the H_2SO_4 to dissipate, making the Earth's atmosphere opaque to sunlight, leading to a reduction of solar transmission to 10-20% of normal for that period.

Southern Hemisphere terrestrial Cretaceous-Paleogene boundary sediments in New Zealand reveal that a diverse Late Cretaceous vegetation was abruptly followed by a short interval dominated by fungi, before the pioneer vegetation of ferns re-conquered the soil. The fern dominated interval, so called fern spike is also evident from Northern Hemisphere Cretaceous-Paleogene boundary sections. The massive depletion in spore-pollen diversity is interpreted to reflect devastation of photosynthetic plant communities, a scenario that agrees well with the atmospheric radiative transfer models. The pattern of vertebrate extinctions revealed by the fossil record accords with the temporary, global devastation of photosynthetic plant communities. Vegetation depletion at high latitudes may also explain the extinction of polar dinosaurs, which apparently were able to withstand relatively cool temperatures and periods of low light intensity: the main reason for their disappearance appears to have been lack of food rather than darkness and cooling.

1 Introduction

Earth history is punctuated by five major mass extinctions (Raup and Sepkoski 1982, 1986; Sepkoski 1990). The latest and perhaps best studied of these mass extinction events is the one that occurred 65 Ma ago as a consequence of a meteorite impact (Alvarez et al. 1980) at the Yucatan peninsula, the so called Chicxulub impact (Hildebrand et al. 1991; Pope et al. 1991). This dramatic event left its signature in the fossil flora, fauna and sedimentological record of the Earth; the biological extinctions defining the Cretaceous-Paleogene boundary. The end Cretaceous mass extinction is also constrained by a sharp geochemical anomaly enabling identification of the boundary commonly at millimetre scale. This provides a robust, independent, calibration point for documenting the biological response to the catastrophe.

A global system approach to mass extinction studies is crucial to gain a greater insight into the environmental mechanisms behind the ecological disruption and biodiversity crisis in the wake of the impact. It may appear that the marine ecosystem was more affected by the mass extinction than its terrestrial counterpart, but perhaps this is partly due to the fact that the terrestrial ecosystem has been less studied and the continental geological and vertebrate record is less complete.

This work aims to review, compare and integrate both published and new results from different fields, such as palynology, vertebrate palaeontology and post-impact sedimentary mechanisms in order to better resolve the processes affecting terrestrial ecosystems and depositional systems at a global scale at the Cretaceous-Paleogene boundary.

2 Geological evidence

The question whether there is a correlation between large impacts craters and extinctions of biota has been dealt with in several studies of impact craters throughout the geological record (Heissig 1986; Aubry et al. 1990;

Dypvik 1996; Kring 1997; Becker et al. 2004). However for the Chicxulub impact there is a clear correlation to a large biotic turnover and mass extinction where the Cretaceous-Paleogene boundary horizon is globally identified as a sharp contact associated with an iridium anomaly in parts per billion (Alvarez 1980). Other features characterizing the boundary horizon is the presence of shocked quartz (Figure 1) and tektites reflecting the geochemical signature of the target rock (Izett 1991; Izett et al. 1991).

The proximal ejecta layers (200-500 km from the crater) consist of vast sediments up to tens of meters in thickness, thinning out with increased distance from the crater. As a comparison the distal boundary layers are only centimetres to millimetres thick in terrestrial and marine settings worldwide (Smit 1999). In these deposits, the first few minutes, hours, days and months after the impact are preserved.

Fig. 1. A. Grain of shocked quartz (~200µm) found in Belize in the Cretaceous-Paleogene fireball layer. **B.** Thin section of an accretionary lapilli with a glass core (pink centre) from Albion Formation, Belize. **C.** Thin section from the diamictite, Albion Formation, Belize. Shard of vesicular impact glass and foraminifera fragment in a calcareous matrix.

The near ejecta sequence dramatically outcrops in Belize and Southern Mexico, consisting of tens of meters in thickness and some of these exposures have been investigated over the last decade (Ocampo et al. 1996; Ocampo 1997, Vega et al. 1997, Fouke et al. 2002; Keller et al. 2003a, 2003b). The Belize and southern Mexico exposures show proximal ejecta of the Chicxulub impact in the form of a sub-aereal ejecta blanket which is characterized by a two member stratigraphical unit, the so called Albion Formation. The Albion Formation was first defined in Northern Belize at about 300 km from the impact site as the most proximal outcrop of the ejecta blanket (Ocampo et al. 1996; Ocampo 1997). The lower unit of the Albion Formation, also called the Spheroid Bed, contains spherules, altered vesicular glass shards, and other impact debris. The upper unit, the so-called Diamictite Bed, overlays the Spheroid Bed and contains abundant shocked quartz, and is also rich in siderophile elements, most notably iridium. The Albion Formation (Spheroid Bed and Diamictite Bed) was formed during the first few minutes following the impact and reflects different depositional processes generated by the impact (Figure 2). Subsequently, the Albion Fm was recognized at several new Cretaceous-Paleocene localities in Southern Quintana Roo, Mexico, exposed at Alvaro Obregon, Ramonal, Agua Dulce and Thompson Quarry, to name a few sites. The formation was also identified in Central Belize, Armenia, at a distance of 500 km from the Chicxulub crater where the contact between the Cretaceous dolomite of Barton Creek Fm and the overlying spheroid bed of Albion Fm is found. Interestingly, at a distance of 800 km from the crater, in Santa Teresa, southern Belize, a tektite layer is exposed, similar to the tektites found in Guayal of southern Mexico (Salge et al. 2000).

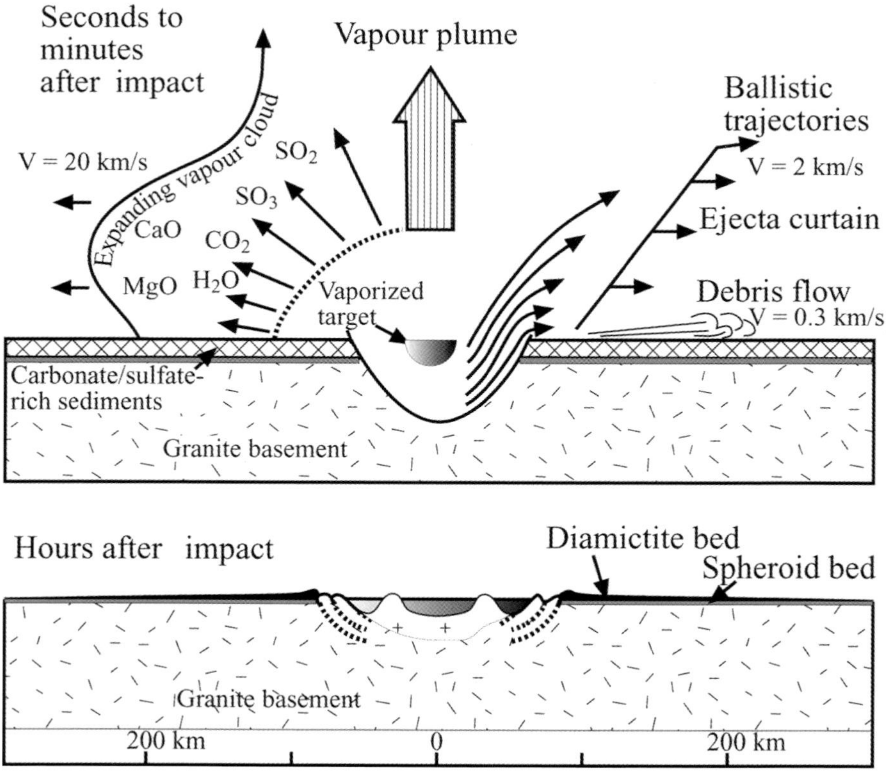

Fig. 2. Scenario immediately following the Chicxulub impact showing the three main transport processes that contributed to the genesis of the Spheroid Bed and Diamictite Bed.

The origin of this two-part stratigraphy is controversial, but at least three major processes were involved in the genesis of these deposits; debris flow, ballistic ejecta and carbonate vapour condensation (Ocampo et al. 1996). The first to be deposited were the spherules which condensed from the cooling vapour plume, expanding at a rate of approximately 20 km/s. These spherules yield a light oxygen isotopic signature that is consistent with genesis within the expanding vapour plume cloud by condensation and coagulation and subsequently mixing with fragmented rock. Precipitation of these millimetres to centimetre-sized spherical pebbles (spherules) of $CaCO_3$ and $(CaMg)(CO_3)_2$ produced the Spheroid Bed, the lower ejecta layer. The Diamictite Bed, which overlies the Spheroid Bed, is mainly composed of larger pebbles and blocks up to 8 m in diameter (many with striations), and was deposited from a debris flow travelling near the ground at a speed of 0.3 km/s. Ballistic material transported at an estimated speed of 2 km/s is also incorporated in the Diamictite Bed (Pope et al. 1994, 1997; Ocampo et al. 1996). The ejecta blanket distribution around the crater remains an important factor in determining the impact angle (Pierazzo and Melosh 2000a,b). The distribution of near ejecta deposits found in southern Mexico and Belize further corroborates the oblique angle of impact. Rock volumes of 300 - 2000 km^3 were vaporized at the impact, releasing 2.7 to 8.2 x 10^{17} grams of sulfur (Pope et al. 1994, 1997; Ivanov et al. 1996). A minor part of the sulfur was released as SO_3 or SO_4^{2-}, but the majority produced sulfuric acid aerosol (Figure 2). The vast quantities of sulfur dioxide released into the atmosphere (Kring et al. 1996), in combination with the atmospheric water vapour, produced H_2SO_4 clouds that were distributed globally in the stratosphere, and attenuated sunlight penetration to the surface to approximately 20% of normal for about 10 years. The sulfuric acid-rich clouds' capability to remain in Earth´s atmosphere for up to a decade was the prime mechanism inhibiting sunlight penetration to the surface producing near freezing temperatures (Pope 2002).

Previous results based on 3D hydrocode simulations have shown that the impact angle affects the strength and distribution of the shock wave generated and that the volume of melt is directly proportional to the volume of the transient crater generated by the impact (Pierazzo and Melosh 2000a,b). For a crater such as Chicxulub, with an impact angle below 45 degrees, the amount of melted target material is less compared to a vertical impact of the same size impactor. However, due to the substantial amount of anhydrite in the target rock, sulfur became more disruptive to the biosphere than impact-generated CO_2 (Pope et al. 1994, 1997; Pierazzo et al. 1998, Pierazzo and Melosh 2000a,b). During the first months to a year after the impact, dust and soot from post impact wild fires played a greater role in blocking sunlight (Wolbach et al. 1988; Kring and Durda 2002), however the sulfuric acid aerosol provided the most damaging longer-term effect to the biosphere (Pope 2002; Toon et al. 1997; Covey et al. 1994; Pollack et al. 1983).

3 Vegetation response

There is now overwhelming evidence for global disruption of vegetation at the Cretaceous-Paleogene boundary. However, there are important regional differences in the signature of vegetation turnover. The data suggest both massive devastation and mass extinction of plants at many Cretaceous-Paleogene boundary sections in North America (Nichols and Johnson 2002; McIver 1999; Hotton 2002; Norris 2001) but mainly mass-kill of vegetation at Southern Hemisphere high latitudes resulting in dramatic but short-term changes in the relative abundance of plant groups (Vajda et al. 2001; Vajda and Raine 2003; Vajda and McLoughlin 2004).

The floristic turnover is evident at centimetre scale in many terrestrial sections, enabling precise positioning of the Cretaceous-Paleogene boundary. The most characteristic palynological feature of the Cretaceous-Paleogene boundary is the sudden disappearance of a diverse Maastrichtian pollen assemblage, usually followed by a low-diversity succession of ferns in the earliest Paleocene. This so-called fern-spike is well-documented in the southern United States where several extinctions and an abrupt decline in angiosperm pollen is followed by an impoverished flora dominated by ferns. The fern-spore spike associated with iridium enrichment in the sediments was first reported by Orth et al. (1981) in material from New Mexico. Subsequently, palynology has been used extensively for high resolution investigations of non-marine mid-continental North American sections, to document the response of terrestrial plants to the this event (Jerzykiewicz and Sweet 1986; Nichols et al. 1986; Wolfe and Upchurch 1986; Bohor et al. 1987; Lerbekmo et al. 1987; Upchurch and Wolfe 1987; Upchurch 1989; Nichols and Fleming 1990; Sweet et al. 1990, 1999; Wolfe 1991; Sweet and Braman 1992, 2001; Nichols et al. 1992; Braman et al. 1993; McIver 1999; Nichols and Johnson 2002; Hotton 2002; Barclay and Johnson 2004). These studies agree that all North American plant communities experienced severe simultaneous disruption at the Cretaceous-Paleogene boundary. The extinction rate of plants based on the palynoflora ranges from 15% in the Raton Basin (Fleming 1985) to 30% at Morgan Creek and southwestern North Dakota (Nichols et al. 1986; Nichols and Johnson 2002). Megafloral turnover has been documented from a vast number of sections in southwestern North Dakota where a loss of 57% of the plant species are correlated to the end-Cretaceous extinction (Wilf and Johnson 2004).

There seems to exist a major discrepancy between the turnover traced in the pollen record, compared to extinction seen in the megaflora. In sediments where palynology and megafloras have been compared, an extinction of 80% of the megaflora corresponds to a turnover rate of 30 % in the pollen-spore record (Johnson et al. 1989; Johnson 1992; Nichols and Johnson 2002). However, the high extinction rate does not appear to be consistent throughout North America as detailed studies of miospores from a Cretaceous-Paleogene boundary section in Saskatchewan, Canada, demonstrate mass kill of standing vegetation but lacks evidence of mass extinction (McIver 1999). Additionally, some sections in western Canada reveal an anomalous rise in angiosperm pollen abundance following the Cretaceous-Paleogene boundary comparable to the fern-spike identified elsewhere (Sweet et al. 1990). In these localities the pioneer vegetation most probably comprised opportunistic, herbaceous angiosperms instead of ferns.

Well-defined, terrestrial, KT sections are non-existent in Europe but a bryophyte spike (moss spores), comparable to the fern-spike, has been reported from marine KT boundary sections in The Netherlands at Curf Quarry (Herngreen et al. 1998) and in the Geulhemmerberg caves (Brinkhuis and Schiøler 1996). The sudden bloom of bryophytes immediately after the KT boundary indicates a major change in the terrestrial ecosystem.

New Zealand, located in the southern hemisphere at the time of impact and far from the Chicxulub crater site, provides an invaluable opportunity to test the global effects of the impact on the biodiversity of the ecosystems. We have recent palynological data from five New Zealand Cretaceous-Paleogene boundary sections (both outcrops and drillcores), all located within the Greymouth Coalfield. The palaeoenvironment consisted of a braided river system covered with swamp vegetation, where silt and mud was sporadically deposited on coal-forming floodplain mires between sandy channel tracts.

The Cretaceous-Paleogene boundary falls within the Paparoa Group, which comprises several coal-bearing intervals (Nathan 1978; Ward 1997). The boundary has been pinpointed by palynology in both marine and terrestrial sediments (Vajda et al. 2001; Vajda and Raine 2003; Vajda and McLoughlin 2004) and is characterized by massive vegetation disruption. The palynological signal through the New Zealand KT sections is nearly identical to those of the southern and central United States. The turnover in the flora is very sudden in the investigated sequences. The latest

Maastrichtian palynoflora includes over 100 species of miospores, but these are suddenly replaced, following the geochemical anomaly, by diversifying fern spores in the earliest Paleocene.

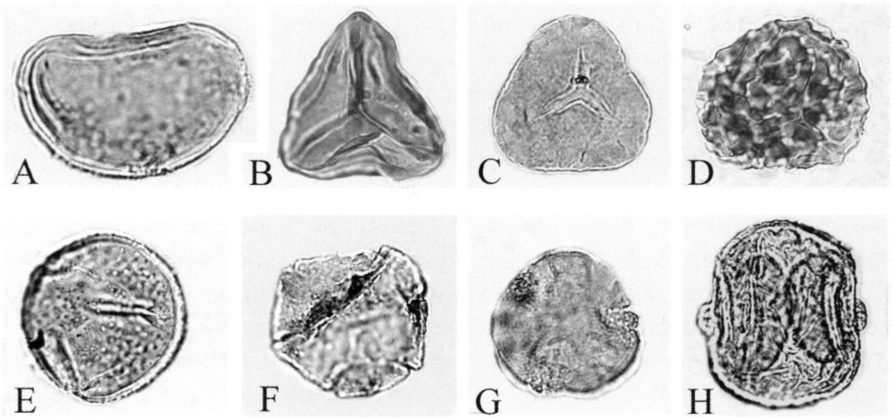

Fig. 3. Spores from the Cretaceous-Paleogene boundary section at Moody Creek Mine, New Zealand, magnification 500x. **(a)** *Laevigatosporites ovatus,* spore related to modern *Blechnum*, a groundfern and first pioneer species recovering after the end-Cretaceous event in Southern Hemisphere, New Zealand sections. **(b)** *Gleicheniidites senonicus*, spore related to modern *Gleichenia,* ground fern and early coloniser **(c)** *Cyathidites minor*, of probable tree fern affinities **(d)** *Cibotiidites tuberculiformis,* another tree fern component and part of recovery succession following the ground ferns. **(e)** *Tricolpites lilliei*, flowering plant pollen and a good stratigraphic marker as it becomes extinct at the KT boundary. **(f)** *Nothofagidites kaitangata*, another key species extinct at the KT boundary. **(g)** *Tricolpites phillipsii*, a flowering plant pollen and one of first new species to evolve after the KT event, encountered several metres above the KT boundary in New Zealand. **(h)** *Phyllocladidites mawsonii* pollen is closely related to those of the modern Huon Pine (*Lagarostrobus franklinii*) and does not attain pre-event abundance until hundreds to thousands of years after the event.

Extinction of only a few Cretaceous key taxa, such as *Tricolpites lilliei* and *Nothofagidites kaitangata* (Figs 3 E-F), at the iridium anomaly horizon, followed by a 4 mm layer only containing fungal spores but barren of plant spores and pollen, typifies the New Zealand KT boundary palynological record (Vajda and McLoughlin 2004). The fungal spores are represented by several genera, e.g., *Monoporisporites* sp. and *Multicellaesporites* sp. and were probably plant saprophytes as the assemblages are of low diversity and are often found associated with fossil tracheids of conifer wood. Interestingly, iridium values are still at their maximum 5 mm above the boundary when the recovery flora of ground ferns return, suggesting rapid re-establishment of ferns in the aftermath of the Chicxulub impact event (Vajda and McLoughlin 2004).

The pioneer recovery vegetation, following the end-Cretaceous catastrophe consists of *Laevigatosporites ovatus* (Figure 3A) succeeded by *Gleicheniidites* (Figure 3B), both representatives of ground ferns. Younger assemblages are dominated by tree fern spores, e.g., *Cyathidites* and *Cibotiidites* (Figures 3C-D).

The period with fern dominance was in New Zealand followed by a stage dominated by the conifer pollen *Phyllocladidites mawsonii* (Figure 3H), closely related to the modern Huon Pine. The evolution of new species was rather slow and only a few new species of pollen appear close above the boundary, e.g., *Tricolpites phillipsii* (Figure 3G) and *Nothofagitides waipawensis,* all belonging to flowering plants.

4
Vertebrate turnover at KT boundary

There has been much debate concerning the extinction of the dinosaurs, and whether their demise was sudden or gradual but there is now substantial support for an abrupt extinction of the non-avian dinosaurs, although doubts are still sometimes expressed about the reality of that mass extinction (Sarjeant and Currie 2001). The argument that dinosaurs did not really become extinct at the KT boundary, because their descendants, the birds, survived, is merely playing on words. The fact that all non-avian dinosaurs disappeared at the KT boundary did have a very profound effect on the composition and structure of terrestrial communities. One of the most striking consequences of that event was the disappearance of all large terrestrial vertebrates. Contrary to what has been claimed (Sarjeant and Currie 2001), the large flightless birds recently reported from the Late Cretaceous of Europe (Buffetaut and Le Loeuff 1998) were archaic forms not related to present-day ratites, or, for that matter, to the Early Paleogene giant flightless Gastornithiformes (Buffetaut 2002), and there is not the slightest evidence that they survived the end-Cretaceous mass-extinction event.

Documentation of faunal diversity from terrestrial ecosystems shows that dinosaurs and pterosaurs were the only major terrestrial vertebrate groups that went completely extinct at the KT boundary event. Most other vertebrate groups were subjected to a mass-kill but no mass extinction is evident (Milner 1998). The main problem for detailed statistical analyses of vertebrate assemblages is the scarcity of fossils and the few terrestrial KT boundary sections containing vertebrate fossils. The sediments of the western interior of North America provide one of the world's best records of dinosaur fossils up to the KT boundary boundary. For example, the Hell Creek Formation of Montana and North Dakota and the Lance Formation of Wyoming, USA, contain assemblages of dinosaur fossils extending up to the Cretaceous-Paleogene boundary, which is well constrained by terrestrial palynology and, in some cases, by geochemistry. Data from as many as 53 vertebrate sites from the Hell Creek Formation strongly support a scenario of sudden extinction of dinosaurs (Pearson et al. 2001, 2002). A much-publicised "3 m gap", supposedly lacking dinosaur fossils at the top of the Hell Creek Formation, was used as evidence that dinosaurs had become extinct (or nearly so) before the KT boundary in North America. Recent research (Sheehan et al. 2000) has shown that in Montana and North Dakota the abundance of dinosaur fossils in the top 3 m of the Hell Creek Formation is comparable to what it is at lower levels in the formation, so that evidence for gradual extinction is absent.

In Europe there are no known dinosaur localities encompassing a well-defined KT boundary. This may be partly a consequence of difficulties in correlating the terrestrial dinosaur-bearing sediments with well-dated marine sediments. However, there are latest Maastrichtian dinosaur localities in the French and Spanish Pyrenees, Aix en Provence (France), Romania, The Netherlands, Belgium, Germany and the Ukraine (López-Martínez et al. 2001). Knowledge of Late Maastrichtian ecosystems in Europe has been greatly improved in recent years by discoveries of stratigraphically well-constrained diverse dinosaur assemblages from the Pyrenées in France (Laurent et al. 2002) and Spain (López-Martínez et al. 2001). Fossil bones of hadrosaurian dinosaurs are most frequently found in the sediments but recent investigations of the Cassagnau locality, southwestern France (Laurent et al. 2002) have revealed, apart from abundant hadrosaur bones, theropod and sauropod dinosaurs, which indicate that at least five dinosaur families inhabited western Europe in the Maastrichtian. The European dinosaur assemblages do not support any gradual decrease in species diversity during the Late Maastrichtian (López-Martínez et al. 2001; Laurent et al. 2002).

On a global scale, a recent examination of dinosaur diversity through the Mesozoic (Fastovsky et al. 2004) shows a steadily increasing rate of diversification. Against this background, fluctuations in known dinosaur diversity during the Campanian and Maastrichtian have little significance, and the data do not suggest a decline in diversity leading to extinction during the last ten million years of the Cretaceous.

Dinosaur fossils are encountered in Maastrichtian sediments from every continent and there have also been sporadic reports of dinosaurs in Paleocene sediments from various parts of the world (e.g., France, United States, Bolivia, China). Some of the older reports are based on misidentifications of non-dinosaurian remains. In all instances when undoubted dinosaur remains have been reported from post-Cretaceous rocks, subsequent studies have shown that either the fossils were reworked, or the purported Paleogene age of the sediments was incorrect. In some instances, accurate placement of the KT boundary is crucial. A recent example of purported Paleocene dinosaurs comes from China. Zhao et al. (2002) claimed evidence of a major extinction at the KT boundary in the Nanxiong Basin, South China, but suggested that dinosaurs in that province survived the KT event by about 250 000 years. The KT boundary is constrained by palynological data. The evidence for dinosaurs surviving into the Paleocene consists of dinosaur nests where the eggshells show an enrichment of iridium. It is suggested that the anomalous trace element concentrations were provided by the food source. Pathological development is traced in the eggs that were produced after the KT event and only a few of them seemed to have hatched due to environmental poisoning. However, the palynological definition of the boundary by Zhao et al. (2002) contradicts previous works, (Zhao 1978, 1993, 1994; Zhao et al. 1991) where the boundary is set higher in the sedimentary succession, at the last occurrence of the eggshells. Both the extinction process by environmental poisoning and the placement of the KT boundary appear highly dubious, and this report from the Nanxiong Basin does not convincingly demonstrate that dinosaurs survived the Cretaceous-Paleogene catastrophe.

Similarly, it has recently been claimed (Fassett et al. 2002) that Paleocene dinosaur remains occur in the Ojo Alamo Sandstone of New Mexico. According to Fassett et al. (2002), persistence of dinosaurs for about one million years after the end-Cretaceous asteroid impact might have resulted from the survival of embryos inside eggs laid just before the disaster. However, this claim is based mainly on palynological evidence for a Paleocene age for the sediments containing the so-called "Lazarus dinosaurs", and renewed sampling has not confirmed it, indicating instead a Maastrichtian age (Sullivan et al. 2002).

Despite various claims to the contrary, there is thus no convincing evidence for the survival of dinosaurs after the KT boundary anywhere in the world.

5 Discussion and summary

Although dinosaurs are the most spectacular victims of the catastrophe, they do not provide the best evidence of the events at that time because their fossil record is much more scanty and discontinuous than that of the marine planktonic organisms or terrestrial palynomorphs. However, there is apparently no significant decline of dinosaur communities prior to the KT boundary (Le Loeuff 2000), and abrupt extinction is likely, albeit difficult to demonstrate because of the nature of the record. It should also be remembered that several groups of vertebrates survived the KT (Buffetaut 1990) with moderate or insignificant damage, among them the ectothermic reptiles, which are known to be sensitive to climate change (Figure 4). A likely scenario involves worldwide food chain collapse (Buffetaut 1984) leading to the extinction of large herbivores and subsequently of the carnivores, which preyed on them, whereas freshwater ecosystems were much less affected because food webs were less immediately dependent on photosynthetic organisms (Sheehan and Fastovsky 1992). Similarly, small continental vertebrates such as mammals and small reptiles, apparently were less affected because they were part of food chains based on organic matter in soils (Sheehan and Fastovsky 1992). Freshwater life may also have been protected from the effects of acid rain due to the formation of larnite, Ca_2SiO_4 (Maruoka and Koeberl 2003). Larnite was formed as a consequence of contact metamorphism of the limestone in the Chicxulub impact area and was globally dispersed via the stratosphere. Larnite accumulating in freshwater bodies may have buffered the low pH conditions created by the sulfuric acid rain making the environment less lethal for aquatic species compared to the land dwelling biota.

Despite the many questions remaining (e.g., why did small carnivorous dinosaurs disappear while birds survived?), the vertebrate fossil record at the KT boundary is generally in good agreement with the scenario of food chain collapse following devastation of plant communities. From that point of view, the importance of the discoveries from New Zealand, revealing a pattern of vegetation devastation followed by recovery very similar to that recorded in North America, lies in the fact that they strongly suggest that the vegetation crisis was global, and not a local North American phenomenon caused by geographic proximity to the Chicxulub impact. The New Zealand record shows that the Southern Hemisphere was just as badly affected as the Northern Hemisphere. A genuinely global crisis is needed to explain the world-wide extinction of dinosaurs, and the palynological record appears to support the idea that general food chain collapse following cessation or reduction of photosynthesis is a valid explanation for the observed patterns of terrestrial vertebrate extinctions. The low extinction rate observed in New Zealand vegetation (10-12%) is, however, in contrast to the North American values of 15-30%. The possible explanation is that the high latitude Southern Hemisphere vegetation at the time was dominated by ferns and conifers in contrast to the more light-dependent flowering plants.

Although there is good agreement between the pattern of vertebrate extinctions revealed by fossils and the scenario of food chain collapse, other aspects of the KT extinction are more problematic. Models involving a drastic and relatively protracted drop in temperatures (Figure 5) are not easily reconciled with the fossil record (Buffetaut 1984, 1990), which unambiguously shows that temperature-sensitive organisms, such as ectothermic reptiles (turtles, lizards, crocodilians) were little affected by KT events (Figure 4). In other words, the hypothesis of a severe cold spell is not supported by palaeontological evidence.

Advances in our knowledge of high latitude vertebrate faunas in the Cretaceous are revealing in this regard (Buffetaut 2004). In Arctic North America, climatic cooling from the Turonian to the Maastrichtian resulted in the disappearance of ectothermic reptiles, whereas it apparently did not affect dinosaurs. The opposite pattern is seen at the KT boundary, when ectothermic reptiles survived while dinosaurs disappeared. The hypothesis according to which dinosaurs fell victim to climatic cooling during the last stages of the Cretaceous is therefore highly unlikely.

How to accommodate darkness resulting in a severe reduction of photosynthesis with a continuation of mild climates is an obvious problem. In any case, there is no evidence that climatic cooling played an important part in KT vertebrate extinctions, food chain collapse being a much more convincing scenario. A study made in order to assess the difference in survival rates between plants subjected to prolonged darkness (10 weeks) and warm conditions (15 degrees) versus plants subjected to prolonged darkness and cold conditions (4 degrees, Read and Francis 1992) revealed that more damage was recorded in the 15 degrees dark treatment than in the 4 degrees treatment among evergreen species due to excess respiration. This may demonstrate that prolonged darkness does not have to be combined with low temperatures to produce massive die-back of vegetation.

Plant-insect associations from the Williston Basin of southwestern North Dakota provide evidence for a major extinction of herbivorous insects (Labandeira et al. 2002). This large loss of specialized insect associations can perhaps explain the higher extinction rate among insect-pollinated angiosperms (flowering plants), compared to conifers and ferns.

Furthermore, it should be emphasized that, according to the scenario outlined above, most of the extinctions at the KT boundary must have taken place within a very short time span. Again the New Zealand palynological and geochemical record provides important clues as the recovery succession of ground ferns appears in the layer where the anomalous trace element concentration is still at its peak, suggesting the first recovery of *Blechnum*-related ferns appeared within a year of the impact, when dust was still settling but light levels were high enough to allow photosynthesis and when temperatures may have been lower.

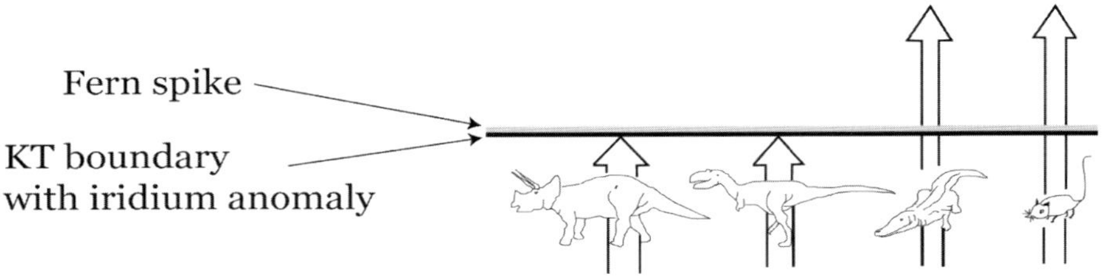

Fig. 4. Scenario of vertebrate extinction at the KT boundary based on food chain collapse. The iridium anomaly (left) illustrates the consequences of the Chicxulub asteroid impact. The fern spore spike is a marker of floristic devastation caused by cessation or reduction of photosynthesis, itself caused by dust and aerosols injected into the atmosphere by the impact. The food chain to which dinosaurs belong collapses when herbivorous dinosaurs (illustrated by *Triceratops*) disappear because of lack of food, followed by carnivorous dinosaurs (illustrated by *Tyrannosaurus*). Members of freshwater ecosystems (crocodile), belonging to different food chains not directly dependent on primary productivity, are not strongly affected. Similarly, some small terrestrial vertebrates (mammal) are able to survive because they belong to food chains based on organic matter in soils. After Buffetaut (1994).

Atmospheric radiative transfer models of sunlight filtration following the Chicxulub impact indicate that the photosynthesis crisis did not last more than a few years (Pope 1994, 1997; Figure 5) and that the longer effect of about 10 years was due to the slow dissipation of the sulfuric acid clouds that enveloped the planet shortly after the impact. The amount of biological evidence is in agreement with this, although more quantitative data are needed: a more protracted crisis would probably have resulted in more extinctions in the plant world, and depletion of organic matter reservoirs in freshwater and soils, which in turn would have led to more severe extinctions among vertebrates. The apparent lowering of oceanic ph agrees with the more devastating effect that took place on planktic foraminifera, where 75% went extinct.The pattern of vertebrate extinction at the KT boundary does suggest a rather brief but significant reduction of photosynthetic activity, not accompanied by a severe temperature drop. Models of KT boundary events must take these palaeontological constraints into consideration.

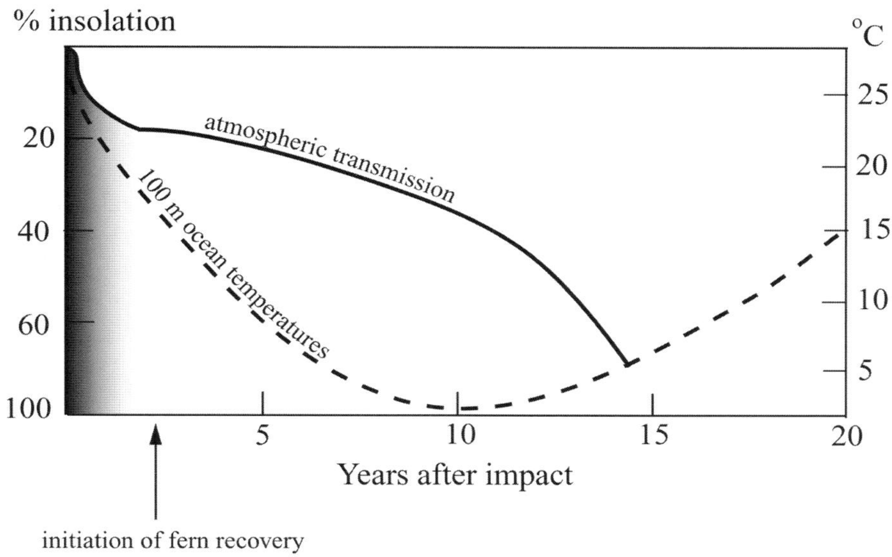

Fig. 5. Models of the changes in insolation and ocean temperature during the 20 years following impact. The shaded area indicates the period in which dust is the major factor in insolation drop - sulfur aerosols are the primary control on insolation inhibition during the remaining interval.

The models show that vaporization of sulfates by the Chicxulub impact, and the subsequent generation of long-lived sulfuric acid aerosol haze, caused major cooling during the decade after the impact. A few years after the impact, surface temperatures may have dropped below freezing in many areas, especially large continental regions. According to the insolation models, these factors played a key role in the mass extinction that marks the KT boundary (Pope 1999). Recent evidence shows that the Chicxulub impactor was chondritic (Kyte 1998; Shukolyukov and Lugmair 1998) in nature, reducing the contribution of the sulfur from the bolide and pointing towards the majority of the sulfur contribution being from the target rock.

The Chicxulub impact event brought rapid and long term effects and the massive quantities of dust and sulfate aerosols released from the target rock were a lethal combination for the flora and fauna of the time.

In summary, the processes that brought about this mass extinction 65 Ma are complex and diverse, and only with a multidisciplinary approach can the enigma of the KT mass extinction be unravelled.

Acknowledgements

This research was supported by The European Space Agency, planetary division, the California Institute of Technology Jet Propulsion Laboratory and The Planetary Society (Ocampo); the Swedish Research Council, Crafoord Foundation and Carl Tryggers Foundation (Vajda) and the ECLIPSE Programme of the Centre National de la Recherche Scientifique (Buffetaut). Stephen McLoughlin (Queensland University of Technology) is gratefully acknowledged for improving this manuscript with comments and suggestions.

References

Alvarez L, Alvarez W, Asaro F, Michel HV (1980) Extraterrestrial cause for the Cretaceous-Tertiary extinction. Science 208: 1095-1108

Aubry MP, Gradstein FM, Jansa LF (1990) The late early Eocene Montagnais bolide: no impact on biotic diversity. Micropaleontology 36: 164-172

Barclay RS, Johnson KR (2004) West Bijou Site Cretaceous-Tertiary boundary, Denver Basin, Colorado. In: Nelson EP, Erslev EA (eds) Field Trips in the Southern Rocky Mountains, USA. The Geological Society of America Field Guide 5: 59-68

Becker L, Poreda RJ, Basu AR, Pope KO, Harrison TM, Nicholson C, Iasky R (2004) Bedout: A possible End-Permian impact crater offshore of Northwestern Australia. Science 304: 1469-1476

Bohor BF, Modreski PJ, Foord EE (1987) Shocked quartz in the Cretaceous-Tertiary boundary clays - evidence for a global disruption. Science 236: 705-709

Braman DR, Sweet AR, Lerbekmo F (1993) Palynofloristic changes across the Cretaceous-Tertiary boundary and contiguous strata [abs.] Canadian Geophysical Union, Joint Annual Meeting 18: 12

Brinkhuis H, Schiøler P (1996) Palynology of the Geulhemmerberg Cretaceous/Tertiary boundary section (Limburg, SE Netherlands). Geologie en Mijnbouw 75: 193-213

Buffetaut E (1984) Selective extinctions and terminal Cretaceous events. Nature 310: 276

Buffetaut E (1990) Vertebrate extinctions and survival across the Cretaceous-Tertiary boundary. Tectonophysics 171: 337-345

Buffetaut E (1994) Les Dinosaures. Presses Universitaires de France, Paris. 128 p

Buffetaut E (2002) Giant ground birds at the Cretaceous-Tertiary boundary: extinction or survival? In: Koeberl C, MacLeod KG (eds) Catastrophic events and mass extinctions; impacts and beyond. Geological Society of America Special Paper 356: 303-306

Buffetaut E (2004) Polar dinosaurs and the question of dinosaur extinction: a brief review. Palaeogeography, Palaeoclimatology, Palaeoecology 214: 225-231.

Buffetaut E, Le Loeuff J (1998) A new giant ground bird from the Upper Cretaceous of southern France. Journal of the Geological Society, London 155: 1-4

Covey C, Thompson SL, Weissman PR, MacCracken MC (1994) Global climatic effects of atmospheric dust from an asteroid or comet impact on earth. Global and Planetary Change 9: 263-273

Dypvik H, Gudlaugsson ST, Tsikalas F, Attrep M, Ferrell RE, Krinsley DH, Mork A, Faleide JI, Nagy J (1996) Mjølnir structure: An impact crater in the Barents sea. Geology 24: 779-782

Fassett JE, Zielinski RA, Budahn JR (2002) Dinosaurs that did not die: Evidence for Paleocene dinosaurs in the Ojo Alamo Sandstone, San Juan Basin, New Mexico. In: Koeberl C, MacLeod KG (eds) Catastrophic events and mass extinctions; impacts and beyond. Geological Society of America Special Paper 356: 307-336

Fastovsky D E, Huang Y, Hsu J, Martin-McNaughton J, Sheehan PM, Weishampel DB (2004) Shape of Mesozoic dinosaur richness. Geology 32: 887-880

Fleming RF (1985) Palynological observations of the Cretaceous-Tertiary boundary in the Raton Formation, New Mexico. In: abstracts of the proceedings of the Seventeenth annual meeting of the American Association of Stratigraphic Palynologists. Palynology 9: 242

Fouke BW, Zerkle AL, Alvarez W, Pope KO, Ocampo AC, Wachtman RJ, Nishimura JMG, Claeys P, Fischer AG (2002) Cathodoluminescence petrography and isotope geochemistry of KT impact ejecta deposited 360 km from the Chicxulub crater, at Albion island, Belize. Sedimentology 49: 117-138

Grieve RAF (1991) Terrestrial impact: the record in the rocks. Meteoritics 26, 175-194

Heissig K (1986) No effect of the Ries impact event on the local mammal fauna. Modern Geology 10: 171-179

Herngreen GFW, Schuurman HAHM, Verbeek JW, Brinkhuis H, Burnett JA, Felder WM, Kedves M (1998) Biostratigraphy of Cretaceous/Tertiary

boundary strata in the Curfs Quarry, the Netherlands. Mededelingen Nederlands Instituut voor Toegepaste Geowetenschappen 61: 57
Hildebrand A, Boyton W (1990) Proximal Cretaceous-Tertiary boundary impact deposits in the Caribbean. Science 248: 843-847
Hildebrand AR, Penfield GT, Kring DA, Pilkington M, Camargo A, Jacobsen SB, Boyton W (1991) Chicxulub Crater - A possible Cretaceous Tertiary boundary impact crater on the Yucatan Peninsula, Mexico. Geology 19: 867-871
Hotton C (2002) Palynology of the Cretaceous-Tertiary boundary in central Montana; evidence extraterrestrial impact as a cause of the terminal Cretaceous extinctions. In: Hartman JH, Johnson KR, Nichols DJ (eds) The Hell Creek Formation and the Cretaceous-Tertiary Boundary in the Northern Great Plains: An Integrated Continental Record of the End of the Cretaceous. Geological Society of America Special Paper 361: 473-501
Ivanov B, Badukov D, Yakolev O, Gerasimov M, Dikov Y, Pope K, Ocampo A (1996) Degassing of sedimentary rocks due to Chicxulub Impact: Hydrocode and physical simulations. In: Ryder G, Fastovsky D, Gartner S (eds) The Cretaceous-Tertiary event and other catastrophes in Earth history. Geological Society of America Special Paper 307: 125-139
Izett GA (1991) Tektites in Cretaceous-Tertiary boundary rocks on Haiti and their bearing on the Alvarez impact extinction hypothesis. Journal of Geophysical Research-Planets 96: 20879-20883
Izett GA, Dalrymple GB, Snee LW (1991) Ar-40/Ar-39 age of Cretaceous-Tertiary boundary tektites from Haiti. Science 252: 1539-1542
Jerzykiewicz T, Sweet AR (1986) The Cretaceous-Tertiary boundary in the central Alberta Foothills: stratigraphy. Canadian Journal of Earth Sciences 23: 1356-1374
Johnson KR (1992) Leaf-fossil evidence for extensive floral extinction at the Cretaceous-Tertiary boundary, North-Dakota, USA. Cretaceous Research 13: 91-117
Johnson KR, Nichols DJ, Attrep Jr M, Orth CJ (1989) High-resolution leaf-fossil record spanning the Cretaceous/Tertiary boundary. Nature 340: 708-711
Keller G, Stinnesbeck W, Adatte T (2003a) Multiple impacts across the Cretaceous-Tertiary boundary. Earth Science Reviews 62: 327-363
Keller G, Stinnesbeck W, Adatte T, Holland B, Stueben D, Harting M, de-Leon C, de la Cruz J (2003b) Spherule deposits in Cretaceous-Tertiary boundary sediments in Belize and Guatemala. Journal of the Geological Society of London 160: 783-795
Kring DA (1997) Air blast produced by the Meteor Crater impact event and a reconstruction of the affected environment. Meteoritics and Planetary Science 32: 517-530
Kring DA (2003) Environmental consequences of impact cratering events as a function of ambient conditions on Earth. Astrobiology 3: 133-152
Kring DA, Durda DD (2002) Trajectories and distribution of material ejected from the Chicxulub Impact Crater: implications for post-impact wildfires. Journal of Geophysical Research-Planets 107: 5062

Kring DA, Melosh HJ, Hunten DM (1996) Impact-induced perturbations of atmospheric sulfur. Earth and Planetary Science Letters 140: 201-212

Kyte FT (1998) A meteorite from the Cretaceous/Tertiary boundary. Science 396: 237-239

Labandeira CC, Johnson KR, Wilf P (2002) Impact of the terminal Cretaceous event on plant-insect associations. Proceedings of the National Academy of Sciences of the USA 99: 2061-2066

Laurent Y, Bilotte M, Loeuff JL (2002) Late Maastrichtian continental vertebrates from southwestern France: Correlation with marine fauna. Palaeogeography, Palaeoclimatology, Palaeoecology 187: 121-135

Le Loeuff J (2000) Les derniers dinosaures. In: La Valse des Espèces, Dossier Pour la Science, Paris: 94-99

Lerbekmo JF, Sweet AR, St Louis RM (1987) The relationship between the iridium anomaly and palynological floral events at three Cretaceous-Tertiary boundary localities in western Canada. Geological Society of America Bulletin 99: 325-330

López-Martínez, Canudo JI, Ardevol L, Suberbiola XP, Orue-Etxebarria X, Cuenca-Bescos G, Ruiz-Omenaca JI, Murelaga X, Feist M (2001) New dinosaur sites correlated with Upper Maastrichtian pelagic deposits in the Spanish Pyrenees: implications for the dinosaur extinction pattern in Europe. Cretaceous Research 22: 41-61

Maruoka T, Koeberl C (2003) Acid-neutralizing scenario after the Cretaceous-Tertiary impact event. Geology 31: 489-492

McIver EE (1999) Paleobotanical evidence for ecosystem disruption at the Cretaceous-Tertiary boundary from Wood Mountain, Saskatchewan, Canada. Canadian Journal of Earth Sciences 36: 775-789

Milner AC (1998) Timing and causes of vertebrate extinction across the Cretaceous- Tertiary boundary. In: Grady MM, Hutchison R, McCall GJH, Rothery DA (eds) Meteorites; flux with time and impact effects. Geological Society, London, Special Publication 140: 247-257

Nathan S (1978) Sheet S44 Greymouth. Geological map of New Zealand 1:63 360. Wellington, N.Z. Department of Scientific and Industrial Research

Nichols DJ, Jarzen DM, Orth CJ, Oliver PQ (1986) Palynological and iridium anomalies at Cretaceous-Tertiary boundary, south-central Saskatchewan. Science 231: 714-717

Nichols DJ, Fleming RF (1990) Plant microfossil record of the terminal Cretaceous event in the western United States and Canada In: Sharpton VL, Ward PD (eds) Global catastrophes in Earth history; an interdisciplinary conference on impacts, volcanism, and mass mortality. Geological Society of America, Special Paper 247: 445-455

Nichols DJ, Brown JL, Attrep Jr M, Orth CJ (1992) A new Cretaceous-Tertiary boundary locality in the western Powder River basin, Wyoming: biological and geological implications. Cretaceous Research 13: 3-30

Nichols DJ, Johnson KR (2002) Palynology and microstratigraphy of Cretaceous-Tertiary boundary sections in southwestern North Dakota. In: Hartman JH, Johnson KR, Nichols DJ (eds) The Hell Creek Formation and the Cretaceous-

Tertiary Boundary in the Northern Great Plains: An Integrated Continental Record of the End of the Cretaceous. The Geological Society of America, Special Paper 361: 95-144

Norris RD (2001) Impact of K-T boundary events on marine life. In: Briggs DEG, Crowther PR (eds) Palaeobiology II. Blackwell Science, Oxford, 229-231

Ocampo A (1997) The Geology of the Chicxulub Impact Ejecta in Belize, Master of Science Thesis, Geology Department, California State University, Northridge. 67 p

Ocampo A, Pope K, Fischer A (1996) Ejecta blanket deposits of the Chicxulub crater from Albion Island, Belize. In: Ryder G, Fastovsky D, Gartner S (eds) The Cretaceous-Tertiary event and other catastrophes in Earth history.Geological Society of America Special Paper 307: 75-88

Orth CJ, Gilmore JS, Knight JD, Pillmore CL, Tschudy RH, Fasett JE (1981) An iridium abundance anomaly at the palynological Cretaceous-Tertiary boundary in Northern New Mexico. Science 214: 1341-1343

Pearson DA, Schaefer T, Johnson KR, Nichols DJ (2001) Palynologically calibrated vertebrate record from North Dakota consistent with abrupt dinosaur extinction at the Cretaceous-Tertiary boundary. Geology 29: 39-42

Pearson DA, Schaefer T, Johnson KR, Nichols DJ, Hunter JP (2002) Vertebrate biostratigraphy of the Hell Creek Formation in southwestern North Dakota and northwestern South Dakota. In: Hartman JH, Johnson KR, Nichols DJ (eds) The Hell Creek Formation and the Cretaceous-Tertiary boundary in the Northern Great Plains; an integrated continental record of the end of the Cretaceous. Geological Society of America, Special Paper 361: 145-167

Pierazzo E, Melosh HJ (2000a) Understanding oblique impacts from experiments, observations, and modelling. Annual Review of Earth and Planetary Sciences 28: 141-167

Pierazzo E, Melosh HJ (2000b) Melt production in oblique impacts. Icarus 145: 252-261

Pierazzo E, Kring DA, Melosh HJ (1998) Hydrocode simulation of the Chicxulub impact event and the production of climatically active gases. Journal of Geophysical Research-Planets 103: 28607-28625

Pierazzo E, Hahmann AN, Sloan LC (2003) Chicxulub and climate: radiative perturbations of impact-produced S-bearing gases. Astrobiology 3: 99-118

Pollack JB, Toon OB, Ackerman TP, McKay CP, Turco RP (1983) Environmental-effects of an impact-generated dust cloud - implications for the Cretaceous-Tertiary extinctions. Science 219: 287-289

Pope K (2002) Impact dust not the cause of Cretaceous-Tertiary mass extinction. Geology 30: 99-102

Pope K, Ocampo A, Duller C (1991) Mexican site for K/T impact crater? Nature 351: 105

Pope K, Baines K, Ocampo A, Ivanov B (1994) Impact winter and the Cretaceous/Tertiary extinctions: Results of a Chicxulub asteroid impact model, Earth and Planetary Science Letters 128: 719-725

Pope K, Baines K, Ocampo A, Ivanov B (1997) Energy, volatile production, and climatic effects of the Chicxulub Cretaceous/Tertiary impact. Journal of Geophysical Research 102: 21645-21664

Pope KO, D'Hondt SL, Marshall CR (1998) Meteorite impact and the mass extinction of species at the Cretaceous/Tertiary boundary. Proceedings of the National Academy of Sciences of the United States of America 95: 11028-11029

Raup D, Sepkoski J (1982) Mass extinction in the marine fossil record. Science 215: 501- 503

Raup D, Sepkoski J (1986) Periodic extinctions of families and genera. Science 231: 833- 836

Read J, Francis J (1992) Responses of some southern-hemisphere tree species to a prolonged dark period and their implications for high-latitude Cretaceous and Tertiary floras. Palaeogeography, Palaeoclimatology, Palaeoecology 99: 271-290

Romein AJT, Smit J (1981) Carbon-oxygen isotope stratigraphy of the Cretaceous – Tertiary boundary interval: data from the Biarritz section (SW France). Geologie en Mijnbouw 60: 541-544

Salge T, Tagle R, Claeys P (2000) Accretionary lapilli from the Cretaceous-Tertiary boundary site of Guayal, Mexico: Preliminary insights into expansion plume formation [abs.]. Meteoritics and Planetary Science 35: A140-A141

Sarjeant WAS, Currie PJ (2001) The "Great Extinction" that never happened: the demise of the dinosaurs considered. Canadian Journal of Earth Sciences 38: 239-247

Sepkoski J (1990) The taxonomic structure of periodic extinction. In: Sharpton VL, Ward PD (eds) Global catastrophes in Earth history; an interdisciplinary conference on impacts, volcanism, and mass mortality. Geological Society of America, Special Paper 247: 33-44

Sheehan PM, Fastovsky DE (1992) Major extinctions of land-dwelling vertebrates at the Cretaceous-Tertiary boundary, eastern Montana. Geology 20: 556-560

Sheehan PM, Fastovsky DE, Barreto C, Hoffmann R.G. (2000) Dinosaur abundance was not declining in a "3 m gap" at the top of the Hell Creek Formation, Montana and North Dakota. Geology 28: 523-526

Shukolyukov A, Lugmair G (1998) Isotopic evidence for the Cretaceous-Tertiary impactor and its type. Science 282: 927-929

Smit J (1999) The global stratigraphy of the Cretaceous-Tertiary boundary impact ejecta. Annual Review of Earth Planetary Science 27: 75-113

Sullivan RM, Lucas SG, Braman D (2002) Paleocene dinosaurs? A critique of the ages assigned to the upper Kirtland Formation, San Juan Basin, New Mexico. Journal of Vertebrate Paleontology 22: 212A.

Sweet AR, Braman DR (2001) Cretaceous-Tertiary palynoflora perturbations and extinctions within the Aquilapollenites phytogeographic province. Canadian Journal of Earth Sciences 8: 249-269

Sweet R, Braman DR (1992) The K-T boundary and contiguous strata in western Canada: interactions between paleoenvironments and palynological assemblages. Cretaceous Research 13: 31-79

Sweet R, Braman DR, Lerbekmo JF (1990) Palynofloral response to K/T boundary events: a transitory interruption within a dynamic system. In: Sharpton VL, Ward PD (eds) Global catastrophes in Earth history; an interdisciplinary conference on impacts, volcanism, and mass mortality. Geological Society of America, Special Paper 247: 457-469

Sweet AR, Braman DR, Lerbekmo JF (1999) Sequential palynological changes across the composite Cretaceous-Tertiary (K-T) boundary claystone and contiguous strata, western Canada and Montana, USA. Canadian Journal of Earth Sciences 36: 743-768

Toon OB, Zahnle K, Morrison D, Turco RP, Covey C (1997) Environmental perturbations caused by the impacts of asteroids and comets. Reviews of Geophysics 35: 41-78

Tschudy RH, Pillmore CL, Orth CJ (1984) Disruption of the terrestrial plant ecosystem at the Cretaceous-Tertiary boundary, Western Interior. Science 225: 1030-1032

Upchurch Jr GR (1989) Terrestrial environmental changes and extinction patterns at the Cretaceous-Tertiary boundary, North America. In: Donovan SK (ed) Mass Extinctions; processes and evidence. Columbia University Press. New York, NY, United States: 195-216

Upchurch Jr GR, Wolfe JA (1987) Plant extinction patterns at the Cretaceous-Tertiary boundary, Raton and Denver basins. Abstracts with Programs, Geological Society of America 19: 874

Vajda V, McLoughlin S (2004) Fungal proliferation at the Cretaceous-Tertiary boundary. Science 303: 1489

Vajda V, Raine JI (2003) Terrestrial palynology of the Cretaceous/Tertiary boundary at mid-Waipara River, North Canterbury, New Zealand. New Zealand Journal of Geology and Geophysics 46: 255-273

Vajda V, Raine JI, Hollis CJ (2001) Indication of global deforestation at the Cretaceous-Tertiary boundary by New Zealand fern spike. Science 294: 1700-1702

Vega FJ, Feldmann RM, Ocampo AC, Pope KO (1997) A new species of Late Cretaceous crab (*Brachyura: Carcineretidae*) from Albion Island, Belize. Journal of Paleontology 71: 615-620

Ward S (1997) Lithostratigraphy, palynostratigraphy and basin analysis of the Late Cretaceous to early Tertiary Paparoa Group, Greymouth Coalfield, New Zealand. Ph.D. thesis, University of Canterbury, 200 pp.

Wilf P, Johnson KR (2004) Land plant extinction at the end of the Cretaceous: A quantitative analysis of the North Dakota megafloral record. Paleobiology 30: 347-368

Wolbach WS, Gilmour I, Anders E, Orth CJ, Brooks RR (1988) Global fire at the Cretaceous-Tertiary Boundary. Nature 334: 665-669

Wolfe JA, Upchurch Jr GR (1986) Vegetation, climatic and floral changes at the Cretaceous-Tertiary boundary. Nature 324: 148-152

Wolfe JA, Russell DA (2001) Impact of K-T boundary events on terrestrial life. In: Briggs DEG, Crowther PR (eds) Palaeobiology II. Blackwell Science, Oxford, pp 232-234

Zhao ZK (1978) A preliminary investigation on the thinning of the dinosaurian eggshells of the Late Cretaceous and some related problems. Vertebrate Palaentology of Asia 16: 213–221

Zhao ZK (1993) Structure, formation, and evolutionary trends of dinosaur eggshell. In: Kobayashi I, Mutvei H, Sahni A (eds) Structure, Formation and Evolution of Fossil Hard Tissues. Tokai University Press, Tokyo: pp 195–212

Zhao ZK (1994) The dinosaur eggs in China: on the structure and evolution of eggshells. In: Carpenter K, Hirsch KF, Horner JR (eds) Dinosaur Eggs and Babies. Cambridge University Press, Cambridge: pp 184–203

Zhao ZK, Ye J, Li H, Zhao Z, Yan Z (1991) Extinction of the dinosaurs across the Cretaceous–Tertiary in Nanxiong Basin, Guangdong Province. Vertebrate Palaeontology of Asia 29: 1–20

Zhao Z, Mao XY, Chai ZF, Yang GC, Kong P, Ebihara M, Zhao ZH (2002) A possible causal relationship between extinction of dinosaurs and K/T iridium enrichment in the Nanxiong Basin, South China: evidence from dinosaur eggshells. Palaeogeography, Palaeoclimatology, Palaeoecology 178: 1-17

Impacts and Wildfires - An Analysis of the K-T Event

C. M. Belcher

Department of Geology, Royal Holloway University of London,
Egham, Surrey, TW20 0EX, UK (c.belcher@gl.rhul.ac.uk)

Abstract. Models of the Cretaceous-Tertiary (K-T) impact at Chicxulub have suggested that thermal radiation would have been sufficient to have ignited extensive or near global wildfires. The discovery of abundant soot, increased levels of polyaromatic hydrocarbons (PAHs), and the possible occurrences of the fullerenes C60 and C70 have been considered to support the wildfire hypothesis. However, the charcoal record from K-T sites stretching the length of the Western Interior of the USA reveals amounts of charcoal below background levels and an abundance of non-charred material in the K-T and earliest Tertiary rocks.

Explanations to account for this disagreement between the charcoal record and the other potential wildfire indicators include: charcoal formation but subsequent oxidation (by acid rains or during diagenesis), and intense thermal radiation converting charcoal directly to CO_2. In both scenarios significant quantities of non-charred material would not be expected to survive. There is no satisfactory hypothesis to explain how charcoal could be transported away from the entire Western Interior, and it is unlikely that conditions prevailed in the Western Interior that prevented the fires.

Following re-analysis of the proposed wildfire evidence, the abundance of soot, PAHs, and fullerenes can be explained without invoking the global wildfire hypothesis. It has been concluded that fullerenes are not a suitable indicator for impact-related palaeo-wildfires. The morphology of the K-T soot is characteristic of soots produced during combustion of hydrocarbons. The uniformity of its carbon isotope signature between sites across the globe is better explained by the vaporisation of one pool of hydrocarbons. The PAH record of the K-T rocks includes compounds that are never formed from the burning of biomass but that are released during the combustion of hydrocarbons. Recent Chicxulub drill cores reveal that the target rocks contain hydrocarbons, the vaporisation of which could produce the soot and PAHs found at the K-T boundary.

1 Introduction

Since the detection of an enrichment of iridium at the K-T boundary (Alvarez et al. 1980), and the discovery of the 65 million year old Chicxulub crater (Hildebrand et al. 1991), it has become generally accepted that an extraterrestrial body collided with the Earth 65 million years ago. The fossil record reveals a considerable biotic turnover across the K-T boundary, which is arguable evidence for a mass extinction (e.g., MacLeod et al. 1996). However, how this impact explains the extinction patterns observed at the K-T boundary remains a topic of hot debate.

Several models have suggested that the energy released by the impact ought to have been sufficient to have ignited wildfires locally, if not globally (Melosh et al. 1990; Kring and Durda 2001). Wolbach et al. (1985, 1988, 1990) reported an enrichment of soot in the K-T rocks that is not only isotopically uniform, suggesting a single source, but also bears an isotopic signature consistent with burning of biomass. Arinbou et al. (1999) investigated the abundance of poly-aromatic hydrocarbons (PAHs) in K-T rocks from Caravaca in Spain, observing a 112 fold increase in some PAHs compared to the Cretaceous background. They suggested that the most likely source for the PAHs was the combustion of terrestrial organic matter. Heymann et al. (1994, 1998) claimed that the presence of the fullerenes C60 and C70 in the K-T rocks was also indicative of global wildfires.

Extensive wildfires could have led to an extreme disruption in the Earth's plant and animal communities 65 million years ago. This paper aims to review, discuss, and re-assess the literature regarding wildfires and the K-T event.

2
Potential effects of extensive wildfires on the global environment

Globally extensive wildfires would undoubtedly have a significant effect on the Earth's plant and animal communities, either through direct contact with the fire itself or via the environmental perturbations associated with massive wildfires. The K-T impact extinction hypothesis of Alvarez et al. (1980) suggested that there was a collapse in the food chain due to the shutdown of photosynthesis by sun-blocking silicate dust that was injected into the stratosphere. 10^{16} g of submicrometer-sized dust is suggested as being the threshold for which light availability will fall below the level required for photosynthesis (Gerstl and Zardecki 1982; Toon et al. 1982). However, Pope (2002) calculated that just 10^{14} g of submicrometer-sized dust was produced by the K-T impact based on the total clastic debris of the "fireball layer" (upper K-T claystone). This suggested that the amount of clastic debris ejected into the atmosphere following the K-T impact would have been insufficient to have caused global darkness.

A widespread fire would produce a vast amount of soot which, combined with clastic debris already in the air, could have had the potential to contribute towards the blocking of sunlight. This would have lead to prolonged darkness and cold following the impact, as soot absorbs sunlight more effectively than rock dust (Gilmour et al. 1990). A shutdown in photosynthesis would have lead to the destruction of plant communities, as recognised by Tschudy et al. (1984) Sweet (2001) Nichols and Johnson (2002). It would also have destroyed marine algae (MacLeod et al. 1996), and in turn the animals that fed on them.

The burning of the total Latest Cretaceous terrestrial biomass has been estimated to have produced 1000 ppm of CO_2 and ~100 ppm of CO. This would have had an enormous greenhouse effect, potentially raising temperatures by as much as 10°C (Wolbach et al. 1990), which may have proven fatal to land-dwelling creatures and surface-dwelling marine organisms. Large amounts of toxins such as polyaromatic hydrocarbons (PAHs) and CO (100 ppm) would have been produced by the fires. Most PAHs are known to be carcinogenic and CO is toxic to oxygen breathing animals (Wolbach et al. 1990). PAHs have been found in marine K-T sequences (Gilmour and Guenther 1988; Venkatesan and Dahl 1989; Arinbou et al. 1999), suggesting that they polluted both the terrestrial and marine realms.

3
K-T impact models and wildfires

Melosh et al. (1990) suggested that the re-entering ejecta from the Chicxulub impact could have delivered enough thermal radiation to ignite forests. Their model produced an average global pulse of around 50 kWm^{-2}, which ought to be sufficient to ignite vegetation, although they noted that the amount delivered to the ground would be near the lower limit required for ignition of solid wood. Hildebrand (1993) argued that the global average suggested by Melosh et al. (1990) for the amount of thermal radiation released from the impact was unrealistic. Hildebrand postulated that close to the impact site, at 1000 to 2500 km distances, the thermal pulse would be three to two orders of magnitude greater, and at distances greater than 10,000 km, the thermal pulse would be at least an order of magnitude less than that calculated by Melosh et al. (1990), implying that the thermal pulse ought to have had a regional effect. This regional effect might have offered places of refuge for organisms at greater distances from the impact (Hildebrand 1993). Hildebrand (1993) suggested that this regional effect implied that the two continents close to the impact site (North and South America) should have been "burnt to the ground", whilst the rest of the globe, with the possible exception of western Europe and Africa, should have remained relatively unscathed.

Kring and Durda (2002) modelled the trajectories of the low and high energy ejecta from the Chicxulub impact. They suggested that whilst 12% of the high energy ejecta would be lost to space, 25% of the material would reaccrete within 2 hours, 55% within 8 hours and 85% within 72 hours. Because the debris does not reaccrete all at once, then shock heating of the atmosphere will be drawn out into a series of pulses. The calculations suggested that the ejecta would have ignited wildfires on several continents around the world (Kring and Durda 2002). Colorado ought to have received a thermal pulse in the order of 150 $kW.m^{-2}$ within the first few hours following the impact (Kring and Durda 2002), more than sufficient to ignite even wet vegetation in the area.

Kring and Durda (2002) hypothesised that a vertical impact would have ignited wildfires more or less across the globe, whereas the distribution of wildfires from an oblique impact would have depended on the trajectory of the projectile. However, Shuvalov and Artemieva (2002) showed that the difference in the total radiation impulse between a vertical and an inclined impact does not exceed a factor of two. Moreover, such variations cannot change the area of forest ignition significantly (Shuvalov and Artemieva

2002). Based on the lack of charcoal across the Western Interior of North America, Belcher et al. (2003) suggested that if the trajectory of the asteroid did have any effect, then a trajectory that did not focus thermal radiation across North America must be assumed.

Shuvalov and Artemieva (2002) suggested that the direct radiation from the Chicxulub impact expansion plume could have been responsible for igniting wildfires over ~3% to 10% of the Earth's surface close to the impact point. However, they argued that the process of ejecta re-entry (Melosh et al. 1990) is probably not responsible for global wildfires. They considered that fires could be ignited across the globe by lightning strikes and that dead forests would be more likely to be ignited, suggesting that the forests were killed first and later burned. Shuvalov and Artemieva (2002) suggested that this would allow the periods of wildfires to continue for several years, and hence large volumes of soot could be produced from a large number of local wildfires in dead forests. They concluded that "the global mortality of forests was not the result of global wildfires, but rather that wildfires and K-T soot could result from the forest mortality".

Belcher et al. (2003) calculated, based on the below background amounts of charcoal found at six K-T sites across the Western Interior, that the K-T impact cannot have delivered a peak irradiance of more than 95 $kW.m^{-2}$ of thermal power to the atmosphere and less than 19 $kW.m^{-2}$ to the ground. They concluded that the thermal power delivered from the impact to North America did not have the destructive potential previously predicted (Belcher et al. 2003).

4 Evidence for a K-T wildfire

If the thermal power released from the impact at Chicxulub was sufficient to ignite wildfires, does the fossil record reveal evidence for these fires?

4.1 Charcoal

Charcoal is perhaps the most distinctive product of wildfire, and is a unique product that is only formed through the combustion of biomass (Belcher et al. 2003).

Jones and Lim (2000) collected charcoal from five marine K-T sites, reporting that all five sites produced charcoal, but that the amounts of charcoal found in the K-T rocks were not significantly higher compared to that routinely found in ancient and modern marine sediments. 53% of the charcoal particles in the K-T rocks were found to be biodegraded prior to being charcoalified (Jones and Lim 2000). It is suggested that either there was some time lag between the plant death, possibly as a result of the K-T events, and their later preservation in wildfires as charcoal, or that the plants died of ordinary causes, started to biodegrade and were preserved as charcoal by the normal regime of wildfires, and simply ended up in the boundary rocks (Jones and Lim 2000).

Kruge et al. (1994, 1996) and Jones (1996) have debated the occurrence of charcoal fragments at Arroyo el Mimbral. The K-T rocks at Arroyo el Mimbral are deep water tsunami deposits and have been reported as containing relatively large charcoal fragments (Kruge et al. 1994); however, the provenance of these is problematic, hence they may represent reworked Cretaceous charcoal.

Sweet and Cameron (1991) studied charcoal from two non-marine K-T sites in Saskatchewan in Canada - Rock Creek West and Wood Mountain Creek. They concluded that fires occurred throughout the sequence and that the total amount of charred material in and around the K-T boundary was low, indeed lower than the amounts recorded at higher levels in both sections.

Scott et al. (2000) studied charcoal throughout the non-marine K-T section at Sugarite and quantified charcoal throughout the section from small orientated polished pieces of rock. Hence their study involved analysis of charred material in situ, and from bulk macerations of the coal. The section at Sugarite revealed that charcoal was abundant below the K/.T boundary (up to 65% charcoal), at the boundary (~25% charcoal), and above it (up to 48% charcoal), indicating that fire was common throughout the section (Scott et al. 2000). (Percentages were measured by a placing a whipple grid of 100 squares in the eye piece. The material that appeared under the cross of each square was recorded for the percentage area it occupied [mineral matter, non charred material, charred material, etc]).

Wildfire was a regular occurrence in the late Cretaceous and early Tertiary at Sugarite, and the K-T layers were found to contain no more than background levels of charcoal (Scott et al. 2000).

An extensive study of charcoals across the K-T boundary was conducted by Belcher et al. (2003). Six non-marine K-T sites stretching from Colorado in the south to Saskatchewan in the North across the Western Interior of North America were studied: Clear Creek South (Colorado), Teapot Dome (Wyoming), Rick's Place (Montana), Mud Buttes (North Dakota), and Rock Creek East and Wood Mountain Creek (Saskatchewan). Both macro- and microscopic charcoal particles occurred throughout the sections, indicating that the charcoal studied not only represents local fires (Innes and Simmons 2000), but that smaller particles, sourced from outside the Western Interior, could also be detected (e.g., charcoal can be transported up to 700 km or more by water [Piperno 1997]). Fires were found to be a typical part of the Late Cretaceous and early Tertiary ecosystem across this Western Interior transect (see Fig. 1). The overall Cretaceous and Tertiary background charcoal levels for these Western Interior sites is 16.3% (Belcher et al. 2003). By contrast, the average percentage of charcoal in the K-T boundary rocks is just 1.75%, making charcoal around 6 times more abundant in the Cretaceous than it is in the K-T boundary rocks (Belcher et al. 2003). A K-T wildfire might also be preserved in the earliest Tertiary (Belcher et al. 2003); however, the earliest Tertiary rocks across the Western Interior (those immediately overlying the K-T boundary layers) contain an average of 11% charcoal (Belcher et al. 2003), which is again less than the background level of charcoal at these sites. The K-T boundary charcoal record from the western interior shows no evidence that a wildfire engulfed North America as part of the K-T events (Belcher et al. 2003).

4.2 Non-charred plant material

The first indirect hint that the K-T layers contained non-charred plant material was made by Izett (1990). This work showed illustrations of thin sections and polished blocks of the "impact layer", the upper K-T claystone layer, from sites in the Raton Basin. Izett (1990) reported that "a chief diagnostic feature of the Impact Layer is its laminated nature, where the laminations display impressions of macerated plant material as planar vitrinite laminae a few tens of millimetres thick that probably formed from leaves and other plant materials". The importance of this statement was underestimated until the study by Belcher et al. (2003), as it is clear that this upper layer of the boundary claystone is laminated with non-charred plant material (vitrinite). Belcher et al. (2003) found that the upper layer of the K-T claystone contains on average 22% non-charred material (ranging from 4% to 60% across their Western Interior transect). The non-charred material commonly occurs as vitrinite laminae (see Fig. 2 of Belcher et al. 2003) and has been found to occur at Clear Creek South (Colorado), Teapot Dome (Wyoming), Rick's Place (Montana), Mud Buttes (North Dakota), and Rock Creek East and Wood Mountain Creek (Saskatchewan) (Belcher et al. 2003), Berwind Canyon, Clear Creek North, and Madrid East, (Colorado) (Izett 1990), Knudsen's Coulee (Alberta), and Frenchman Valley and Rock Creek West, (Saskatchewan) (Sweet et al. 1999). The non-charred plant remains occur as elongate fibrous material (as long as 4 cm), that are also visible on bedding surfaces in the outcrop at Madrid East South, Clear Creek North, and Berwind Canyon (field observations of Belcher and Hildebrand, unpublished data). The earliest Tertiary rocks contain an average of 61% non-charred material (range from 21% to 87%). These notable occurrences of non-charred material in the K-T and earliest Tertiary rocks from more than 12 K-T sites across North America are particularly significant (as much as the lack of charcoal). Belcher et al. (2003) concluded that it was hard to imagine how so much organic matter could have remained non-charred if wildfire consumed the vegetation across the North American continent.

4.3 Soot

Wolbach et al. (1985, 1988, 1990) and Hildebrand and Wolbach (1989) presented evidence that soot occurs at K-T sites in Europe, New Zealand and North America and concluded that this suggested that worldwide forests fires were ignited by the impact. Wolbach et al. (1990) reported that the boundary layers contain up to 0.69% graphitic carbon, mainly as fluffy aggregates of soot spherules, 0.1 to 0.5 μm in size, with a characteristic "chained cluster" morphology. The amount of soot present in the K-T rocks was calculated to be 10^3 times the mean abundance of the Late Cretaceous (Wolbach et al. 1985). The peak in soot abundance coincides with the Iridium layer suggesting that the fires were ignited promptly following the impact as the soot accompanies the Iridium, even in the first fraction of the fallout. Apparently, the fires started well before all the ejecta from the impact had settled (Wolbach et al. 1988). The carbon isotopic value of the soot is remarkably constant ($\sim\delta^{13}C = -25.36$), leading Wolbach et al. (1988; 1990) to suggest that this constancy supports the interpretation that the carbon is global fallout from the K-T event from a single source. The carbon isotopic value of the soot is consistent with it being from a biomass source, as it resembles values for natural charcoal from forest fires (Wolbach et al. 1990). However, the authors note that at least some fossil carbon sources would have the same isotopic range, and therefore the soot could also be sourced from combustion of hydrocarbons (e.g., coal or oil).

The amount of elemental carbon (soot) found at the K-T boundary (0.012 g cm^{-2}) is considered to be very large and requires either that most of the Cretaceous biomass burned down, or that the soot yield was higher than in small fires (i.e. > 2%) (Gilmour et al. 1990).

4.4 Polyaromatic hydrocarbons

Wolbach et al. (1988, 1990) and Gilmour et al. (1990) suggested that a strong argument that the K-T soot is sourced from wildfires comes from the presence of polyaromatic hydrocarbons. Polyaromatic hydrocarbons (PAH's) are typical products of partial combustion and are known to be formed in forest fires (Masclet et al. 1995). Gilmour and Guenther (1988) reported the presence of retene (1-methyl-7-isopropyl phenanthrene) in the boundary clay at Woodside Creek (New Zealand). Wolbach et al. (1990) suggested that retene is diagnostic of resinous wood fires and probably forms by low-T pyrolysis of abietic acid which is a common constituent of plant resins in conifers and some angiosperms (Ramdahl 1983; Gilmour and Guenther 1988), and therefore that such trees provided part of the fuel for the K-T wildfire.

Venkatesan and Dahl (1989) reported enhanced PAH contents in K-T samples from Stevns Klint (Denmark), Gubbio (Italy) and Woodside Creek (New Zealand). The K-T rocks are enriched in phenanthrene, 2 methylfluorene, coronene, benzo(b + k) fluoranthene, chrysene/ triphenylene, pyrene, fluoranthene, benzopyrene and benzo(g,h,i)perylene (Venkatesan and Dahl, 1989). These compounds are formed from pyrolysis of organic matter (Blumer 1975; Simoneit 2002), although some are also formed by combustion of hydrocarbon materials (Oros and Simoneit 2000). Venkatesan and Dahl (1989) argued that the PAH distributions at the K-T boundary from Woodside Creek, Stevns Klint and Gubbio are characteristic of a combustion origin, based on the dominance of parent PAHs over alkylated species (which are suggested as being more characteristic of petroleums), concluding that the abundance of the PAHs is consistent with the suggestion of massive global wildfires.

Arinbou et al. (1999) reported that a 112 to 154-fold enrichment of typical pyrosynthetic PAHs such as, coronene, benzo(g,h,i)perylene and benzo(e)pyrene occurs in the boundary clay at Caravaca in Spain and estimated that the geologically instantaneous combustion of ~18 - 24% of the terrestrial above ground biomass occurred at the K-T boundary.

4.5 Fullerenes

Fullerenes have been reported from five K-T boundary sites from geographically separated areas, leading to the suggestion that the Earth's surface was covered by C60 and C70 (Heymann et al. 1996). The estimated amount of C60 present at the K-T boundary has been suggested as being consistent with the hypothesis that the fullerenes are from a wildfire source (Heymann et al. 1996). The reports that fullerenes occur in outcrops of 65 million-year-old rocks conflicts with their known ease of rapid oxidation (Taylor and Abdul Sada 2000). Fullerenes are unstable towards high temperatures and measurable decomposition can be observed for C60 above 750°C (Taylor 1999). Decomposition is more rapid in the presence of UV light (Taylor 1999). These facts suggest that unlike the highly inert forms of carbon, such as the charcoals and soots, fullerenic carbon is unlikely to be able to survive burial and diagenesis, and perhaps not even reach the sediment surface before being oxidised. Taylor and Abdul Sada (2000) carefully re-examined K-T boundary samples from Woodside Creek and Flaxbourne River and found no traces of either C60 or C70 (using a sensitivity four times greater than that used by Heymann et al. 1996), concluding that there are no fullerenes whatsoever in the K-T rocks.

Furthermore, fullerenes have not been detected in modern-day wildfires (Becker et al. 1995). However, they could potentially be synthesized during the impact (Becker et al. 2000), or sourced from the asteroid itself (Ehrenfreund and Foing 1995). Heymann et al. (1996) calculated that if the fullerenes were sourced via impact processes the C60 content of the carbon ought to be 2500 times larger than that found, suggesting that the wildfire scenario provides a better explanation. However, Heymann et al. (1996) estimated that the amount of biomass that would need to be burned to produce the quantity of C60 found in the K-T sediments is about twice that estimated to have been present at the time of the K-T. This suggests that at least some of the C60 at the K-T boundary may be due to the burning of fossil fuel.

Overall, the evidence suggests that fullerenes cannot provide a reliable indicator for impact related palaeo-wildfires. Not only have fullerenes never been detected in modern wildfires but there is reasonable doubt about the preservation of fullerenes in the fossil record. Furthermore, an impact-related source for any fullerenes found in K-T or other impact related rocks cannot be ruled out.

5 Discussion

5.1 Conflicting evidence

It is clear that the wildfire indicators outlined above show some disagreement. The soot and PAH record has been suggested to provide evidence that extensive or global wildfires occurred as part of the K-T events. Yet the charcoal record and the abundance of non-charred material across the K-T boundary suggests otherwise. Can this be explained?

5.1.1 Re-consideration of the charcoal record

If the thermal radiation released from the impact was as severe as predicted by Hildebrand (1993), then temperatures ought to have been sufficient even for spontaneous ignition of wet biomass, with temperatures so high it could be suggested that the charcoal was converted directly to CO_2. According to this scenario, not only would the charcoal be converted to CO_2, but such a severe post-impact conflagration would also tend to convert any soot to CO_2. However, this hypothesis would fail to explain how significant quantities of non-charred material in the K-T rocks managed to survive.

Another possibility is that charcoal was formed, but was subsequently oxidized away, either by acid rains following the impact (Hildebrand and Boynton 1989) or later during diagenesis. Not only do the K-T layers contain a small portion of charred material (see Fig. 1), implying that the depositional and diagnentic conditions were favourable to the preservation of charcoal (Belcher et al. 2003), but non-charred material (which is reactive) would be oxidized away much more rapidly than charcoal (which is highly inert). As significant non-charred material remains, it is hard to imagine a situation that would oxidize charred fragments and not non-charred material.

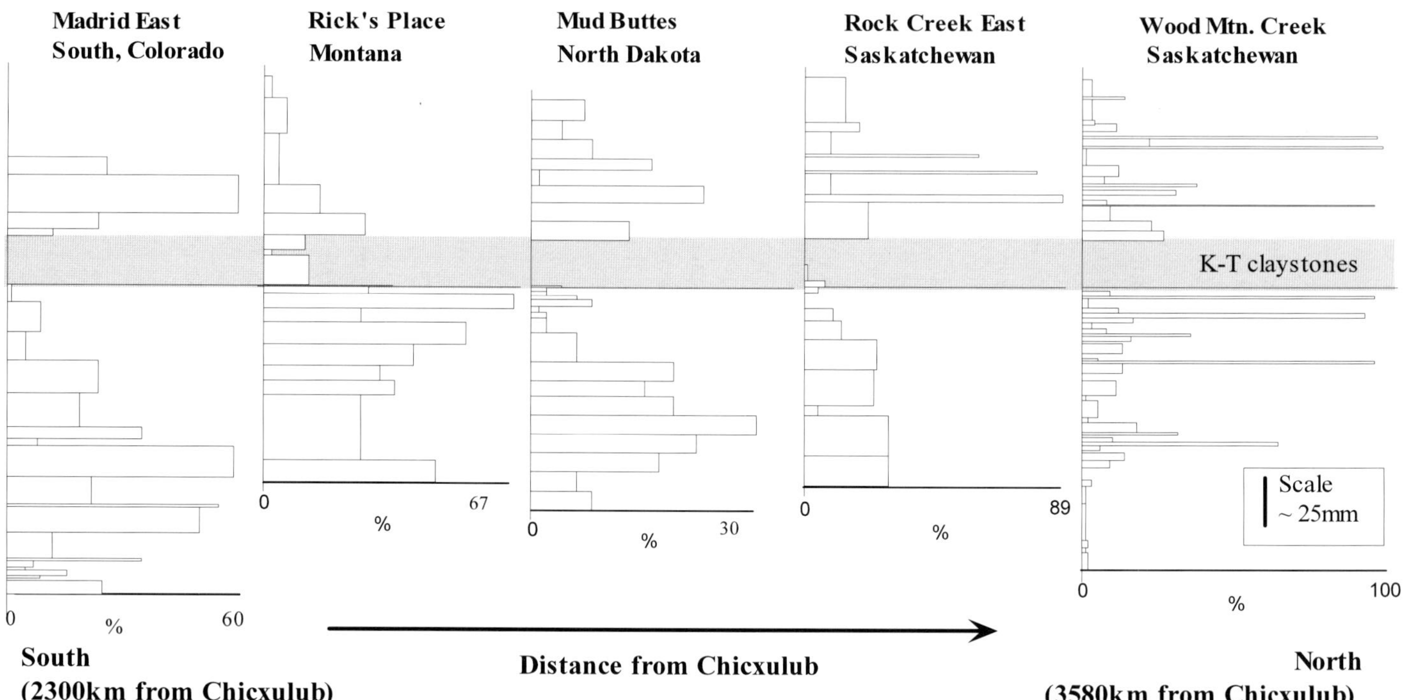

Fig 1. Percentage of charred material across the K-T boundary along a transect of sites across North America (modified from Belcher et al. 2003).

A further possibility is that the majority of the charcoal was transported away from the entire Western Interior, either by water and/ or wind. The late Cretaceous and Early Tertiary charcoal assemblages across the Western Interior contain both macro- and microscopic charcoal particles (tens to hundreds of micrometers in size). This indicates that the charcoal not only represents local fires (Innes and Simmons 2000), but that charred particles sourced from burning outside the Western Interior could also be included in the assemblages (Belcher et al. 2003).

The Western Interior at this time was dominated by fluvio-deltaic systems with forests growing around swamps, oxbow lakes and ponds (Pillmore and Flores 1990). Charcoal would have been washed into and concentrated in the ponds, increasing the apparent abundance of charcoal at these sites. The particles in the lower K-T claystone fine upwards, and the upper K-T claystone, form laminated layers. This suggests that the particles settled in tranquil conditions, with minimal currents available to wash any charcoal away. The charcoal could have been carried away by strong winds (which could only remove particles $<100\mu m$ in size [Patterson et al. 1987]). To remove charcoal in this way the vegetation would have had to have been flash burned into tiny fragments and then picked up by the wind instantly, before any could be washed or blown into the ponds. Again, the significant amount of non-charred material argues against rapid flash burning of all the vegetation. Even allowing for the fact that perhaps not all the vegetation was burnt, it seems unlikely that the non-charred material was preferentially deposited and the charcoal carried away.

Charcoal was not transported away from the Western Interior throughout the late Cretaceous and earliest Tertiary (Fig. 1). It is implausible that charcoal but not non-charred material could be transported away from this area only during the K-T event. Moreover, there is no evidence of vast deposits of charcoal in the marine K-T record of France, Haiti, USA, Tunisia and Spain (Jones and Lim 2000), suggesting that charcoal was not washed or blown into the sea.

It could be argued that the fluvio-deltaic and swamp environment of the Western Interior was sufficiently wet to prevent major fires across this region, whilst wildfires raged across the rest of the U.S and the globe. The latest Cretaceous and early Tertiary charcoal record (Fig. 1) shows that fires were frequent throughout this ecosystem, ruling out the possibility that this area was simply not fire-prone. Moreover, 60,000 km^2 of peat swamp in Indonesia have been burnt in recent years, indicating that fuel availability is far more important than moisture content (Page et al. 2002). If huge amounts of thermal energy had been released by the impact, then whether the area was fire-prone or not ought not to have been a factor.

There is no reason the Western Interior should have escaped burning compared to any other area in the USA. The only way to prevent fires in the Western Interior would be if rains fell on this area, and suddenly following the impact, or if for some reason extensive cloud cover shielded the area from the intense radiation (Melosh et al. 1990). Allowing for the chance that fires did not occur in the Western Interior, Belcher et al. (2003) argued that if fires burned across the rest of the continent, charcoal ought to have been carried into this region, and would have been identifiable in their samples.

5.1.2 Re-assessment of the soot record

Wolbach et al. (1990) observed that the coarse K-T carbon contained no particles that were diagnostic of wood. Moreover, the soot particles from the K-T rocks are described as fluffy aggregates of soot spherules, with a characteristic "chained cluster" morphology (Wolbach et al. 1990), typical of aciniform carbon agglomerates. Fernandes et al. (2003) characterised carbonaceous combustion residues of diesel soot, urban dust, carbon black, chimney soot, vegetation fire residues, wood and straw charcoals via morphological, elemental and spectroscopic features. They concluded that: a) Diesel soot is composed of spherical particles, essentially of aciniform carbon, and b) wood and straw charcoals exhibit virtually no aciniform carbon and consist of layered structures with hard-edge boundaries. Stoffyn-Egli et al. (1997) used an analytical SEM to identify soot particles from oil, coal and biomass combustion. They concluded that: a) Oil and coal soot particles were often spherical, whereas biomass soots are not, b) Only biomass soots and unburnt coal show plant structures, and c) that biomass burning generates mostly non-spheroidal particles. Belcher et al. (*in press*) collected soot from the combustion of coal, oil (petroleum) and biomass (Hayman Wildfire, Colorado U.S.A, 2002) and studied it's morphology under the scanning electron microscope. They observed that: a) soot from the Hayman wildfire is characterised by numerous tiny plant fragments, generally smaller than 30 μm in length, b) soot produced from the combustion of coal and oil is characterised by fluffy aggregates of spherical to sub-spherical particles welded into chains and clusters. Therefore, the morphology of the K-T soot appears to be inconsistent with it being sourced from combustion of biomass, and is morphologically similar to that produced by burning oil or coal.

Chicxulub drill cores reveal that the target rocks contain hydrocarbons (Gilmour et al. 2003), the vaporisation of which could produce soot. The

K-T soot is isotopically uniform between locations; however, the carbon isotopic value of soot produced from the burning of the global biomass might be expected to be more heterogeneous (i.e., different between Europe, New Zealand, and the USA) unless the soot became thoroughly and globally mixed before fallout. Soot produced from vaporisation of one pool of hydrocarbons might be expected to be more isotopically homogeneous. Moreover, the large amount of soot in the K-T rocks requires that most of the Cretaceous biomass burned down (Gilmour et al. 1990). This amount might be accounted for more easily by assuming that fossil carbon rather than biomass was the major fuel source (Gilmour et al. 1990).

5.1.3 Re-assessment of the PAH record

Wolbach et al. (1988, 1990), Gilmour and Guenther (1988) and Gilmour et al. (1990) argued that the presence of the PAH retene in the K-T rocks was strong evidence that the K-T soot was sourced from wildfires. However, retene is not exclusively formed by biomass burning, but is also derived through diagenesis of pimaric and abietic acids from organic matter (Killops and Killops 1993).

Oros and Simoneit (2000, 2001a, b) and Simoneit (2002) published extensive literature on the organic tracers (including PAHs) from incomplete combustion of biomass and coals. There are no PAHs produced exclusively by biomass burning. However, there are PAHs that are formed only during the combustion of hydrocarbon material. Therefore, should any of these PAHs be found along with those that may also be produced by biomass burning, an alternative source might be suggested.

Both Venkatesan and Dahl (1989) and Arinbou et al. (1999) found coronene to be one of the dominant PAHs in the K-T rocks. Coronene is formed from the combustion of coals, and not biomass (Oros and Simoneit 2000, 2001a, b; Simoneit 2002), whilst the other PAHs found in the K-T sediments (phenanthrene, 2 methylfluorene, coronene, benzo(b + k) fluoranthene, chrysene/ triphenylene, pyrene, fluoranthene, benzopyrene, benzo(g,h,i)perylene [Venkatesan and Dahl 1989], benzo(g,h,i)perylene and benzo(e)pyrene [Arinbou et al. 1999]) can be formed both through combustion of biomass and hydrocarbons (Oros and Simoneit 2000, 2001a, b; Simoneit 2002). The presence of a significant quantity of coronene suggests that the PAHs might be sourced from combustion of hydrocarbons, rather than biomass material.

Venkatesan and Dahl (1989) concluded that the PAH distributions in the K-T rocks were consistent with the suggestions of massive global fires.

However, they considered that the PAHs at Woodside Creek and Gubbio could be characteristic of wood or kerosene whereas the Stevens Klint profile was more consistent with being from the combustion of coal (Venkatesan and Dahl 1989). This would accord with the interpretation that the PAHs were sourced from the combustion of hydrocarbon material rather than biomass burning.

5.1.4 Were there wildfires associated with the K-T event?

Hypotheses to explain the lack of charred material in the K-T and earliest Tertiary sediments across the Western Interior involve scenarios that require intense charcoal-destructive wildfires. The presence of uncharred organic material argues against these explanations. The area was fire-prone before and after the K-T event, with abundant charcoal preserved in the rocks. Minor amounts of charcoal are found in the K-T rocks. Therefore, there seems to be no taphonomic bias against charcoal preservation or recognition. The abundance of soot and certain PAHs in the K-T rocks are consistent with them being from a non-biomass source, such as from the vaporisation of hydrocarbons in the Chicxulub target rocks. This explanation would be more consistent with the record of charcoal and non-charred material found in the K-T rocks. There is no irrefutable evidence to suggest that extensive wildfires were ignited as part of the K-T events.

6 How does the lack of wildfires affect the proposed extinction mechanisms of the K-T event?

It is recognised that there was a major disruption to plant communities (Tschudy et al. 1984; Sweet 2001; Nichols and Johnson 2002) across the K-T boundary. However, the analysis of Belcher et al. (2003) suggested that extensive wildfires were not the cause. Plant and animal communities may not have been burned alive, but the lack of wildfires does not mean that they did not suffer the effects of mild thermal radiation, global darkness, cold, and poisoning.

The fact that there were not extensive wildfires as part of the K-T events allowed Belcher et al. (2005) to place an upper limit on the ground temperatures that could have resulted from the K-T impact event (based on the fact that temperatures greater than 325°C are required to initiate smouldering of vegetation). The lack of wildfires implies that temperatures did not exceed 325°C (much lower than any model has previously predicted e.g., Melosh et al. 1990), but it does not rule out the possibility that relatively high temperatures (of the order of a couple of hundred degrees centigrade) did not play a role in some of the extinctions seen at this time.

Irrespective of the source, the large amounts of soot found in the K-T rocks (Wolbach et al. 1985, 1988, 1990) combined with a substantial amount of silicate dust (Pope 2002) would still have had the potential to produce initial global cold and darkness. The combustion of fossil carbon produces significant quantities of pyrotoxins, including PAHs (e.g., Oros and Simoneit 2000). The distribution of PAHs in the K-T rocks (Venkatesan and Dahl 1989) suggests that these may have been released in sufficient quantities to have proven toxic to animal life. The fact that the PAH's were not formed by a wildfire would not have effected their toxic potential.

Global wildfires are not the only way in which vast quantities of CO_2 could have been released into the atmosphere. The impact at Chicxulub into a sequence dominated by carbonate rocks would have lead to shock devolatisation of CO_2 in vast amounts (O'Keefe and Ahrens 1989). The impact would have produced a CO_2 pulse of $\sim 10^{20}$ g (O'Keefe and Ahrens 1989), which is likely to have remained resident in the atmosphere on a timescale of 10^3 to 10^5 years (e.g., Broecker and Peng 1982; Berner et al. 1983). Such a quantity of CO_2 is ~50 times the current atmospheric level and would have been capable of producing greenhouse warming by as much as ~15°C (Hildebrand 1993), following the initial cold and darkness created by the impact. The suggestion that global warming was responsible for the extinctions seen across the K-T boundary (O'Keefe and Ahrens 1989) does not require that the global biomass was burned down. It is better explained by devolatisation of CO_2 in the carbonate platform at Chicxulub.

7 Summary

Several models have predicted the occurrence of extensive wildfires associated with the K-T impact event (Melosh et al. 1990; Kring and Durda 2002). The abundance of soot (Wolbach et al. 1985, 1988, 1990; Gilmour et al. 1990), and increased concentrations of PAHs (Venkatesan and Dahl 1989; Arinbou et al. 1999) in the K-T rocks, have been considered to support the wildfire hypothesis. Taylor and Abdul Sada (2000) rejected the claim made by Heymann et al. (1996) that the fullerenes C60 and C70 could be found in the K-T rocks. It is concluded that fullerenes are not a suitable indicator of impact-related palaeo-wildfires based on their occurrence in extraterrestrial material (Ehrenfreund and Foing 1995) and the possibility that they might be a synthesized during the impact (Becker et al. 2000). The charcoal record across K-T sites stretching the length of the Western Interior of the USA reveals below background amounts of charcoal (Sweet and Cameron 1991; Scott et al. 2001; Belcher et al. 2003) and an abundance of non-charred material in the K-T and earliest Tertiary rocks (Belcher et al. 2003).

Explanations to account for the disagreement between the charcoal record and the other potential wildfire indicators include: charcoal formation but subsequent oxidation (by acid rains or during diagenesis), or intense thermal radiation converting charcoal directly to CO_2. However, in both scenarios significant quantities of non-charred material would not be expected to survive. There is no satisfactory hypothesis to explain how charcoal could be formed and transported away from the entire Western Interior, and it is extremely unlikely that conditions prevailed in the Western Interior alone that prevented the fires. Furthermore, the record of soot and PAHs can be explained without invoking the global wildfire hypothesis.

The morphology of the K-T soot is inconsistent with it being from a biomass source, and is more characteristic of soots produced during combustion of hydrocarbons (Stoffyn-Egli et al. 1997; Fernandes et al. 2003). The K-T soot is isotopically uniform between sites across the globe. This might not be expected if the entire late Cretaceous biomass had been burnt, however the vaporisation of one pool of hydrocarbons might be expected to give a more homogenous result. The PAH signature of the K-T rocks includes compounds that are not formed from the burning of biomass, but that are released during the combustion of hydrocarbons (Oros and Simoneit 2000, 2001a, b; Simoneit 2002), or formed during diagenesis (Killops and Killops 1993). Chicxulub drill cores reveal that the target rocks contain hydrocarbons (Gilmour et al. 2003), the vaporisation of which could produce the soot and PAHs found at the K-T boundary. It seems that wildfires were not a major factor in the K-T event and that they were not responsible for the environmental perturbations or extinctions observed at this time.

Acknowledgments

I thank the reviewers for constructive reviews and Margaret E. Collinson, Paul Finch, Alan R. Hildebrand, Andrew, C. Scott, Arthur R. Sweet, Roger Taylor, and Wendy S. Wolbach for helpful discussions and exchange of ideas on the K-T event and/or wildfires. I thank Margaret E. Collinson, Jana Hanova, Alan R. Hildebrand, Kirk Johnson, Dean Pearson, Arthur R. Sweet, the staff from Grasslands National Park, Canada, and the Land owners (Miles Anderson and Ritchie Hordenchuck) for help with fieldwork throughout my studies. I also thank Neil Holloway for expert preparation of polished blocks of sediment, and Sharon Gibbons for lab assistance. Funding through a NERC studentship (NER/S/A /2001/06342) and CASE support from Royal Botanic Gardens, Kew, is gratefully acknowledged.

References

Alvarez LW, Alvarez W, Asaro F, Michel HV (1980) Extraterrestrial cause for the Cretaceous-Tertiary extinction. Science 208: 1095 –1108

Arinbou T, Ishiwatari R, Kaiho K, Lamolda MA (1999) Spike of pyrosynthetic polycyclic aromatic hydrocarbons associated with an abrupt decrease in $\delta^{13}C$ of a terrestrial biomarker at the Cretaceous-Tertiary boundary at Caravaca Spain. Geology 27: 723-726

Becker L, Bada J, Bunch TE (1995) Fullerenes in the K-T boundary: Are they a result of global fires [abs]. Lunar and Planetary Science 26: 85-86

Becker L, Poreda RJ, Bunch TE (2000) Fullerenes: An extraterrestrial carbon carrier phase for noble gases. Proceedings of the National Academy of Sciences of the USA 97: 2979-2983

Belcher CM, Collinson ME, Sweet AR, Hildebrand AR, Scott AC (2003) "Fireball passes and nothing burns"- The role of thermal radiation in the K-T event: Evidence from the charcoal record of North America. Geology 31: 1061–1064

Belcher CM, Collinson ME, Sweet AR, Hildebrand AR, Scott AC (2005) Constraints on the thermal energy released from the Chixculub impactor: new evidence from multi-method charcoal analysis. Journal of the Geological Society of London, 162: 591-602

Berner RA, Lasaga AC, Garrels RM (1983) The carbonate-silicate geochemical cycle and its effect on atmospheric carbon dioxide over the past 100 million years. American Journal of Science 283: 641-683

Blumer M (1975) Curtisite, Irdrialite and pendletonite, polyaromatic hydrocarbon minerals: their composition and origin. Chemical Geology 16: 245-256

Broecker WS, Peng TH (1982) Tracers in the Sea. Eldigo, New York, 690 pp

Ehrenfreund P, Foing BH (1995) Search for fullerenes and PAHs in diffuse interstellar medium. Planetary and Space Science 43: 1183-1187

Fernandes MB, Skjemstad JO, Johnson BB, Wells JD, Brooks P (2003) Characterization of carbonaceous combustion residues. I. Morphological, elemental and spectroscopic features. Chemosphere 51: 785-795

Gerstl SA, Zaerdecki A (1982) Reduction of photosynthetically active radiation under extreme stratospheric aerosol loads. In: Silver LT, Schultz PH (eds) Geological implications of impacts of large asteroids and comets on Earth. Geological Society of America Special Paper 247: 367-382

Gilmour I, Guenther F (1988) The global Cretaceous-Tertiary fire: biomass or fossil carbon? [abs.]. In: Global Catastrophes in Earth History, Lunar and Planetary Institute Contribution 673: 60-61

Gilmour I, Wolbach WS, Anders E (1990) Major wildfires at the Cretaceous-Tertiary boundary. In: Clube SVM (eds) Catastrophes and Evolution: Astronomical Foundations, Cambridge University Press, Cambridge, pp 195-213

Gilmour I, Sephton MA, Morgan JV (2003) Organic geochemistry of a hydrocarbon-rich calcarenite from the Chicxulub scientific drilling program [abs.]. Lunar and Planetary Science XXXIV [abs.] #1771 (CD-ROM)

Heymann D, Chibante LPF, Brooks RR, Wolbach WS, Smalley RE (1994) Fullerenes in the Cretaceous-Tertiary boundary layer. Science 265: 645-647

Heymann D, Chibante LPF, Brooks RR, Wolbach WS, Smit J, Korochantsev A, Nazarov MA, Smalley RE (1996) Fullerenes of possible wildfire origin in Cretaceous-Tertiary boundary sediments. In: Ryder G, Fastovsky D, Gartner S (eds) The Cretaceous-Tertiary event and other catastophes in earth history. Geological Sociecty of America Special Paper 307: 453-464

Heymann D, Yancey TE, Wolbach WS, Thiemens MH, Johnson EA, Roach D, Moecker S (1998) Geochemical markers of the Cretaceous-Tertiary boundary event at Brazos River, Texas, USA. Geochimica et Cosmochimica Acta 62: 173-181

Hildebrand AR (1993) The Cretaceous/Tertiary boundary impact (or the dinosaurs didn't have a chance). Journal of the Royal Astronomical Society of Canada 87: 77–118

Hildebrand AR, Boynton WV (1989) Hg anomalies at the K/T boundary-evidence for acid rain [abs.] Meteoritics 24: 277–278

Hildebrand AR, Wolbach WS (1989) Carbon and chalcophiles at a non marine K/T boundary: Joint investigations of the Raton section, New Mexico [abs.] Lunar and Planetary Science 20: 414–415

Hildebrand AR, Penfield GT, Kring DA, Pilkington M, Camargo A, Jacobsen SB, Boynton WV (1991) Chicxulub crater: A possible Cretaceous-Tertiary boundary impact crater on the Yucatán Peninsula, Mexico. Geology 19: 867–871

Innes JB, Simmons IG (2000) Mid Holocene charcoal stratigraphy, fire history and palaeoecology at North Gill, North York Moors, UK. Palaeogeography, Palaeoclimatology, Palaeoecology 164: 151–166

Izett GA (1990) The Cretaceous/Tertiary boundary interval, Raton Basin, Colorado and New Mexico, and its content of shock-metamorphosed minerals: Evidence relevant to the K/T boundary impact-extinction theory. Geological Society of America Special Paper 249: 100 pp

Jones TP (1996) Comment on "Fossil charcoal in Cretaceous-Tertiary boundary strata: evidence for catastrophic firestorm and megawave". Geochimica et Cosmochimica Acta 60: 719-720
Jones TP, Lim B (2000) Extra terrestrial impacts and wildfires. Palaeogeography, Palaeoclimatology, Palaeoecology 164: 57- 66
Killops SD, Killops VJ (1993) An Introduction to Organic Geochemistry. Geochemistry Series, Longman Scientific and Technical, John Wiley and Sons, New York, 265 pp
Kring DA, Durda, DD (2002) Trajectories and distribution of material ejected from the Chicxulub impact crater: Implications for post impact wildfires. Journal of Geophysical Research 107: 6–22 DOI:10.1029/2001JE001532
Kruge MA, Stankiewicz BA, Creilling JC, Montanari A, Bensley DF (1994) Fossil charcoal in Cretaceous-Tertiary boundary strata: evidence for catastrophic firestorm and megawave. Geochimica et Cosmochimica Acta 58: 1393-1397
Kruge MA, Stankiewicz BA, Creilling JC, Montanari A, Bensley DF (1996) Reply to the Comment by TP Jones on "Fossil charcoal in Cretaceous-Tertiary boundary strata: evidence for catastrophic firestorm and megawave". Geochimica et Cosmochimica Acta 60: 721-722
MacLeod N, and 21 others (1996) The Cretaceous-Tertiary biotic transition. Journal of the Geological Society, London 153: 1-28
Melosh HJ, Schneider NM, Zahnle KJ, Latham D (1990) Ignition of global wildfires at the Cretaceous/Tertiary boundary. Nature 343: 251–254
Masclet A, Liousse C, Wortham H (1995) Emissions of polycyclic hydrocarbons by savanna fires. Journal of Atmospheric Chemistry 22: 41-54
Nichols DJ, Johnson KR (2002) Palynology and microstratigraphy of Cretaceous-Tertiary boundary sections in southwestern North Dakota. In: Hartman JH, Johnson KR, Nichols DJ (eds) The Hell Creek Formation of the Cretaceous-Tertiary boundary in the northern great plains. Geological Society of America Special Paper 361, pp 95–143
O'Keefe JD, Ahrens TJ (1989) Impact production of CO_2 by the Cretaceous-Tertiary extinction bolide and the resultant heating of the Earth. Nature 338: 247-249
Oros DR, Simoneit BRT (2000) Identification and emission rates of molecular tracers in coal smoke particulate matter. Fuel 79: 515-536
Oros DR, Simoneit BRT (2001a) Identification and emission factors of molecular tracers in organic aerosols from biomass burning. Part 1. Temperate climate conifers. Applied Geochemistry 16: 1513-1544
Oros DR, Simoneit BRT (2001b) Identification and emission factors of molecular tracers in organic aerosols from biomass burning. Part 2. Decidous trees. Applied Geochemistry 16: 1545-1565
Page SE, Siegert F, O. Rieley J, Boehm HDV, Jaya A, Limin S (2002) The amount of carbon released from peat and forest fires in Indonesia during 1997. Nature 420: 61-65
Patterson III, WA, Edwards KJ, Maguire DJ (1987) Microscopic charcoal as a fossil indicator of fire. Quaternary Science Reviews 6: 3-23

Pillmore CL, Flores RM (1990) Stratigraphy and depositional environments of the Cretaceous-Tertiary boundary clay and associated rocks, Raton basin, New Mexico and Colorado. In: Fassett JE, Rigby JK (eds) The Cretaceous-Tertiary boundary in the San Juan and Raton Basins, New Mexico and Colorado. Geological Society of America Special Paper 209: 111–129

Piperno DR (1997) Proceedings of the Ocean Drilling Program, Scientific Results, Vol. 155: College Station, Texas, Ocean Drilling Program, pp 411-418

Pope KO (2002) Impact dust not the cause of the Cretaceous-Tertiary mass extinction. Geology 30: 99–102

Ramdahl T (1983) Retene a molecular marker of wood combustion in ambient air. Nature 306: 580-582

Scott AC, Lomax B, Collinson ME, Upchurch G, Beerling D (2000) Fire across the K-T boundary: Initial results from the Sugarite coal, New Mexico, USA. Palaeogeography, Palaeoclimatology, Palaeoecology 164: 381–395

Shuvalov VV, Artemieva NA (2002) Atmospheric erosion and radiation impulse induced by impacts. In: Koeberl C, MacLeod KG (eds) Catastrophic events and mass extinctions: Impacts and beyond. Geological Society of America Special Paper 356: 695–703

Simoneit BRT (2002) Biomass burning - a review of organic tracers for smoke from incomplete combustion. Applied Geochemistry 17: 129-162

Stoffyn-Egli P, Potter TM, Leonard JD, Pocklington R (1997) The identification of black carbon particles with the analytical scanning electron microscope: methods and initial results. The Science of the Total Environment 198: 211-223

Sweet AR (2001) Plants, a yardstick for measuring the environmental consequences of the Cretaceous-Tertiary boundary event. Geoscience Canada 28: 127–138

Sweet AR, Cameron AR (1991) Palynofacies, coal petrographic facies and depositional environments: Amphitheatre Fornation (Eocene to Oligocene) and Ravenscrag Formation (Maastrichtian to Paleocene). Canada International Journal of Coal Geology 19: 121-144

Sweet AR, Braman DR, Lerbekmo JF (1999) Sequential palynological changes across the composite Cretaceous-Tertiary (K-T) boundary claystone and contiguous strata, western Canada and Montana, USA. Canadian Journal of Earth Sciences 36: 743–768

Taylor R (1999) Lecture Notes on Fullerene Chemistry - a Handbook for Chemists, Imperial College Press, London, 268 pp

Taylor R, Abdul-Sada AK (2000) There are no fullerenes in the K-T boundary layer. Fullerene Science and Technology 8: 47-54

Toon OB, Pollack JB, Ackerman TP, Turco RP, McKay CP, Liu MS (1982) Evolution of an impact generated dust cloud and its effects on the atmosphere. In: Silver LT, Schultz PH (eds) Geological implications of impacts of large asteroids and comets on the Earth. Geological Society of America Special Paper 190: 187–201

Tschudy RH, Pillmore CL, Orth CJ, Gillmore JS, Knight JD (1984) Disruption of the terrestrial plant ecosystem at the Cretaceous-Tertiary boundary, Western Interior. Science 225: 1030-1032

Venkatesan MI, Dahl J (1989) Further geochemical evidence for global fires at the Cretaceous-Tertiary boundary. Nature 338: 57-60

Wolbach WS, Lewis RS, Anders E (1985) Cretaceous extinctions: Evidence for wildfires and search for meteoritic material. Science 230: 167-230

Wolbach WS, Gilmour I, Anders E, Orth CJ, Brooks RR (1988) Global fire at the Cretaceous-Tertiary boundary. Nature 334: 665-669

Wolbach WS, Gilmour I, Anders E (1990) Major wildfires at the Cretaceous/Tertiary boundary. In: Sharpton VL, Ward PD (eds) Global catastrophes in Earth history. Geological Society of America Special Paper 247: 391–400

Continental Vertebrate Extinctions at the Triassic-Jurassic and Cretaceous-Tertiary Boundaries: a Comparison

Eric Buffetaut

CNRS, 16 cour du Liégat, 75013 Paris, France
(eric.buffetaut@wanadoo.fr)

Abstract. Although similar causes (such as asteroid impact or flood basalt volcanism) have been suggested for both the Triassic-Jurassic (Tr/J) and Cretaceous-Tertiary (K/T) mass extinctions, a comparison of extinction patterns among continental vertebrates reveals significant differences between these two events. There is no discernible size factor at the Tr/J boundary, whereas the K/T extinction is marked by the disappearance of all large land vertebrates. Unlike what happened at the end of the Cretaceous, freshwater ecosystems were significantly affected at the end of the Triassic, with the extinction of the crocodile-like phytosaurs. Both dinosaurs and mammals survived the Tr/J event, but the large adaptive radiation of mammals had to wait until dinosaurs disappeared at the end of the Cretaceous. All these differences in pattern suggest that different processes were involved, and that what happened at the K/T boundary cannot be simply extrapolated to the Tr/J boundary. The need for closer studies of extinction patterns during major biotic crises is emphasized.

1 Introduction

While it is now widely accepted that the mass extinction at the end of the Cretaceous was caused by the Chicxulub meteorite impact, the causes of the biotic crisis at the end of the Triassic are not so well understood. Although an impact has repeatedly been invoked (see reviews in Benton 1997, and Hallam and Wignall 1997), it is only recently that detailed evidence linking possible impact signatures (iridium anomaly, fern spike) with significant changes in vertebrate communities has been presented (Olsen *et al.* 2002a,b), on the basis of observations made in eastern North

America (mainly in the Newark Basin). There, according to Olsen *et al.*, the fossil record (including both body fossils and footprints) suggests that large theropod dinosaurs appeared less than 10,000 years after the boundary, which is marked by a mass extinction coincident with an iridium anomaly and a fern spore spike, suggesting a bolide impact.

Although the evidence linking the extinctions of the Tr/J boundary with an impact is still rather tenuous and needs confirmation, the matter clearly deserves consideration. An interesting aspect of the problem is that the pattern of vertebrate extinction at the Tr/J boundary appears to be significantly different from that at the K/T boundary, which obviously raises questions as to a similar cause for both events. The purpose of the present paper is to review some of the major differences between the two mass extinctions in terms of their effects on continental vertebrates, and to discuss the possible significance of these differences.

A detailed review of the evidence for mass extinctions of land vertebrates at the Cretaceous-Tertiary and Triassic-Jurassic boundaries is beyond the scope of this paper. Numerous reviews of vertebrate extinctions at the end of the Cretaceous are available (e.g., Buffetaut 1990, Archibald 1996). Terminal Triassic extinctions have received less attention, but they have been addressed by several authors (see Benton 1997, and Hallam and Wignall 1997, for reviews).

Despite continuing debate about the role of additional factors (such as flood basalt volcanism: see Courtillot and Renne 2003), it is now widely accepted that the main cause of the mass extinction at the Cretaceous-Tertiary boundary was the Chicxulub asteroid impact – with food chain collapse induced by darkness as the most likely direct cause of extinction. The causal factors involved in the Triassic-Jurassic mass extinction are by no means so well understood, and various explanations, ranging from asteroid impact to volcanism and to sea-level change, have been put forward (Benton 1997, Hallam and Wignall 1997).

The present paper will not discuss in any detail the possible *processes* involved in these mass extinctions, but will concentrate instead on a comparison between the observable *patterns* of extinction. One of the interesting characters of all mass extinctions is their selectivity, some groups of organisms being exterminated while others survive without apparently being much affected (Buffetaut 1984). The extinctions at the Tr/J and K/T boundaries are no exception, but, at least among continental vertebrates, the groups which became extinct and those which survived were not the same. As a result, very different patterns of extinction emerge.

2
The size factor

The Cretaceous-Tertiary mass extinction is notorious for having selectively eliminated large land vertebrates, whereas small forms fared better. As a result, the largest land vertebrates at the beginning of the Palaeocene did not exceed an approximate weight of 25 kg. This of course largely reflects the disappearance of the dinosaurs, which were the largest terrestrial animals of the Late Cretaceous. Whatever happened at the K/T boundary, being a small animal undoubtedly conferred a selective advantage. This is understandable within the framework of the hypothesis of food chain disruption (Sheehan and Fastovsky 1992): small terrestrial animals, such as mammals or small lizards, which had not the same food requirements as the larger dinosaurs, may have survived on invertebrates, themselves feeding on organic matter contained in soil. On the other hand, larger animals, which were part of food chains based on living plants, apparently did not survive the well-attested crisis in the plant world (itself probably a result of impact-induced darkness).

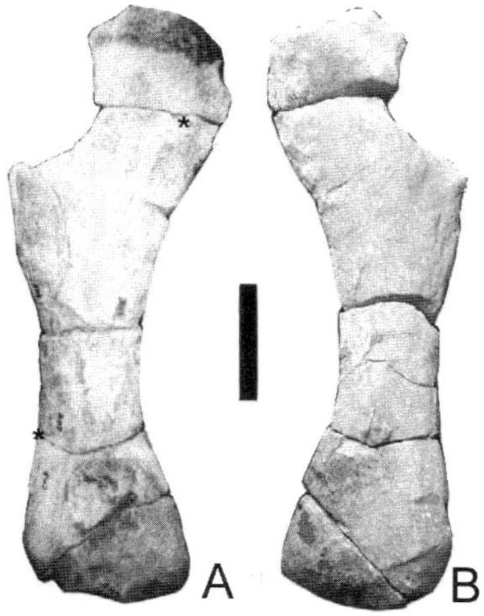

Fig. 1. The largest known Triassic terrestrial vertebrate: a sauropod humerus (possibly *Isanosaurus*), from the Nam Phong Formation (Rhaetian) of Thailand, in cranial (a) and caudal (b) views (after Buffetaut *et al.* 2002). Scale bar: 250 mm. The bone is slightly over one metre in length, suggesting that the animal was 13 to 15 m long. Large sauropod dinosaurs were present in the Late Triassic, apparently were not affected by Tr/J boundary events, and expanded during the Jurassic.

The pattern observed at the Triassic-Jurassic boundary is significantly different, because extinction and survival do not appear to be size-related. Some large animals did disappear, including phytosaurs (see below) and large rauisuchid pseudosuchians. However, by terminal Triassic times, the largest land animals were dinosaurs, following the expansion of the group in the Norian. Dinosaurs were not significantly affected by the events of the Triassic-Jurassic boundary (see below), and as a result the drastic reduction in the size of terrestrial vertebrates which is observed at the K/T boundary has no equivalent at the Tr/J boundary.

Olsen *et al.* (2002a) have reported an *increase* in the size of theropod footprints immediately above the Tr/J boundary in eastern North America, which again indicates a quite different pattern from that seen at the K/T boundary. According to Olsen *et al.* (2002a), this increase in footprint size may reflect the rapid rise of large theropods during a period of ecological release following the extinction event at the Tr/J boundary – a hypothesis which could be falsified by the discovery of large theropod bones or footprints in Triassic strata. Courtillot and Renne (2003) have considered that the recent description of large sauropod dinosaurs from the Late Triassic of Thailand (Buffetaut *et al.* 2002) falsified the hypothesis of Olsen *et al.* (2002a), but it clearly does not, since the hypothesis is based on theropods, not sauropods (Buffetaut 2003). Most Triassic theropods were indeed small, although *Liliensternus*, from the Late Triassic of Germany, could reach a length of five metres (Rauhut and Hungerbühler 1998). The expansion, if not the appearance, of large theropods during the Early Jurassic, may be linked to the demise of the large carnivorous pseudosuchians, as part of the extinction event at the Tr/J boundary (Olsen *et al.* 2002b).

The case of sauropodomorph dinosaurs is different, since they had already become quite large by the Late Triassic. Large prosauropods, reaching a length of nine metres, such as *Plateosaurus*, have long been known from the Norian of various parts of the world. Recently, it has become clear that sauropods were also present during the Triassic (Buffetaut *et al.* 2000; Yates and Kitching 2003), and that some of them had already reached a considerable size, comparable to that of Jurassic forms (Buffetaut *et al.* 2002).

It is thus clear that there is no evidence of size decline among land vertebrates at the Tr/J boundary. The evidence rather points to size *increase* in some groups. This is obviously extremely different from what is observed at the K/T boundary.

3 Extinction and survival of freshwater vertebrates

One of the striking aspects of the mass extinction at the K/T boundary is its limited impact on freshwater ecosystems (Buffetaut 1990, Archibald and Bryant 1990, Sheehan and Fastovsky 1992, Archibald 1996). This is reflected by the high rate of survival among such groups as freshwater fishes (Cavin 2002), turtles and crocodilians. The main reason for this preferential survival is probably that freshwater food chains are based on particles of organic matter in suspension in the water rather than on fresh plants, so that a disruption of photosynthetic activity (because of darkness) probably exerted only a limited influence on such ecosystems. As a result, fairly large freshwater vertebrates, such as crocodilians reaching several metres in length, were able to survive the K/T boundary crisis.

An interesting aspect of Tr/J boundary events is that they led to the extinction of a significant group of freshwater vertebrates, the phytosaurs, which had played an important part in Late Triassic ecosystems. Phytosaurs, or parasuchians, were large aquatic archosaurs which were superficially very similar to crocodilians (Williston 1914, Parrish 1999), having elongated jaws provided with numerous conical teeth. A basic difference with crocodiles, however, was that the external nares were located far posteriorly, slightly anterior and dorsal to the orbits, rather than at the tip of the snout as in crocodilians. A number of phytosaur genera have been described from the Late Triassic, differing in the length and robustness of the jaws, and thus paralleling the various dietary adaptations known in crocodilians, with slender-jawed forms feeding mainly on fish, whereas those with more robust jaws could probably prey on large terrestrial animals. Phytosaurs thus were among the aquatic top predators of the Late Triassic. Although phytosaur fossils have been reported from a few localities thought to be Early Jurassic in age (Buffetaut 1993), these reports are based on doubtful or scanty material, which in some cases may have been reworked from older deposits, and it seems well established that phytosaurs did not survive the Triassic-Jurassic boundary. Although true crocodilians appeared during the Late Triassic, it is unlikely that the extinction of the phytosaurs could have been caused by competition with them, since early crocodilians were mostly small, short-jawed, terrestrial forms showing few adaptations to freshwater habitats (Brochu 1999), which are unlikely to have occupied the same niches as phytosaurs. Long-jawed crocodilians appeared during the Early Jurassic, after the disappearance of the phytosaurs – which may have freed previously occupied ecological space.

Although the exact cause of the demise of the phytosaurs remains unknown, the fact that these large aquatic predators disappeared at the Tr/J boundary, whereas crocodilians survived at the K/T boundary, is an importance difference, in terms of ecological consequences, between the two mass extinctions. Whatever happened at the Tr/J boundary, it seems that it had more negative effects on freshwater ecosystems than the K/T impact. The reason for this particular difference is extremely unclear, but it should be kept in mind when discussing Tr/J boundary events.

4
Dinosaurs and mammals: different fates

A striking difference between the Tr/J and K/T mass extinctions is their effect on dinosaurs. The disappearance of all non-avian dinosaurs at the Cretaceous-Tertiary boundary is the most spectacular aspect of that particular event. Things were quite different at the time of the Tr/J events. Not only did dinosaurs survive with apparently very few extinctions, but they actually seem to have benefited from whatever happened at the Tr/J boundary. As noted by Olsen *et al.* (2002a,b), ichnological evidence suggests that large theropods flourished a short time after the extinction event. As noted above, this expansion of large theropods may well be linked to the disappearance of large pseudosuchians, which removed probable competitors. Be that as it may, the mass extinction of the Tr/J boundary clearly had no adverse effects on dinosaur evolution, and may have been beneficial for the expansion of the group. This is in sharp contrast to what happened at the end of the Cretaceous.

There is no evidence either for significant extinctions among mammals at the Tr/J boundary, and in fact the group as a whole may have suffered more at the Cretaceous-Tertiary boundary, when, for instance, high extinction rates are observed among North American marsupials (Archibald 1996). The main difference, however, lies in what happened to mammals *after* the mass extinction. Although mammals did diversify during the Jurassic, their Mesozoic, post-Triassic diversification (Hopson 1999) is different from their great Cenozoic radiation which enabled them to conquer a very wide range of environments, and in many cases to reach a much larger size than in the Mesozoic. The post-Cretaceous radiation of mammals is frequently seen – probably with right - as a consequence of the disappearance of the dinosaurs, which provided a wide range of ecological opportunities for mammals. Seen in this light, the more modest radiation of Jurassic mammals may be explained by the continuing evolutionary success of dinosaurs across the Tr/J boundary, which restricted the opportunities available to other groups of land vertebrates. Had dinosaurs been wiped out by the Tr/J mass extinction, the major radiation of mammals might have taken place in the Jurassic rather than in the Cenozoic. There is no doubt, in any case, that the events at the Tr/J and K/T boundaries had vastly different consequences for those two important groups of terrestrial vertebrates.

Olsen *et al.* (2002b, p.517) have claimed that "with the exception of the appearance of the large dinosaurian ichnospecies *Eubrontes giganteus*, the earliest Jurassic floral and tetrapod assemblages consist entirely of survivor taxa with no origination and no apparent replacement of Triassic forms by Jurassic ecological vicars". This may be true (at least in the area they have investigated in the Newark Basin), but their claim that the biotic pattern at the Tr/J boundary is "very similar" to that at the K/T boundary seems exaggerated, because earliest Jurassic terrestrial ecosystems had not suffered the same kind of devastation, and were not as depauperate as those of the earliest Palaeocene: dinosaurs were already abundant, diverse, and in some cases large, during the latest Triassic, and suffered no decline at the Tr/J transition. The replacement of terrestrial ecosystems dominated by large dinosaurs by assemblages of small mammals, which is such a striking feature of the mass extinction at the K/T boundary, had no equivalent at the Tr/J boundary.

5
From pattern to process

What can these differences in pattern, as outlined above, tell us about possible differences in processes between the Tr/J and K/T extinctions? Although there is still no complete consensus about the causes of the terminal Cretaceous extinctions, few authors now doubt that the Chicxulub asteroid impact played a major part in it. Things are much less clear concerning the Tr/J boundary, and the real extent of the extinction event has even been disputed (see Hallam and Wignall 1997, for a review). Partly because more research work has been devoted to it, the mass extinction at the K/T boundary is thus better understood than the Tr/J event.

The likeliest cause for the K/T mass extinction appears to be selective food chain collapse caused by a disruption of photosynthesis, itself brought about by diminished light levels following the introduction in the atmosphere of enormous amounts of particles ejected by the impact (Buffetaut 1984, Sheehan and Fastovsky 1992). This hypothesis accounts, for instance, for the better resistance of freshwater ecosystems, which were less directly dependent on living plants than purely terrestrial ones. It also explains why large terrestrial animals were much more affected than small ones, which had different food requirements and were part of food chains based on organic matter in soils rather than on fresh plants. This hypothesis is generally in good agreement with patterns of extinction revealed by the fossil record, although some problems subsist. For instance, why did birds survive, while small non-avian theropods did not ?

As shown above, the pattern of extinction seen at the Tr/J boundary is markedly different from that at the K/T boundary, and this may suggest different processes. Beyond the very different fates of the dinosaurs, the apparent lack of a size factor and the extinction of at least one significant group of freshwater vertebrates at the Tr/J boundary certainly suggest different processes, despite the occurrence of a fern spike and a modest iridium anomaly in eastern North America. It is therefore tempting to assume that the ultimate causes of the Tr/J and K/T extinctions must be different, and to doubt that a meteorite impact played a key role in the terminal Triassic mass extinction. However, it should be stressed that this also applies to other extinction hypotheses. If one assumes, for instance, that both the TR/J and K/T extinctions were caused by flood basalt volcanism (see Palfy *et al.* 2002, and Courtillot and Renne 2003, and references therein, for possible evidence of a link between flood basalt volcanism and Tr/J mass extinction), it should still be explained how the same cause could have produced so widely different effects.

6
Patterns and the search for causes

The search for a common cause for all mass extinctions may go back to Cuvier's "revolutions of the globe", by which he apparently meant sudden changes of the distribution of land and sea (Cuvier 1812). In recent years, many attempts have been made to build up general theories of mass extinctions, based on a recurring cause, whether it be asteroid impacts, flood basalt volcanism, or sea level changes. Most of these attempts are unconvincing because they fail to take into account the widely different patterns that can be observed at each mass extinction, which can hardly be accounted for by simplistic explanations. Obviously, each episode of large scale extinction must be studied in detail in its own right, in order to investigate all aspects of the question. The K/T boundary has been investigated in greater detail than any other mass extinction event, but this does not mean that conclusions about the cause of that particular mass extinction can be extrapolated to other major crises in the history of life.

This being said, the fact that the Tr/J mass extinction shows a very different pattern from that of the K/T boundary does not necessarily imply that a meteorite impact was not involved. We do not know enough about the biological consequences of large impacts to be certain that the only pattern of extinction they can produce is that seen at the end of the Cretaceous. Comparison between the extinction patterns at the Tr/J and K/T boundaries certainly does not suggest that they were caused by exactly the same kind of event. However, if further studies were to produce additional evidence of a large meteorite impact at the Tr/J boundary, coincident with extinctions, it would then become necessary to think of processes different from those suggested for the K/T boundary to account for the kinds of extinctions observed at the end of the Triassic. The environmental and biological consequences of an extra-terrestrial impact may be different according to the type of impactor, the nature, composition and location of the target, etc.

7 Conclusions

A comparison of some aspects of the Tr/J and K/T extinctions among terrestrial vertebrates reveals significant differences in pattern: the size factor so obviously apparent at the K/T boundary does not seem to be involved at the Tr/J boundary, fresh water assemblages were more severely affected at the end of the Triassic (with the disappearance of the phytosaurs) than at the end of the Cretaceous, and the effects on dinosaurs and mammals of the Tr/J and K/T extinctions were widely different. This may suggest that different processes were involved, whether one puts the blame for those mass extinctions on extra-terrestrial impacts or on flood basalt volcanism, or on other causes.

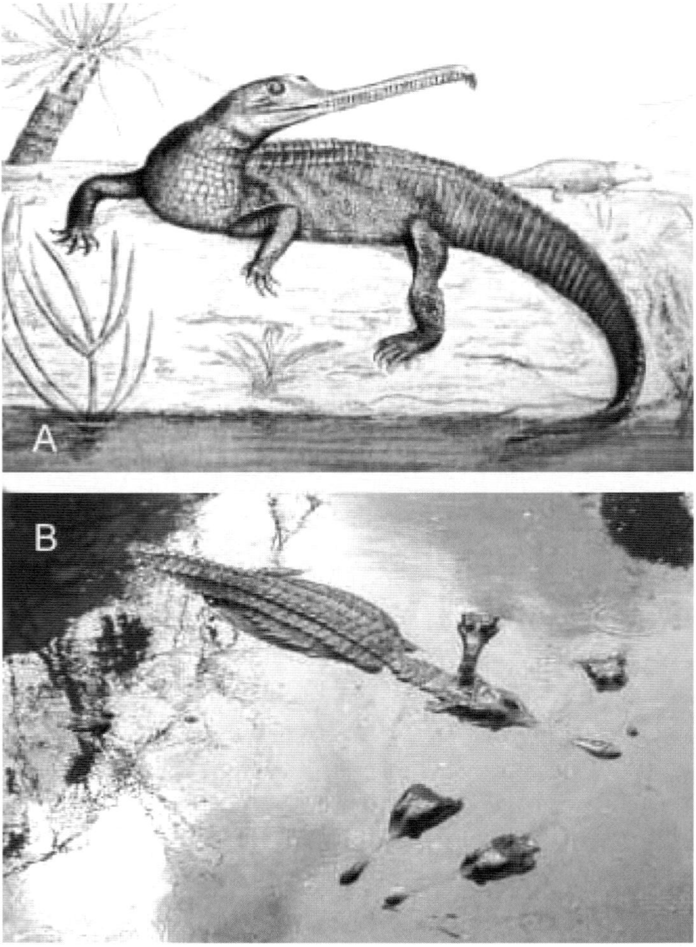

Fig.2. Phytosaurs (A: *Mystriosuchus*, after Williston 1914) were Late Triassic freshwater archosaurs which closely paralleled crocodilians (B: *Tomistoma schlegeli* in a Thai crocodile farm) in their adaptations as freshwater predators, one the main differences being the position of the external nares, in front of the eyes in phytosaurs, at the tip of the snout in crocodilians. While phytosaurs became extinct at the Tr/J boundary, the similarly adapted crocodilians survived the K/T boundary without being much affected.

Whatever the exact cause of the Tr/J mass extinction, the fact that the observed pattern (at least among continental vertebrates) is so distinct from that of the K/T boundary can serve as a useful reminder of the fact that patterns of extinction are not necessarily the same at each major crisis in the history of life. A mass extinction should not be investigated only in quantitative terms (how many taxa became extinct), it is equally or more important to determine which taxa became extinct, in which environments they lived, and (if possible) what their ecological requirements were. Comparisons between the effects of mass extinctions, which may reveal differences or similarities, may turn out to be revealing and to point to inconsistencies in attempts to extrapolate from one mass extinction to another. Only by paying sufficient attention to patterns can it be hoped to obtain really useful information about mass extinction processes (and their ultimate causes) from the fossil record.

Acknowledgments

I thank the referees who reviewed this paper for their useful comments. This work was supported by the ECLIPSE programme of CNRS.

References

Archibald JD (1996) Dinosaur extinction and the end of an Era. Columbia University Press, New York, 237 pp.

Archibald JD, Bryant LJ (1990) Differential Cretaceous/Tertiary extinctions of non-marine vertebrates; evidence from northeastern Montana. In: Sharpton VL, Ward PD (eds) Global catastrophes in Earth history; an interdisciplinary conference on impacts, volcanism and mass mortality. Geological Society of America Special Paper 247: 549-562

Benton MJ (1997) Extinction, Triassic. In: Currie PJ, Padian K (eds) Encyclopedia of dinosaurs. Academic Press, San Diego: 230-236

Brochu CA (1999) Crocodylomorphs. In: Singer R (ed) Encyclopedia of Paleontology, Fitzroy Dearborn, Chicago & London: 322-328

Buffetaut E (1984) Selective extinctions and terminal Cretaceous events. Nature 310: 276

Buffetaut E (1990) Vertebrate extinctions and survival across the Cretaceous-Tertiary boundary. Tectonophysics 171: 337-345

Buffetaut E (1993) Phytosaurs in time and space. Paleontologia Lombarda 2: 39-44

Buffetaut E (2003) Commentary on the note by Vincent Courtillot and Paul R. Renne, On the ages of flood basalt events. Comptes Rendus Geoscience 335: 773-774

Buffetaut E, Suteethorn V, Cuny G, Tong H, Le Loeuff J, Khansubha S, Jongautchariyakul S (2000) The earliest known sauropod dinosaur. Nature 407: 72-74

Buffetaut E, Suteethorn V, Le Loeuff J, Cuny G, Tong H, Khansubha S (2002) The first giant dinosaurs: a large sauropod from the Late Triassic of Thailand. Comptes Rendus Palevol 1 : 103-109

Cavin L (2002) Effects of the Cretaceous-Tertiary boundary event on bony fishes. In: Buffetaut E, Koeberl C (eds) Geological and biological effects of impact events. Springer, Berlin: 140-158

Courtillot VE, Renne PR (2003) On the age of flood basalt events. Comptes Rendus Geoscience 335: 113-140

Cuvier G (1812) Recherches sur les ossemens fossiles de quadrupedes, où l'on rétablit les caractères de plusieurs espèces d'animaux que les révolutions du globe paroissent avoir détruites. Deterville, Paris.

Hallam A, Wignall PB (1997) Mass extinctions and their aftermath. Oxford University Press, Oxford, 320 pp.

Hopson JA (1999) Mammals, Mesozoic and non-therian. In: Singer R (ed) Encyclopedia of Paleontology, Fitzroy Dearborn, Chicago & London: 691-701

Olsen PE, Kent DV, Sues HD, Koeberl C, Huber H, Montanari A, Ramforth EC, Fowell SJ, Szajas MJ, Hartline HW (2002a) Ascent of dinosaur linked to an iridium anomaly at the Triassic-Jurassic boundary. Science 296: 1305-1307

Olsen PE, Koeberl C, Huber H, Montanari A, Fowell SJ, Et-Touhami M, Kent DV (2002b) Continental Triassic-Jurassic boundary in central Pangea: Recent progress and discussion of an Ir anomaly. In: Koeberl C, MacLeod KG (eds) Catastrophic events and mass extinctions: impacts and beyond. Geological Society of America Special Paper 356: 505-522

Pálfy J, Smith PL, Mortensen JK (2002) Dating the end-Triassic and Early Jurassic mass extinctions, correlative large igneous provinces, and isotopic events. In: Koeberl C, MacLeod KG (eds) Catastrophic events and mass extinctions: impacts and beyond. Geological Society of America Special Paper 356: 523-532

Parrish JM (1999) Archosauromorphs. In: Singer R (ed) Encyclopedia of Paleontology, Fitzroy Dearborn, Chicago & London: 102-111

Rauhut OWM, Hungerbühler A (1998) A review of European Triassic theropods. Gaia 15: 76-88

Sheehan PM, Fastovsky DE (1992) Major extinctions of land-dwelling vertebrates at the Cretaceous-Tertiary boundary, eastern Montana. Geology 20: 556-560

Williston SW (1914) Water reptiles of the past and present. University of Chicago Press, Chicago, 251 pp.

Yates AM, Kitching JW (2003) The earliest known sauropod dinosaur and the first steps towards sauropod locomotion. Proceedings of the Royal Society of London B 270: 1753-1758

Geochemical Search for Impact Signatures in Possible Impact-generated Units Associated with the Jurassic-Cretaceous Boundary in southern England and northern France

Iain McDonald[1], Gordon J. Irvine[1], Eveline de Vos[1], Andrew S. Gale[2], and Wolf Uwe Reimold[3].

[1] School of Earth, Ocean & Planetary Sciences, Cardiff University, P.O. Box 914, Cardiff CF10 3YE, U.K. (iain@earth.cf.ac.uk)
[2] Department of Earth & Environmental Sciences, Medway School of Science, University of Greenwich, Chatham Maritime, Kent ME4 4TB, U.K.
[3] Impact Cratering Research Group, School of Geosciences, University of the Witwatersrand, P.O. Wits 2050, Johannesburg, South Africa.

Abstract. Understanding of the extinction event at the Jurassic-Cretaceous (J/K) boundary is hampered by problems of faunal isolation and lack of comparable magneto- and biostratigraphy between the Boreal and Tethyan faunal realms. Three impact craters (Gosses Bluff, Mjølnir, and Morokweng) are known to have formed in the late Jurassic or early Cretaceous. The question of whether these impacts damaged the environment at the time of the J/K boundary sufficiently to affect extinction patterns remains unresolved but the impacts may also provide a tool to help resolve J/K stratigraphy. If an ejecta deposit from at least one of the craters can be located in both faunal realms, this would provide an independent time marker with which to compare the different Boreal and Tethyan biostratigraphies. This study reports on the search for geochemical markers of impact, notably the platinum-group elements (PGE), in two limestone-mudstone sections at Pointe aux Oises, near Boulogne sur Mer in France, and at Durlston Bay near Swanage in England that have been suggested to include possible impact layers. Both sections show zones with PGE enrichment, where individual PGE behave remarkably coherently, and which are invariably coincident with increased concentrations of clay. The highest concentrations of Ir and Ru (up to 0.08 ppb) are not particularly elevated when compared with available data for early Cretaceous clays in Southeast England not linked with any impact event, and the chondrite-normalized PGE patterns of these samples are consistently fractionated relative to chondrite. The PGE in these rocks are probably terrestrial and do not reveal any evidence for impact associated with the two layers. Nevertheless, the data provide one of the most comprehensive sets of baseline data for the PGE in such rocks and may be of value in helping to resolve impact signatures from natural PGE background in future studies.

1 Introduction

In comparison with stratigraphic boundaries, such as the Permian-Triassic (P/T), or the Cretaceous-Tertiary (K/T) - boundaries that are associated with dramatic biological extinction events - evolutionary and environmental changes at the Jurassic-Cretaceous (J/K) boundary are less well known. Statistical analysis of extinction patterns suggests that the J/K boundary records the 5th most significant extinction event – primarily affecting marine genera and reptiles - throughout the Phanerozoic (Raup and Sepkoski 1986; Sepkoski 1995; Rampino 1999). However, the questions of whether this was a sudden or relatively prolonged extinction and whether it was strongly influenced by regional rather than global factors (Hallam 1986; Hallam et al. 1991) remains open. Part of this uncertainty arises from the fact that late Jurassic and early Cretaceous sequences are poorly correlated globally in terms of biostratigraphy (Hallam et al. 1991; Remane 1991) and magnetostratigraphy (Ogg and Lowrie 1986; Ogg et al. 1991). During this time animals were separated into distinct faunal realms – the Boreal, which covered high northern latitudes, and the Tethyan which covered the equatorial northern hemisphere and the southern hemisphere (e.g., Hallam 1976; Remane 1991). The lack of comparable fossils makes correlation of the J/K transition in the two realms very difficult. In the Tethyan realm and some southern Boreal (sometimes termed sub-Boreal) sites in France and England, the J/K boundary is taken as the boundary between the Tithonian and Berriasian stages. This marker has an estimated age of 144 ± 2.6 Ma (Gradstein et al. 1994). At Boreal sites at higher latitudes, the J/K boundary is more commonly taken as the Volgian-Ryazanian stage boundary related to an age of 140-142 Ma (Gradstein et al. 1994; Jacquin et al. 1998; Fig. 1a). It has been suggested that the J/K boundary as currently recognised may be diachronous between the two faunal realms and separated by a period of time from a few hundred thousand, up to 2 million years (Ogg and Lowrie 1986; Smelror et al. 2001).

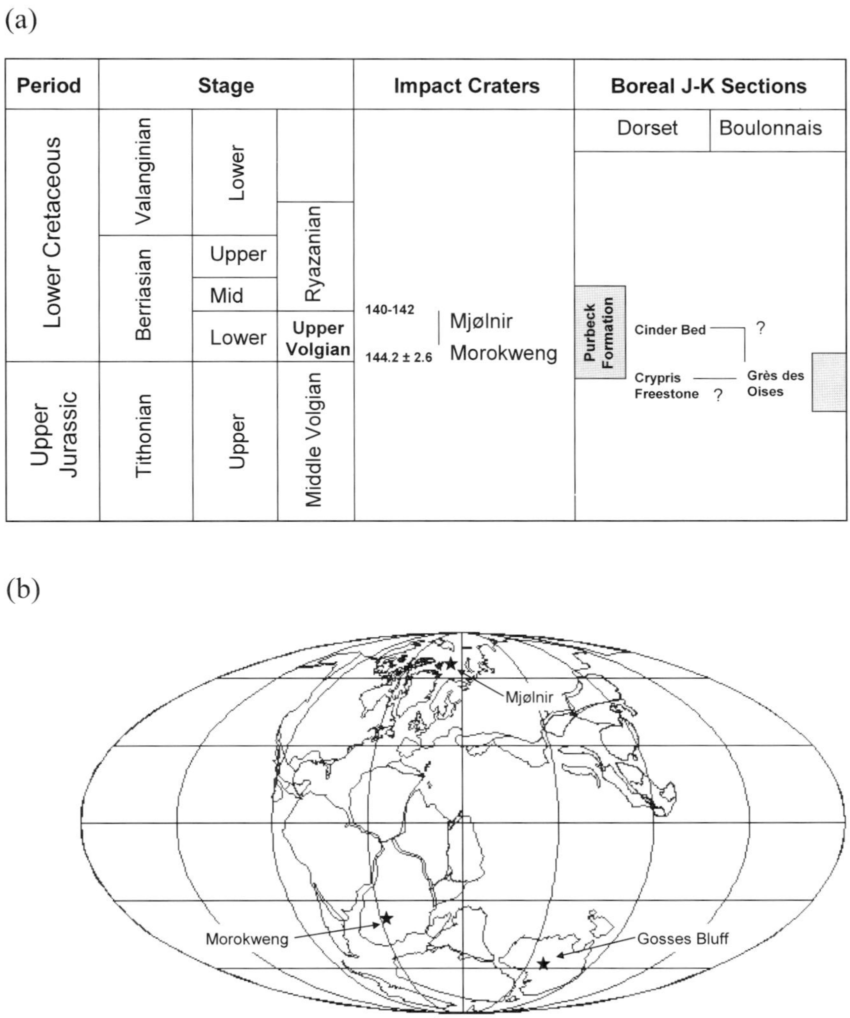

Fig. 1. (a) Upper Jurassic and lower Cretaceous stratigraphy with estimated ages for the Tithonian-Berriasian boundary and the Volgian-Ryazanian boundary (from Gradstein et al. 1999). Relative ages of the Morokweng and Mjølnir craters and the sections in this study are indicated. (b) Geographical locations of the impact craters superimposed on a continental reconstruction of the early Cretaceous (modified after Scotese 2002).

Despite the paucity of comparable fossils between faunal realms and the uncertainty over where exactly to place the J/K boundary, the period between 146 and 141 million years ago records at least three impact events. By analogy with the Chixculub crater and its associated ejecta at the Cretaceous-Tertiary boundary (Smit 1999), one or more of these might have produced an ejecta deposit recognisable in both J/K faunal realms that could provide an independent time marker with which to match the biostratigraphy. The three craters known from late Jurassic and early Cretaceous times are: the 40-km-diameter Mjølnir submarine crater in the Barents Sea (Gudlaugsson 1993; Dypvik et al. 1996; Tsikalas et al. 1998; Dypvik and Ferrell 1998); the >70 km diameter Morokweng impact crater in South Africa (Hart et al. 1997; Andreoli et al. 1999; Koeberl et al. 1997; Henkel et al. 2002; Reimold et al. 2002); and the 24 km diameter Gosses Bluff crater in Australia (Grieve 1991; Milton et al. 1996a;b).

The relatively small size and high palaeolatitude of both the Mjølnir and Gosses Bluff craters in the late Jurassic (65-70°N and 55-60°S respectively; Fig. 1b) restricts the potential for global ejecta deposits dramatically. In both cases, Mjølnir or Gosses Bluff probably released ejecta over a significant area of the host hemisphere (north and south respectively), but not globally. This would imply that Gosses Bluff ejecta are probably restricted to Tethyan sequences in the southern hemisphere whereas most Mjølnir ejecta were deposited in the Boreal northern hemisphere. However, the position of the Tethyan-Boreal transition in Europe at around 40°N in the latest Jurassic (Hallam et al. 1991) makes consideration of the Mjølnir ejecta less clear-cut. Depending on the speed and angle of impact, there is a chance that some Mjølnir-type ejecta could also have been distributed into sub-Boreal basins in England and France, or even northern Tethyan basins in central and southern Europe.

Nevertheless, the most likely source of global ejecta at the J/K boundary however is the Morokweng crater in South Africa. Morokweng is a large structure; at least 70 km in diameter (Reimold et al. 2002; Henkel et al. 2002) based on geophysical modelling and petrographic analysis for shocked minerals in basement rocks sampled by drill cores located external to the main geophysical anomaly. However, possibly larger diameters in excess of 200 km have been suggested by Andreoli et al. (1999) and Hart et al. (2002). The structure has a thick melt sheet, over 800 meters thick in one drill hole (Hart et al. 2002) that contains high concentrations of platinum-group elements (PGE) derived from the chondritic projectile (Koeberl et al. 1997; McDonald et al. 2001; Maier et al. 2003; Koeberl and Reimold 2003). U-Pb ages on zircons constrain the age of the melt sheet and by inference the impact, to 144 ± 1 Ma (Hart et al. 1997; Koeberl et al. 1997). The large size of the structure, coupled with an intermediate palaeolatitude at 40-45°S in the very latest Jurassic (Reimold et al. 1999; Fig. 1b) means that, dependant upon impact angle and velocity, Morokweng could potentially have released large amounts of ejecta across both hemispheres and into both faunal realms.

Prior to the discovery of the Mjølnir and Morokweng craters, Zakharov et al. (1993) reported a PGE anomaly associated with the Boreal J/K boundary on the Nordvik Peninsula in northern Siberia. At the time of this discovery, there were no likely craters known from that period in the terrestrial cratering record and Zakharov et al. (1993) ascribed the PGE anomaly to a prolonged accumulation of interstellar dust in a layer that developed during a period of greatly reduced sedimentation.

Dypvik et al. (1996) found shocked quartz and an Ir anomaly associated with the ejecta blanket on the periphery of the Mjølnir crater. The geochemistry of the impact ejecta led to Dypvik and Attrepp (1999) to suggest that the impactor was an iron asteroid. In subsequent studies, Dypvik et al. (2000) were apparently able to trace the same Ir anomaly in a calcite-cemented silt unit within a shale sequence in the uppermost Jurassic on Janusfjellet Mountain on Svalbard. There are no radiometric age determinations to constrain the age of the Mjølnir impact however, but analysis of micro- and macrofossil assemblages associated with pre-impact and post-impact sediments constrains the age of this impact to between the upper Volgian and lower Ryazanian. This boundary is estimated to correspond to an age of 142 ± 2.6 Ma (Smelror et al. 2001). Smelror et al. (2001) have further suggested that the ejecta layer is the same unit as the PGE-rich shale found by Zakharov et al. (1993) and, as it is apparently recognisable across the Arctic, that it can be used as a marker for the Volgian-Ryazanian boundary in the region.

Impact of the Mjølnir projectile would also have triggered tsunami that radiated out from the structure (Tsikalas et al. 1998). Deconinck et al. (2000) and Schnyder et al. (2001) ascribed coarse conglomerates and sandstones at the top of the late Jurassic Grès des Oises formation in the Boulonnais of northern France to the effects of a tsunami that travelled south down the palaeo-North Sea from the Mjølnir crater. Deconinck et al. (2000) have further speculated that a sudden marine transgression (developed as an oyster lumachelle, known as the Cinder Bed) in similar rocks in southern England might also have been caused as a consequence of the destruction of a lagoonal barrier by a tsunami arising from the Mjølnir event.

The most recent published geochemical investigation of the Tethyan J/K boundary for impact markers is by Kudielka et al. (2001). They studied 40 meters of limestone-dominated sediments from the Bosso River Gorge section in the Umbria-Marche region of central Italy. An anomalous enrichment in iridium (recorded in a single sample) is broadly coincident with enhanced levels of Cr close to the Tithonian-Berriasian boundary defined by micropalaeontology, but the effect is not significant enough to draw any firm conclusions.

In this study we report on our investigations into anomalous units (principally clays and conglomerates) in Boreal and Tethyan J/K boundary sequences in the northern hemisphere to see whether these preserve any geochemical and/or mineralogical evidence for impact(s) associated with this boundary. In this paper, we present data – incorporating the first complete PGE spectra for the sub-Boreal sequences at Points des Oises (France) and the Cinder Bed marine transgression at Durlston Bay in Dorset (England) that Deconinck et al. (2000) have suggested to be linked to the Mjølnir crater.

2
Sample localities

2.1
Points des Oises, Boulogne sur Mer

The late Jurassic-early Cretaceous Purbeckian and Wealden sediments of the Boulonnais in northern France were deposited in shallow marine and marginal marine basins immediately south of the London-Brabant massif (Hallam et al. 1991; Deconinck et al. 2000). Reliable biostratigraphic markers are rare within the sequence and the J/K boundary has not been located precisely but is thought to lie within the Purbeckian interval that encompasses the Tithonian open marine Grès des Oises formation and the overlying continental margin sediments (e.g., Jaquin et al. 1998). At Pointe aux Oises, 6 km north of Boulogne-sur-Mer, Deconinck et al. (2000) described an unusual unit near the top of the Grès des Oises formation. This consists of conglomerate and hummocky cross-stratified sandstone that shows a sharp base with the underlying bioturbated shales and fine-grained sandstones (Fig. 2a). The hummocky, cross- stratified sandstone fines upwards and is overlain by a channel fill that contains abundant pebbles and wood fragments embedded in a clay/silt matrix. In addition to wood, the conglomerate also contains many other exotic bioclastic fragments (Deconinck et al. 2000). The coarse material becomes progressively less abundant in the upper 50 cm of the clay, and the top of the unit contains large calcareous nodules. The clay is capped by a one - meter thick unit of calcrete. Using the base of the conglomerate as a zero reference, samples were collected from 10 cm below the conglomerate, up to 140 cm above the conglomerate (Fig. 2b).

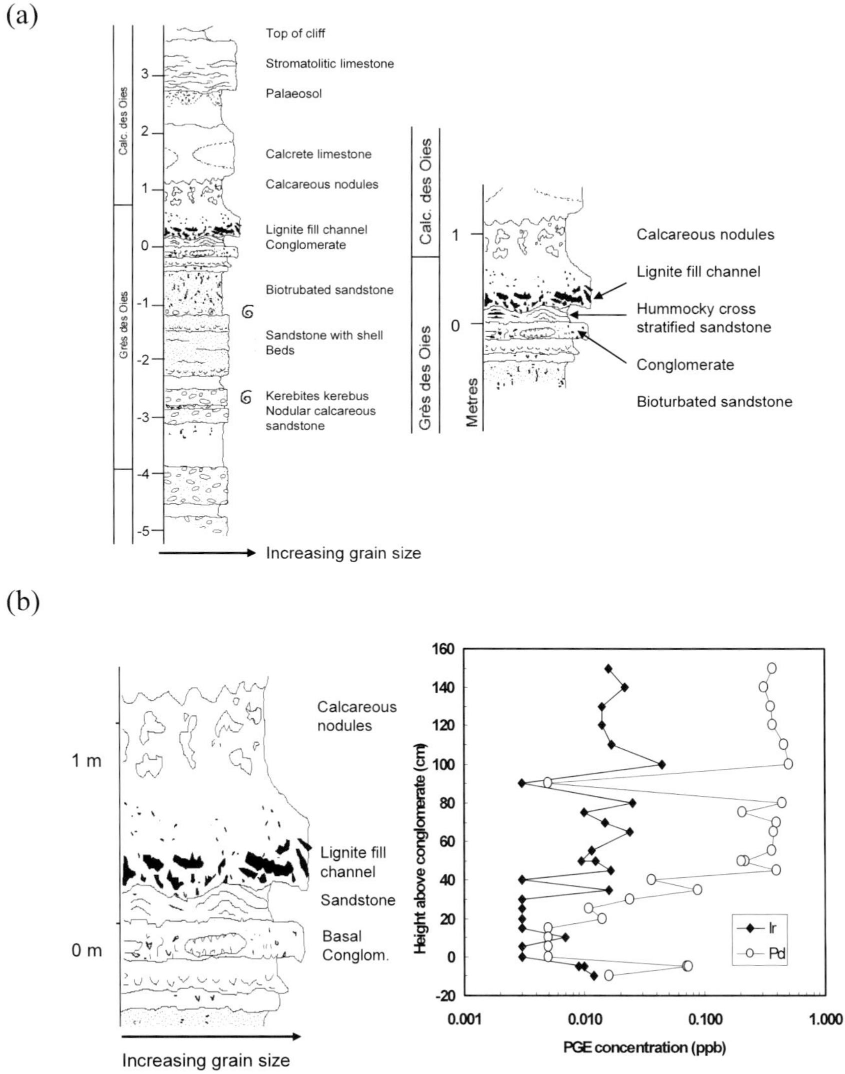

Fig. 2. (a) Summary log of the Purbeckian succession at Pointe aux Oises (this work). (b) Expanded view of the anomalous conglomerate unit with a summary of the Ir and Pd concentrations (see Table 6).

2.2 Durlston Bay, Dorset

The type section for the late Jurassic-early Cretaceous Purbeck Limestone Group (or Purbeckian facies) is at Durlston Bay, south of the town in Swanage in Dorset. Casey (1963) divided the succession into two formations: the lower Lulworth Formation and the overlying Durlston Formation. The boundary between the two formations is the Cinder Bed, an anomalous oyster-rich bed that marks an apparently sudden marine transgression into a sequence of otherwise lagoonal or marginal marine sediments. The most detailed description of the section has been compiled by Clements (1992), and the beds sampled in this study are based on his nomenclature and sequencing. Casey (1963) proposed that the Cinder Bed marked the J/K boundary in England, but this view has been challenged by more recent work (Wimbledon and Hunt 1983; Allen and Wimbledon 1991; Horne 1995).

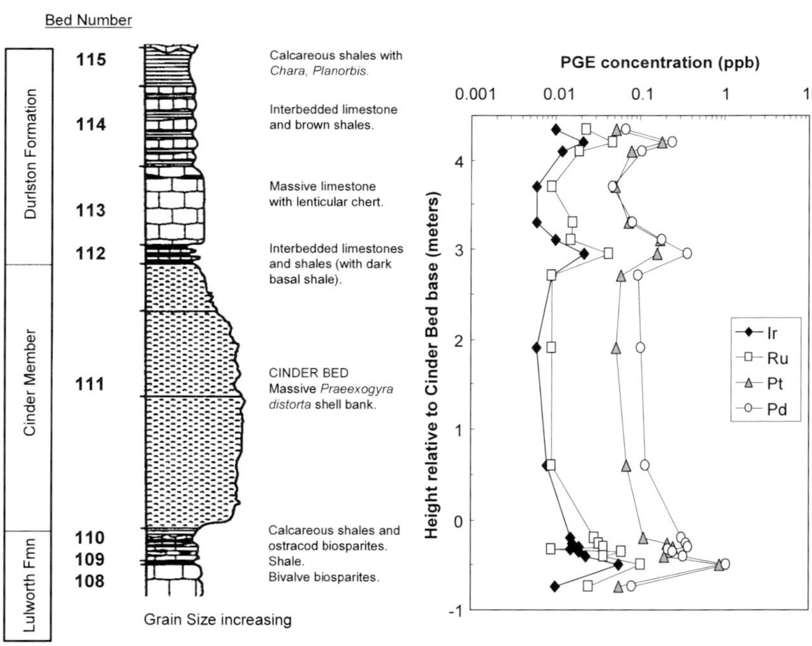

Fig. 3. Summary log of the Cinder bed and units immediately above and below (after Clements 1992). Corresponding concentrations of Ir, Ru, Pt and Pd found in these samples (see Table 5) are shown opposite.

In particular, the most recent study of charophyte biostratigraphy by Feist et al. (1995) suggests that the whole of the sequence at Durlston Bay is Berriasian and, therefore, could overlap in time with the either the Mjølnir or the Morokweng impacts (Fig. 1a). However, the number of beds sampled for charophytes by Feist et al. (1995) was limited, and Deconinck et al. (2000) preferred to keep open the possibility that the Cinder Bed might correlate with the conglomerate at the top of the Grès des Oises formation in France.

A preliminary chemostratigraphic study of the Cinder Bed was carried out by Fleming (2001). This indicated that a 30 cm thick unit of shales and calcareous shales immediately below the Cinder Bed contained anomalously high concentrations of Ni, Cr and Ir (up to 50 ppt) that might have a volcanic or impact source. In this study, we carried out a systematic program of sampling through Clements' beds 109-114 (Fig. 3). This incorporates the Cinder Bed (bed number 111) as well as 0.8 meters of underlying and 1.4 meters of overlying sediments.

3 Analytical methods

3.1 Sampling and crushing

Samples were collected through the sections according to the procedures outlined in Montanari and Koeberl (2000) and stored in sealable plastic bags for transport. No jewellery was worn during sample collection to minimise the possibility of contamination. Samples were dried in air for several days and the dry pieces were cut using a diamond saw to remove any weathering products or surface material. The clean pieces were reduced to chips in a jaw crusher and then reduced to powder in an agate ring mill. Analysis of crushed high purity silica crystals shows that the external contribution of siderophiles (Ni, Co, Cr and PGE) during crushing is negligible. Following crushing, samples were prepared for analysis of major and trace elements by inductively coupled plasma-optical emission spectrometry (ICP-OES) and analysis of PGE by inductively coupled plasma-mass spectrometry (ICP-MS) at Cardiff University. The preparation procedure for ICP-OES uses reagents and components that contain Pt, Rh and Au, so great care was taken to keep this separate from samples prepared for PGE analysis.

3.2 Preparation and analysis of elements other than PGE

Approximately 2.0 g of each sample was accurately weighed and then heated in a muffle furnace at 900°C to release H_2O, CO_2 and other volatiles. The mass of the ignited residue was then determined and the loss on ignition (LOI) calculated. LOIs of up to 40 wt% were obtained from some of the most carbonate-rich samples. A mass of 0.100 g of ignited residue was then accurately weighed and mixed with 0.400 g of Li metaborate flux (Alfa Aesar Spectroflux 100B) and a wetting agent (20% vol LiI) in a Pt-Rh crucible. The mixture was fused over a propane burner on a Claisse FLUXY automated fusion system. After fusion, the melt is automatically poured into a 250 ml Teflon beaker containing 50 ml of 4% HNO_3 and the solution was stirred using a magnetic stirrer until all of the glass fragments had dissolved. After dissolution, the solution was spiked with 1 ml of a 100 ppm Rh spike solution as an internal standard and made up to 100 ml with 18.2 MΩ deionised water, producing a spiked rock solution in 2% HNO_3. Solutions were then analysed for Si, Ti, Al, Fe, Mg, Mn, Ca, K, Na and P (reported as oxides) and Ni, Cu, Co, Cr, Ba, Sr, Zr, Y, Sc and V (reported as elements in ppm) using a JY Horiba ULTIMA2 ICP-OES system. Instrumental parameters are given in Table 1. Blanks were prepared in the same manner as above, but omitting the sample material. Calibration was performed using a reagent blank and solutions of the international certified reference materials DTS-1, PCC1, W2, BIR1, MRG1, JA2 and STM-1 and JG3, prepared as above, and using the 1 ppm Rh spike as an internal standard to correct for drift over the course of the run. Analysis of the W2 calibration solution was repeated every 6 unknowns as an external check on instrumental drift. Accuracy was assessed by routine analysis of certified reference materials GSP1, BHVO1, SARM40 and JB1a as unknowns (Gonvindaraju 1994).

Analyses for rare earth elements (REE), Ti, V, Cr, Mn, Co, Ni, Ga, Sr, Y, Zr, Nb, Ba, Hf, Ta, Pb, Th and U were carried out by ICP-MS using the same solutions as prepared for ICP-OES, except that the initial rock solution was diluted by 20 times with 2% HNO_3 and spiked with 5 ppb of Tl to produce 20 ppb Rh and 5 ppb Tl internal standards to correct for instrumental drift at low–mid and high masses, respectively. Analyses were performed using a Thermo Elemental X Series (X7) ICP-MS system.

Calibration and accuracy checks were performed using the same CRM's as for ICP-OES. Instrumental parameters and limits of detection and quantification for ICP-MS analysis are given in Table 2.

3.3 Sample preparation and analysis of PGE

Samples were prepared for PGE analysis by nickel sulphide fire assay preconcentration and tellurium co-precipitation using procedures as described in Huber et al. (2001). The only modifications to this are as follows: the filtered residue was spiked with 0.1 ml of a 2500 ppb In and Tl spike solution and digested using 3 ml of concentrated HNO_3 and 4 ml of concentrated HCl in a sealed 15 ml Savillex screw-top Teflon vials. Once the residue had dissolved, each sealed vial was cooled with running water before it was opened and the liquid contents quantitatively transferred to a 50 ml volumetric flask and made up to volume with 18.2 MΩ deionised water. Solutions were analysed for PGE and Au on a Thermo X Series (X7) ICP-MS system.

Detection and quantification limits and reagent blank contributions are given in Table 3. Detection limits and quantification limits for PGE (expressed as equivalent ppb of element in the sample) obtainable with the X Series are 8-20 times better than those achieved with earlier generations of Thermo ICP-MS systems (Table 2). This improvement in sensitivity is essential for quantifying the very low PGE concentrations present in the samples analysed in this study. Accuracy was assessed by multiple analyses of the PGE reference materials WITS1 and GP13 (Tredoux and McDonald 1996; Pearson and Woodland 2000; Krogh Jensen et al. 2003) and these data are presented in Table 2.

4 Results

Geochemical data for the two sections under investigation are presented in Tables 4 to 7.

4.1
Points des Oises

The basal conglomerate unit and the hummocky, cross-stratified sandstone contain almost no PGE above the limits of the detection. The bioclastic sand and clay 5 cm below the conglomerate contains detectable Ir and low concentrations of the other PGE, but significant PGE concentrations only appear in the clay unit above the hummocky, cross- stratified sandstone (samples PO+45 and above). Figure 2b shows that Ir and Pd concentrations move in tandem through the section, rising to maxima at 65, 80, 100 and 140 cm above the base of the conglomerate. The highest Ir concentration, 0.044 ppb in sample PO+100, is only 0.049 ppb on a volatile- free basis and not particularly high compared with the available PGE data for stratigraphically higher Wealden (lower Cretaceous) clays and mudstones in Southeast England not linked with any potential impact event (de Vos et al. 2002). PGE show no strong correlations with Ni, Cr, or Co, but all PGE, and especially Pt and Pd, show positive correlations with Ga (which is a proxy for Al). Where both metals are quantified, Pd/Ir ratios vary between 3 and 27. This is consistently above the chondritic ratio (1.17; Jochum 1996), and chondrite- normalized patterns from the most PGE-enriched layers indicate that all of them show consistently smooth and fractionated patterns from Ir to Pd (Fig. 4). Both of these features are consistent with absolute PGE concentrations influenced by the input of crust-derived clay materials. There is no evidence for any meteoritic signature in the section.

Plots of REE abundances normalized to North American Shale Composite (NASC; Gromet et al. 1984) reveal two contrasting sets of patterns in the lower and upper portions of the section (Fig. 5). Below the lignite fill channel (samples PO+50 and below), samples show patterns with negative Ce anomalies and pronounced MREE (Sm-Tb) enrichment. Sample PO+55 and those above show sloping patterns with LREE enrichment, as well as negative Ce anomalies, but no enrichment in the MREE, apart from a very small positive Eu anomaly in samples PO+65, PO+100 and PO+110. MREE enrichment is normally associated with development of phosphates during diagenesis (Cruse et al. 2000), and this is supported by the higher P_2O_5 concentrations in samples PO+35 to PO+50 compared with the other samples.

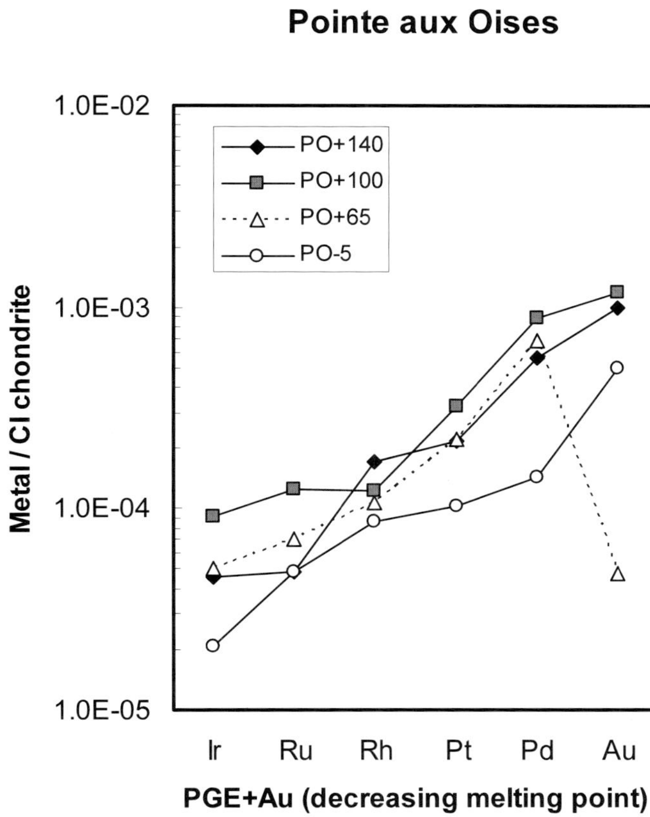

Fig. 4. Chondrite-normalized PGE and Au concentrations for some of the most PGE-rich units at Pointe aux Oises. Chondrite values used for normalization are from Jochum (1996).

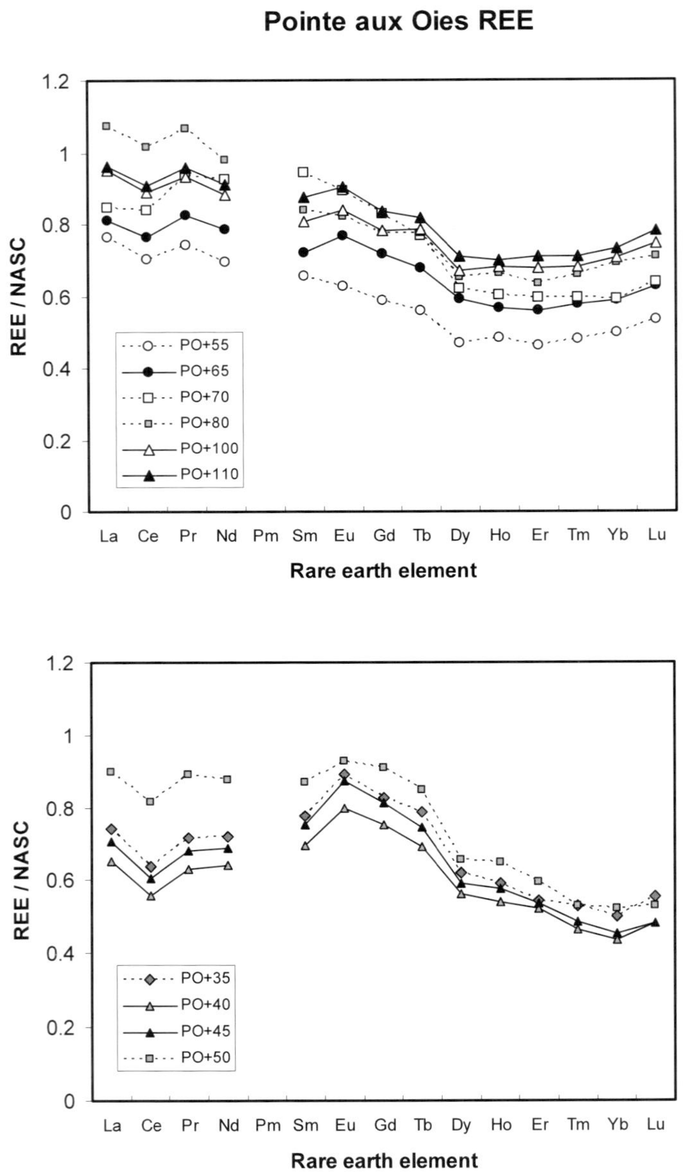

Fig. 5. North American Shale Composite (NASC) normalized rare earth element patterns for different units at Pointe aux Oises (see text for discussion).

4.2 Durlston Bay

The PGE trends with stratigraphic height for Ir, Ru, Pt and Pd move in tandem and show obvious positive anomalies associated with the clay-rich units, with concentrations falling back to close to the limit of detection in the carbonates (Fig. 3b). The highest PGE concentrations occur in the dark shale (bed 109) beneath the Cinder Bed, confirming the results of the preliminary study by Fleming (2001). Iridium reaches a maximum of 0.056 ppb (0.084 ppb on a volatile-free basis) in this layer, but the Pd/Ir ratio is 18.7, significantly higher than the chondrite ratio. Similarly, supra-chondritic Pd/Ir ratios are present in all of the layers where Ir and Pd are quantified, and chondrite-normalized plots show that clay-rich layers produce smooth, quite strongly fractionated patterns (Fig. 6).

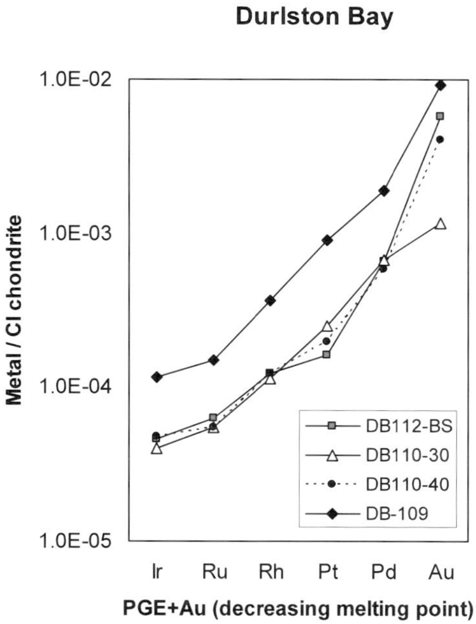

Fig. 6. Chondrite-normalized PGE and Au concentrations for some of the most PGE-rich units at Durlston Bay. Chondrite values used for normalization are from Jochum (1996).

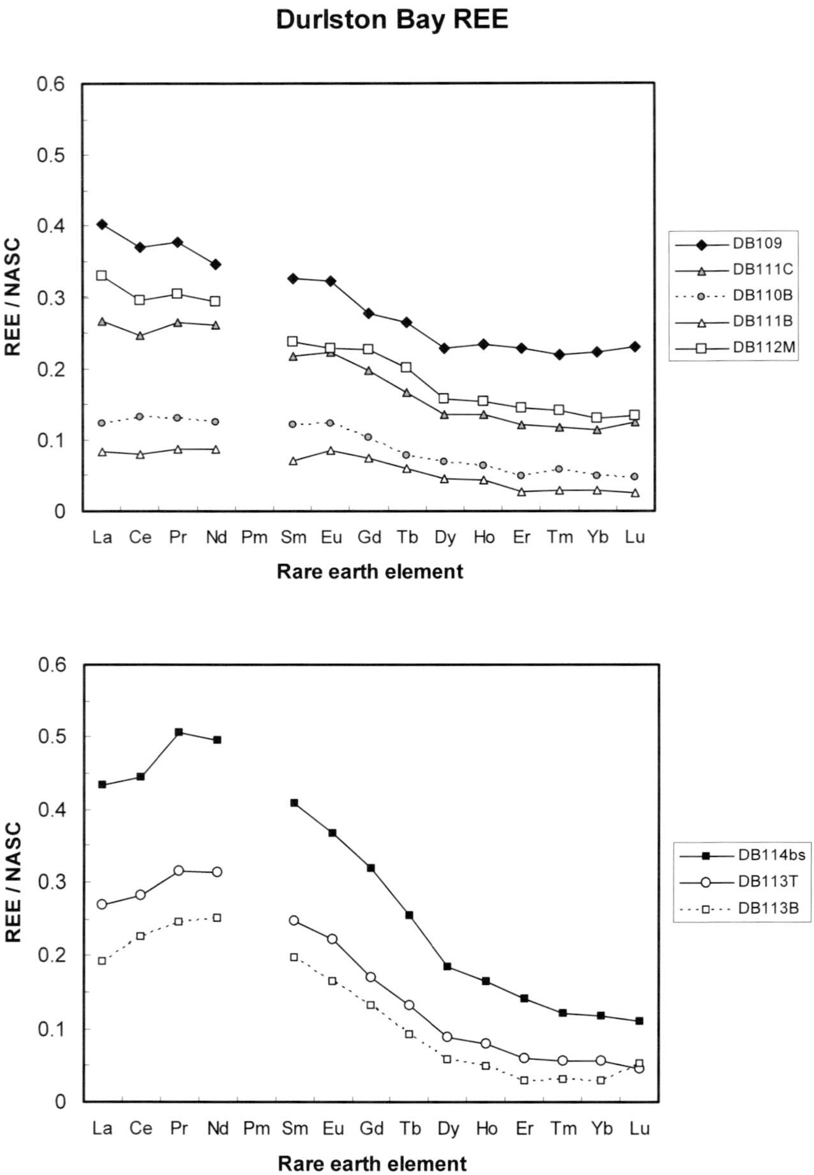

Fig. 7. North American Shale Composite (NASC) normalized rare earth element patterns for different units at Durlston Bay (see text for discussion).

Shale-normalized REE patterns show small changes associated with the development of the Cinder Bed. REE patterns for beds 109 and 110 slope gently from LREE to HREE and have no, or slightly positive, Ce anomalies (Fig. 7). Samples from the Cinder Bed and the overlying limestone and shale (bed 112) show small negative Ce anomalies. This feature is typical of rocks formed in fully marine conditions and is consistent with the abundance of oysters in the Cinder Bed, but nevertheless the Ce anomalies are small and suggest that either the basin did not become fully marine or the seawater geochemistry was buffered by mixing with the water present in the original lagoon.

In contrast, beds 113 and 114 show a consistent series of REE patterns that are significantly different from the underlying units. The limestone and shales of beds 113 and 114 are characterised by LREE enrichment, no Ce anomaly, and a pronounced positive Pr and Nd anomaly (Fig. 7). This change may be linked to a change in the input of sediment from another source after the marine incursion.

5 Conclusions

Geochemical analysis of the rock immediately below, through and above the anomalous conglomerate at Pointe aux Oises and the Cinder Bed marine transgression at Durlston Bay show no compelling evidence for an impact event, be it Morokweng or Mjølnir, or a still unknown event, associated with deposition of either unit. At both sites, the anomalous unit (conglomerate/sandstone or oyster lumachelle) contains very low concentrations of siderophile elements, and any siderophile anomalies present are found in units that visibly contain the highest percentage of clay. Chondrite-normalized plots show that the enhanced concentrations of PGE are consistently fractionated relative to chondrite and could be derived from a number of potential crustal sources. The data presented here provide no independent support for the proposal by Deconinck et al. (2000) that these units are related to late Jurassic or early Cretaceous impacts.

Another potential source of elevated levels of PGE is volcanic ash, and firmly established ash beds are potentially important correlative markers on scales of 100s to 1000s of km. Basalts have higher PGE concentrations than andesites or rhyolites (Pearson and Woodland 2000). However there are no major subaerial basaltic provinces known to have been active in Europe or adjacent regions during the late Jurassic. Furthermore, concentrations of Ir and Ru (the most immobile PGE and the least likely to be lost during diagenesis) in the most clay-rich layers studied here are uniformly less than 0.08 ppb. This is significantly lower than virtually all continental or oceanic island plateau basalts (cf. Greenough and Owen 1992; Vogel and Keays 1997; Tatsumi et al. 1999), and suggests that a significant basaltic input into these layers appears unlikely.

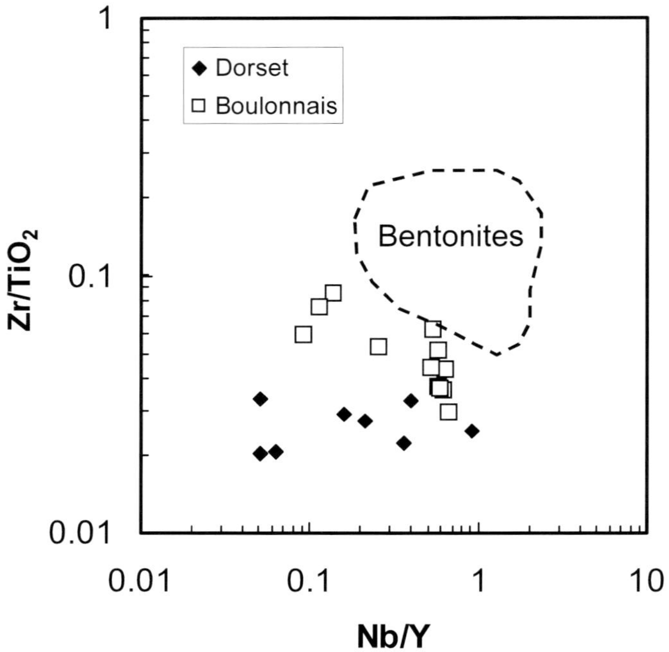

Fig. 8. Boulonnais and Durlston Bay samples plotted on a discrimination diagram for bentonitic (volcanic) and detritial clays (after Wray 1999).

The Cretaceous chalk sequences of northern Europe contain numerous bentonites derived form the alteration of volcanic ash and these have proved useful in regional correlation (Wray 1999). Geochemically, these bentonites are characterised by negative Eu anomalies on shale normalized REE plots and enrichment in immobile elements such as Zr and Nb. Both of these features are typical of explosive acidic volcanism and such events are the assumed source of most of the Cretaceous bentonites. The clay-rich units at Pointe aux Oises and Durlston Bay lack negative Eu anomalies and on the Zr/TiO_2 versus Nb/Y plot used by Wray (1999) to separate bentonitic clays from detrital material, they fall firmly into the detrital field

(Fig. 8). It therefore seems unlikely that any of the clay-rich units is the product of volcanism and that the clay and PGE were supplied by normal erosion and surface run-off. If this is the case then this study provides one of the most complete datasets yet available for the natural PGE background in sedimentary sequences dominated by limestones and calcareous sands and clays. It provides an important baseline to help future studies of potential impact horizons in similar rocks.

Smelror and Dypvik (2003) and Shuvalov and Dypvik (2004) have recently suggested that the distribution of Mjølnir ejecta close to the crater in the Barents Sea indicates that the impact was oblique and that principal transport and deposition of ejecta took place downrange to the northeast of the crater. Their modelling predicts little or no ejecta south of the impact site. The data presented here do not in any way contradict this proposal, but unfortunately nor do they offer any clear genetic or time linkage between the anomalous units in the Boulonnais and Durlston Bay and tsunamis that may still have radiated out south from the Mjølnir crater. In the case of both units, non-impact origins related to local tectonic factors are more probable.

Acknowledgements

This study was supported by the Leverhulme Trust through grant F/00407/K to Iain McDonald. The Cardiff ICP laboratory is supported by the NERC, through Joint Infrastructure Fund award NER/H/S/2000/00862 to J.A. Pearce and others. This paper is University of the Witwatersrand Impact Cratering Research Group Contribution No. 70. We thank Tony Ramsay and John Cope for helpful discussions during the early stages of this project. The paper benefited from reviews and comments by Frank Kyte, Henning Dypvik and Christian Koeberl.

References

Allen P, Wimbledon WA (1991) Correlation of the NW European Purbeck-Wealden (non-marine lower Cretaceous) as seen from the English type areas. Cretaceous Research 12: 511-526

Andreoli MAG, Ashwal LD, Hart RJ, Huizenga JM (1999) Petrology and geochemistry of nickel and iridium-rich quartz norite from the late Jurassic Morokweng impact structure, South Africa. In: Large Meteorite Impacts and

Planetary Evolution II, Dressler BO, Sharpton VL (eds), Geological Society of America Special Paper 339: 91-108
Casey R (1963) The dawn of the Cretaceous period in Britain. Bulletin of South East Union of Science Societies 117: 1-15
Clements RG (1992) Type section of the Purbeck Limestone Group, Durlston Bay, Awanage, Dorset. Proceedings of the Dorset Natural History and Archaeological Society 114: 181-206
Cruse AM, Lyons TW, Kidder DL (2000) Rare-earth element behaviour in phosphates and organic-rich host shales: an example from the Upper Carboniferous of midcontinent North America. In Marine Authigenesis: From Global to Microbial, Society for Sedimentary Geology Special Publication 66: 445-453.
Deconinck J-F, Baudin F, Tribovillard N (2000) The Purbeckian facies of the Boulonnais: a tsunami deposit hypothesis (Jurassic-Cretaceous boundary, northern France). Comptes Rendus Academy of Sciences, Paris (Earth and Planetary Sciences) 330: 527-532
de Vos E, Edwards SJ, McDonald I, Wray DS, Carey PJ (2002) A baseline survey of the distribution and origin of platinum-group elements in contemporary fluvial sediments of the Kentish Stour, England. Applied Geochemistry 17: 1115-1121
Dypvik H, Attrep Jr. M (1999) Geochemical signals of the late Jurassic, marine Mjølnir impact. Meteoritics and Planetary Science 34: 393-406
Dypvik H, Gudlaugsson ST, Tsikalas F, Attrep Jr. M, Ferrel Jr. RE, Krinsley DH, Mørk A, Faleide JI, Nagy J (1996) Mjølnir structure: an impact crater in the Barents Sea. Geology 24: 779-782
Dypvik H, Kyte FT, Smelror M (2000) Iridium peaks and algal blooms – the Mjølnir impact. Lunar and Planetary Science 31: abstract #1538 (CD-ROM)
Feist M, Lake RD, Wood CJ (1995) Charophyte biostratigraphy of the Purbeck and Wealden of England. Cretaceous Research 16: 639-663
Fleming E (2001) Chemostratigraphy of the Cinder bed member and surrounding units at Durlston Bay, Dorset. BSc. Dissertation, University of Greenwich, Chatham Maritime, 52pp (unpublished)
Govindaraju K (1994) 1994 Compilation of working values and sample descriptions for 383 geostandards. Geostandards Newsletter 18: 1-158
Gradstein FM, Agterberg FP, Ogg JG, Hardenbol J, van Veen P, Thierry J, Huang Z (1994) A Mesozoic time scale. Journal of Geophysical Research 99: 24051–24074
Greenough JD, Owen JV (1992) Platinum-group element geochemistry of continental tholeiites: analysis of the Long Range dyke swarm, Newfoundland, Canada. Chemical Geology 98: 203-219
Grieve RAF (1991) Terrestrial impact: the record in the rocks. Meteoritics 26: 175-194
Gromet LP, Dymk RF, Haskin LA, Korotev RL (1984) The "North American Shale Composite": its compilation, major and trace element characteristics. Geochimica et Cosmochimica Acta 48: 2469-2482

Gudlaugsson ST (1993) Large impact crater in the Barents Sea. Geology 21: 291-294

Hallam A (1975) Jurassic environments. Cambridge University Press, Cambridge, 218 pp.

Hallam A (1986) The Pliensbachian and Tithonian extinction events. Nature 319: 765-768

Hallam A, Grose JA, Ruffell AH (1991) Palaeoclimatic significance of changes in clay mineralogy across the Jurassic-Cretaceous boundary in England and France. Palaeogeography, Palaeoclimatology, Palaeoecology 81: 173-187

Hart RJ, Andreoli MAG, Tredoux M, Moser D, Ashwal LD, Eide EA, Webb SJ, Brandt D (1997) Late Jurassic age for the Morokweng impact structure, southern Africa. Earth and Planetary Science Letters 147: 25-35

Hart RJ, Cloete M, McDonald I, Carlson RW, Andreoli MAG (2002) Siderophile-rich inclusions from the Morokweng impact melt sheet, South Africa: possible fragments of a chondritic meteorite. Earth and Planetary Science Letters 198: 49-62

Henkel H, Reimold WU, Koeberl C (2002) Magnetic and gravity model of the Morokweng impact structure. Journal of Applied Geophysics 49: 129-147

Horne DJ (1995) A revised ostracod biostratigraphy for the Purbeck-Wealden of England. Cretaceous Research 16: 639-663

Huber H, Koeberl C, McDonald I, Reimold WU (2001) Geochemistry and petrology of Witwatersrand and Dwyka diamictites from South Africa: search for an extraterrestrial component. Geochimica et Cosmochimica Acta 65: 2007-2016

Jacquin T, Dardeau G, Durlet C, de Graciansky P-C, Hantzpergue P (1998) The North Sea cycle: an overview of 2nd order transgressive/regressive facies cycles in Western Europe. In: Mesozoic and Cenozoic Stratigraphy of European Basins. SEPM (Society for Sedimentary Geology) Special Publication 60: 445-466

Jochum K (1996) Rhodium and other platinum-group elements in carbonaceous chondrites. Geochimica et Cosmochimica Acta 60: 3353-3357

Koeberl C, Reimold WU (2003) Geochemistry and petrography of impact breccias and target rocks from the 145 Ma Morokweng impact structure, South Africa. Geochimica et Cosmochimica Acta 67: 1837-1862

Koeberl C, Armstrong RA, Reimold WU (1997) Morokweng, South Africa, A large impact structure of Jurassic-Cretaceous boundary age. Geology 25: 731-734

Krogh Jensen K, Baker J, Waight T, Frei R, Peate DW (2003) High precision Ru, Pd, Ir, Pt Re and REE determinations in the Stevns Klint Cretaceous-Tertiary boundary reference material (FC-1) by Isotope Dilution Multiple Collector Inductively Coupled Plasma-Mass Spectrometry. Geostandards Newsletter 27: 59-66

Kudielka G, Koeberl C, Montanari A, Newton J, Reimold WU (2001) Stable-isotope and trace element stratigraphy of the Jurassic/Cretaceous boundary, Bosso River Gorge, Italy. In: Geological and Biological Effects of Impact

Events, Buffetaut E, Koeberl C (eds), Impact Studies vol. 1, Springer Verlag, Heidelberg-Berlin: 25-68

Maier WD, Andreoli MAG, McDonald I, Irvine GJ, Ashwal LD, Prevec S (2003) The Morokweng impact melt sheet, South Africa: a reconnaissance study with implications for Ni-Cu-PGE mineralization. Applied Earth Science (Transactions Institution of Mining and Metallurgy B) 112: B150-B152

McDonald I, Andreoli MAG, Hart RJ, Tredoux M (2001): Platinum-group elements in the Morokweng impact structure, South Africa: evidence for the impact of a large ordinary chondrite projectile at the Jurassic-Cretaceous boundary. Geochimica et Cosmochimica Acta 65: 113-123

Milton DJ, Glikson AY, Brett R (1996a) Gosses Bluff – a latest Jurassic impact structure, central Australia. Part 1: geological structure, stratigraphy, and origin. AGSO Journal of Australian Geology and Geophysics 16: 453-486

Milton DJ, Barlow BC, Brown AR, Moss FJ, Manwaring EA, Sedmik ECE, Young GA, Van Son J (1996b) Gosses Bluff – a latest Jurassic impact structure, central Australia. Part 2: seismic, magnetic, and gravity studies. AGSO Journal of Australian Geology and Geophysics 16: 487-527

Montanari A, Koeberl C (2000) Impact Stratigraphy – The Italian record. Lecture Notes in Earth Sciences, vol. 93. Springer Verlag, Heidelberg, 364 pp

Ogg JG, Lowrie W (1986) Magnetostratigraphy of the Jurassic-Cretaceous. Geology 14: 547-550

Ogg JG, Hasenyager RW, Wimbledon WA, Channell JET, Bralower B (1991) Magnetostratigraphy of the Jurassic-Cretaceous boundary interval, Tethyan and English faunal realms. Cretaceous Research 12: 455-482

Pearson DG, and Woodland SJ (2000) Solvent extraction/anion exchange separation and determination of PGEs (Os, Ir, Pt, Pd Ru) and Re-Os isotopes in geological samples by isotope dilution ICP-MS. Chemical Geology 165: 87-107

Rampino MR (1999) Impact crises, mass extinctions and galactic dynamics: the case for a unified theory. In Large Meteorite Impacts and Planetary Evolution II Dressler, BO, Sharpton VL (eds), Geological Society of America Special Paper 339: 241-248

Raup DM, Sepkoski JJ (1986) Periodic extinction of families and genera. Science 231: 833-836

Reimold WU, Koeberl C, Brandstätter F, Kruger FJ, Armstrong RA, and Bootsman C (1999) Morokweng impact structure, South Africa: geologic, petrographic and isotopic results, and implications for the size of the structure. In Large Meteorite Impacts and Planetary Evolution II Dressler BO and Sharpton VL (eds), Geological Society of America Special Paper 339: 61-90

Reimold WU, Armstrong RA, Koeberl C (2002) A deep drillcore from the Morokweng impact structure, South Africa: petrography, geochemistry, and constraints on the crater size. Earth and Planetary Science Letters 201: 221-232

Remane J (1991) The Jurassic-Cretaceous boundary: problems of definition and procedure. Cretaceous Research 12: 447-453

Schnyder J, Baudin F, Deconinck J-F (2001) A possible tsunami deposit in the Purbeckian facies of the Boulonnais (France, Jurassic-Cretaceous boundary). 21st International Association of Sedimentologists Meeting, Davos, Switzerland, 3-5th September 2001: abstract# 188

Scotese CR (2002) http://www.scotese.com (PALEOMAP website)

Sepkoski JJ (1995) Patterns of Phanerozoic extinction: a perspective from global databases. In Global Events and Event Stratigraphy in the Phanerozoic (ed. O.H. Walliser), Springer-Verlag, Berlin: 35-51

Shuvalov V, Dypvik H. (2004) Ejecta formation and crater development of the Mjølnir impact. Meteoritics and Planetary Science 39: 467-480

Smelror M, Dypvik H (2003) The sweet aftermath: environmental changes and biotic restoration following the marine Mjølnir impact (Volgian-Ryazanian boundary, Barents Sea). In Biological Processes Associated with Impact Events, ESF IMPACT Workshop, Cambridge: p 50 (abstract)

Smelror M, Kelly SRA, Dypvik H, Mørk A, Nagy J, Tsikalas F (2001) Mjølnir (Barents Sea) meteorite impact ejecta offers a Volgian-Ryazanian boundary marker. Newsletter of Stratigraphy 38: 129-140

Smit J (1999) The global stratigraphy of the Cretaceous-Tertiary boundary impact ejecta. Annual Review of Earth and Planetary Sciences 27: 75-113

Tatsumi Y, Oguri K, Shimoda G (1999) The behaviour of the platinum-group elements during magmatic differentiation of Hawaiian tholeiites. Geochemical Journal 33: 237-248

Tredoux M, McDonald I (1996) Komatiite Wits-1: a low concentration noble metal standard for the analysis of non-mineralized samples. Geostandards Newsletter 20: 267-276

Tsikalas F, Gudlaugsson ST, Eldholme O, Faleide JI (1998) Integrated geophysical analysis supporting the impact origin of the Mjølnir structure, Barents Sea. Tectonophysics 289: 257-280

Vogel DC, Keays RR (1997) The petrogenesis and platinum-group element geochemistry of the Newer Volcanic Province, Victoria, Australia. Chemical Geology 136: 181-204

Wimbledon WA, Hunt CO (1983) The Portland-Purbeck junction (Portlandian-Berriasian) in the Weald, and correlation of the latest Jurassic-early Cretaceous rocks in southern England. Geological Magazine 120: 267-280

Wray DS (1999) Identification and long-range correlation of bentonites in Turonian-Coniacian (Upper Cretaceous) chalks of northwest Europe. Geological Magazine 136: 361-371

Zakharov VA, Lapukhov AS, Shenfil OV (1993) Iridium anomaly at the Jurassic-Cretaceous boundary in northern Siberia. Russian Journal of Geology and Geophysics 34: 83-90.

Table 1. Details of ICP-OES analysis using the JY Horiba ULTIMA2 spectrometer. Data are given for an analysis of certified reference material SARM40 (carbonatite) and its certified values (Govindaraju 1994) to illustrate the accuracy of the analysis. This CRM has the closest composition to many of the unknown samples.

Element	Line (nm)	Limit of detection in sample	SARM40	SARM40 cert
Si	251.61	0.0119 (as wt% oxide in rock)	3.53	3.08
Ti	334.94	0.0002	0.06	0.05
Al	308.21	0.0055	0.52	0.41
Fe	259.94	0.0044	2.67	2.75
Mg	279.55	0.0004	1.78	1.97
Mn	257.61	0.0194	0.19	0.18
Ca	317.93	0.0029	48.96	49.77
Na	588.99	0.0029	0.05	0.03
K	766.49	0.0169	0.04	0.05
P	214.94	0.0044	1.98	2.05
Ba	233.53	0.95 (as ppm element in rock)	317	310
Co	228.62	0.64	22.1	20
Cr	267.72	1.45	30.6	35
Ni	216.56	3.41	28.9	25
Sc	361.38	0.31	26.1	no data
Sr	407.77	0.32	1635	1600
V	310.23	18.1	32.0	27
Y	371.03	0.67	31.6	33
Zr	343.82	1.85	90.2	87
Rh	369.23	0.064 (as ppm Rh in solution)		

Instrumental parameters: Plasma gas - Argon; Forward power = 1000 W; Ar purge gas flow rate, 2 liters min^{-1}; Auxiliary Ar gas flow rate, 12 liters min^{-1}; Sheath gas flow rate (for K analysis), 8 liters min^{-1}; Nebuliser – Meinhard. Amplifier gainage, G1 (G3 for K analysis).

Table 2. Details of the ICP-MS parameters for analysis of major and trace elements in a dilute Li metaborate fusion solution using the Thermo X Series ICPMS. Detection and quantitation limits are expressed as ppm in the original sample unless stated otherwise. Data are given for a typical analysis of certified reference material JB1a (basalt) and its certified values (Govindaraju 1994) to illustrate the accuracy of the analyses.

Element	Isotope	Detect. Limit	Quant. Limit	JB1a	JB1a cert
Ti	^{49}Ti	0.0001*	0.0003*	1.28*	1.30*
V	^{51}V	0.07	0.22	206	220
Cr	^{52}Cr	0.20	0.69	411	415
Mn	^{55}Mn	0.0005*	0.0017*	0.15*	0.15*
Co	^{59}Co	0.03	0.10	38.5	39.5
Ni	^{60}Ni	0.24	0.80	146.1	140
Ga	^{71}Ga	0.022	0.076	18.2	18
Rb	^{85}Rb	0.031	0.103	35.7	41
Sr	^{88}Sr	0.29	0.99	448	443
Y	^{89}Y	0.02	0.07	24.1	24
Zr	^{90}Zr	0.05	0.17	144.7	146
Nb	^{93}Nb	0.09	0.29	27.6	27
Ba	^{137}Ba	0.41	1.36	504.1	497
La	^{139}La	0.011	0.036	36.3	38.1
Ce	^{140}Ce	0.006	0.020	65.7	66.1
Pr	^{141}Pr	0.003	0.010	7.28	7.3
Nd	^{146}Nd	0.006	0.021	25.7	25.5
Sm	^{147}Sm	0.005	0.018	5.03	5.07
Eu	^{153}Eu	0.002	0.007	1.42	1.47
Gd	^{157}Gd	0.028	0.095	4.49	4.54
Tb	^{159}Tb	0.009	0.033	0.68	0.69
Dy	^{163}Dy	0.003	0.010	4.11	4.19
Ho	^{165}Ho	0.001	0.004	0.69	0.64
Er	^{166}Er	0.003	0.010	2.11	2.18
Tm	^{169}Tm	0.001	0.003	0.31	0.31
Yb	^{172}Yb	0.003	0.010	2.16	2.10
Lu	^{175}Lu	0.004	0.013	0.31	0.32
Hf	^{187}Hf	0.002	0.007	3.44	3.48
Ta	^{181}Ta	0.001	0.003	1.62	2
Th	^{232}Th	0.002	0.007	8.72	8.8
U	^{238}U	0.004	0.013	1.61	1.6

* as wt% TiO_2 or MnO

Instrumental parameters: Plasma gas - Argon; Forward power = 1200 W; Nebuliser – Meinhard with impact bead spray chamber; Pump speed = 15 rpm; Sample uptake ~ 0.5 ml min^{-1}; Nebuliser gas flow rate = 0.95 l min^{-1}; Auxiliary gas flow rate = 0.7 l min^{-1}; Coolant flow rate = 13.0 l min^{-1}; Cones – nickel. Lens parameters optimized to achieve >50,000 cps per ppb for ^{103}Rh and ^{115}In and to achieve <1% CeO/Ce. Analysis mode = peak jumping. Dwell times from 1 ms (for Ti and Mn) to 20ms (for REE, Hf, Ta, Th and U).

Table 3. Limits of detection and quantitation for PGE analysis using the X Series ICP-MS and an assessment of the accuracy of the analyses using certified reference materials WITS-1 and GP13. Also figures are in parts per billion. For more information, see text.

	Ir	Ru	Rh	Pt	Pd	Au
Detection Limit (X Series)	0.003	0.009	0.004	0.010	0.005	0.007
Quant. Limit (X series)	0.010	0.031	0.013	0.032	0.016	0.024
Quant. Limit (PQ2+)	0.060	0.160	0.060	0.460	0.330	0.065
Reagent Blank 1	<0.006	0.195	bdl	0.089	0.031	0.353
Reagent Blank 2	<0.008	0.190	bdl	0.091	0.050	0.283
Reagent Blank 3	<0.006	0.171	bdl	0.115	0.035	0.271
Reagent Blank 4	<0.009	0.189	bdl	0.086	0.042	0.245
WITS1-1	1.22	4.02	1.19	6.51	4.47	8.36
WITS1-2	1.29	4.23	1.11	6.34	6.06	6.45
WITS1-3	1.43	4.08	1.32	6.55	5.53	6.07
WITS1-4	1.21	3.79	1.05	4.97	5.42	8.01
WITS1-5	1.34	4.71	1.11	6.84	6.23	5.96
Wits1 Recommended (±2σ)	1.4±0.3	3.9±0.8	1.1±0.2	5.7±1.4	4.9±1.2	4.9±2.6
GP13-2	3.37	7.01	1.13	6.62	5.42	1.91
GP13-3	3.20	6.67	1.11	6.10	5.33	1.98
GP13-4	3.38	6.42	1.16	6.22	5.31	2.29
GP13 Recommended (±2σ)	3.4±0.8	6.9±0.6	no data	6.9±1.0	5.6±0.6	no data

Instrumental parameters: Plasma gas - Argon; Forward power = 1200 W; Nebuliser – Meinhard with impact bead spray chamber; Pump speed = 15 rpm; Sample uptake ~ 0.5 ml min^{-1}; Nebuliser gas flow rate = 0.95 l min^{-1}; Auxiliary gas flow rate = 0.7 l min^{-1}; Coolant flow rate = 13.0 l min^{-1}; Cones, nickel; Lens parameters optimized to achieve >50,000 cps per ppb for ^{232}Th and ^{115}In and to achieve <1% CeO/Ce. Analysis mode = peak jumping. Dwell times, 20 ms for all PGE and Au isotopes.

Table 4. Major and trace element data for selected Pointe aux Oises (PO) samples determined by ICP-OES and ICP-MS.

	PO+35	PO+40	PO+45	PO+50	PO+55	PO+65	PO+70	PO+80	PO+100	PO+110	PO+120	PO+140
SiO2 (wt%)	68.55	71.22	70.15	69.21	60.93	59.61	68.99	65.84	67.09	67.88	66.64	65.43
TiO2	0.21	0.14	0.16	0.36	0.49	0.49	0.43	0.65	0.56	0.58	0.58	0.59
Al2O3	7.99	6.67	7.04	8.21	8.62	8.20	7.22	9.91	12.11	11.98	11.91	10.68
Fe2O3	8.21	7.12	7.49	7.78	9.57	10.88	9.18	8.71	3.48	3.05	3.45	6.21
MgO	1.03	0.98	1.07	0.98	1.16	1.01	0.91	1.20	1.58	1.46	1.65	1.30
MnO	0.07	0.04	0.05	0.04	0.04	0.01	0.01	0.01	0.02	0.01	0.02	0.01
CaO	4.71	3.88	3.39	3.32	3.02	4.54	3.79	3.21	6.27	4.16	5.36	3.04
K2O	0.83	1.03	0.99	0.88	0.93	1.29	1.08	1.43	0.24	0.16	0.22	1.47
Na2O	0.39	0.69	0.52	0.45	0.40	0.27	0.29	0.37	0.77	0.21	0.93	0.25
P2O5	0.24	0.34	0.33	0.18	0.03	0.02	0.01	0.02	0.28	0.19	0.24	0.20
LOI	6.22	5.99	6.98	8.53	12.93	10.77	5.70	7.40	9.38	8.38	8.52	9.21
Total	98.45	98.10	98.17	99.94	98.12	97.09	97.61	98.74	101.78	98.06	99.53	98.38
V (ppm)	24.1	17.9	19.7	34.6	70.6	72.5	64.9	88.8	100.3	104.9	93.0	98.1
Cr	20.1	10.5	20.2	36.7	34.9	20.2	29.9	9.4	26.8	17.2	6.7	8.8
Co	7.4	2.0	17.0	23.5	21.6	8.9	17.7	16.3	10.2	5.1	3.5	4.3
Ni	23.5	8.8	18.3	29.4	38.2	16.2	36.1	30.1	28.7	17.9	10.9	11.3
Ga	1.4	0.3	0.5	3.9	11.8	12.1	10.5	15.6	18.0	16.6	16.6	17.3
Rb	10.0	9.0	10.8	32.0	45.6	62.8	52.3	69.5	62.1	63.2	86.8	61.0
Sr	340	342	458	184	119	170	205	210	185	155	161	171
Y	25.4	23.3	24.5	25.5	17.2	21.0	21.8	24.0	24.3	24.8	26.4	28.6
Zr	164.3	76.5	122.6	183.6	226.7	217.4	288.3	230.4	196.6	233.2	238.5	362.4
Nb	3.5	2.2	2.8	6.7	11.2	11.0	11.7	14.1	16.3	15.6	15.7	16.6
Ba	149.5	156.4	167.1	145.1	196.9	191.5	186.0	274.7	232.9	211.5	226.9	236.8
La	23.0	20.2	22.0	28.0	23.7	25.3	26.4	33.4	29.6	30.0	30.1	35.1
Ce	42.8	37.4	40.6	54.7	47.0	51.3	56.2	68.1	59.7	60.9	60.8	75.2
Pr	5.66	4.97	5.38	7.04	5.85	6.51	7.40	8.42	7.37	7.56	7.52	9.35
Nd	21.9	19.5	20.9	26.7	21.1	23.9	28.2	29.7	26.8	27.7	27.5	34.0
Sm	4.65	4.16	4.51	5.21	3.93	4.31	5.64	5.02	4.83	5.23	5.07	6.16
Eu	1.12	1.00	1.10	1.16	0.78	0.93	1.14	1.03	1.02	1.13	1.08	1.24
Gd	4.55	4.15	4.48	5.01	3.22	3.94	4.55	4.26	4.30	4.59	4.49	5.06
Tb	0.67	0.59	0.63	0.72	0.47	0.58	0.65	0.66	0.67	0.70	0.70	0.78
Dy	3.43	3.11	3.28	3.63	2.60	3.27	3.42	3.61	3.72	3.93	3.98	4.38
Ho	0.61	0.56	0.60	0.67	0.50	0.59	0.63	0.69	0.71	0.73	0.75	0.83
Er	1.78	1.72	1.76	1.94	1.51	1.83	1.95	2.08	2.22	2.32	2.32	2.64
Tm	0.26	0.23	0.24	0.26	0.24	0.29	0.30	0.33	0.34	0.35	0.37	0.39
Yb	1.57	1.36	1.41	1.62	1.55	1.83	1.84	2.15	2.20	2.28	2.33	2.62
Lu	0.25	0.22	0.22	0.24	0.44	0.29	0.29	0.32	0.34	0.36	0.35	0.41
Hf	3.67	1.73	2.75	4.27	5.55	5.20	6.74	5.55	4.69	5.66	5.67	8.27
Ta	0.30	0.21	0.27	0.49	0.77	0.77	0.71	0.91	0.99	0.97	1.00	1.01
Th	6.46	5.31	5.46	6.30	8.70	8.44	8.29	10.46	10.91	11.23	10.87	11.87
U	2.25	1.66	1.87	2.61	2.59	2.06	1.79	2.13	2.28	1.78	1.95	2.00

Table 5. Major and trace element data for selected Durlston Bay (DB) samples determined by ICP-OES and ICP-MS.

	DB109	DB110-33	DB111A	DB111B	DB111C	DB112M	DB113B	DB113M	DB113T	DB114B
SiO2 (wt%)	22.21	2.51	5.33	4.86	4.95	4.84	5.12	2.27	2.14	4.99
TiO2	0.25	0.03	0.06	0.03	0.05	0.07	0.02	0.02	0.03	0.05
Al2O3	7.12	0.49	1.35	0.88	1.38	1.63	0.34	0.33	0.67	1.39
Fe2O3	3.39	0.30	0.53	0.52	0.54	0.90	0.24	0.02	0.32	0.88
MgO	2.14	0.91	0.85	0.74	0.80	0.86	0.64	0.69	0.70	0.74
MnO	0.04	0.03	0.02	0.02	0.04	0.04	0.02	0.01	0.03	0.04
CaO	29.33	51.59	49.99	50.47	49.82	49.73	48.79	53.32	52.96	47.90
K2O	0.14	0.11	0.06	0.05	0.13	0.07	0.06	0.03	0.06	0.08
Na2O	1.93	0.03	0.11	0.02	0.06	0.11	0.04	0.05	0.05	0.09
P2O5	0.07	0.02	0.08	0.02	0.05	0.09	0.04	0.01	0.04	0.06
LOI	33.55	42.22	39.95	40.28	40.29	39.88	43.17	43.33	42.54	41.96
Total	100.16	98.24	98.34	97.89	98.11	98.22	98.48	100.07	99.54	98.19
V (ppm)	88.6	15.5	21.2	14.3	17.1	23.1	10.3	10.4	12.8	17.8
Cr	29.1	10.8	3.8	3.5	3.9	9.8	2.7	2.9	3.7	19.5
Co	9.4	3.7	1.6	1.0	2.7	4.2	0.3	0.1	1.5	7.2
Ni	26.6	10.6	5.6	4.2	2.9	6.7	3.7	4.1	4.2	17.1
Ga	9.6	0.1	1.4	0.1	0.9	1.7	0.1	0.1	0.1	1.4
Rb	4.5	4.0	3.5	3.2	3.1	3.9	3.5	3.2	3.0	2.9
Sr	758	1597	350	475	492	436	516	584	507	458
Y	8.1	2.3	3.5	1.6	4.8	5.9	1.7	0.8	2.9	5.8
Zr	70.2	8.9	22.4	3.7	14.9	17.6	1.7	0.6	5.5	13.1
Nb	7.4	0.1	1.4	0.1	1.0	2.1	0.1	0.1	0.2	0.9
Ba	180.5	71.4	51.7	115.2	43.4	47.0	27.4	44.5	58.3	41.7
La	12.5	3.8	6.0	2.6	8.3	10.3	5.9	4.6	8.4	13.5
Ce	24.8	8.9	11.7	5.3	16.5	19.8	15.2	9.7	18.9	29.8
Pr	2.97	1.03	1.50	0.68	2.09	2.41	1.95	1.10	2.48	3.99
Nd	10.5	3.8	5.5	2.6	7.9	8.9	7.6	4.3	9.5	15.1
Sm	1.95	0.73	0.86	0.43	1.30	1.42	1.18	0.80	1.48	2.44
Eu	0.41	0.15	0.18	0.11	0.28	0.29	0.21	0.18	0.28	0.46
Gd	1.52	0.57	0.74	0.41	1.08	1.25	0.73	0.60	0.94	1.75
Tb	0.23	0.07	0.09	0.05	0.14	0.17	0.08	0.06	0.11	0.22
Dy	1.27	0.38	0.51	0.25	0.75	0.87	0.32	0.20	0.49	1.02
Ho	0.24	0.07	0.10	0.05	0.14	0.16	0.05	0.02	0.08	0.17
Er	0.75	0.16	0.28	0.09	0.40	0.47	0.10	0.02	0.20	0.46
Tm	0.12	0.03	0.05	0.01	0.06	0.07	0.02	0.01	0.03	0.06
Yb	0.81	0.15	0.28	0.09	0.35	0.41	0.09	0.02	0.18	0.37
Lu	0.13	0.02	0.06	0.01	0.06	0.06	0.02	0.00	0.02	0.05
Hf	1.80	0.22	0.62	0.13	0.42	0.45	0.05	0.01	0.19	0.35
Ta	0.44	0.04	0.08	0.03	0.07	0.12	0.02	0.02	0.04	0.08
Th	5.56	0.44	1.16	0.40	0.99	1.53	0.50	0.26	0.88	1.47
U	2.60	0.55	0.41	0.28	0.24	0.39	0.23	0.17	0.27	0.37

Table 6. PGE and Au concentrations (in ppb) in Pointe aux Oises samples. Height expressed in meters relative to the base of the conglomerate. "n.d." indicates below the detection limit. Values between the detection and quantification limits are reported as <x, where x is the next greatest integer. Samples indicated A and B are repeat analyses.

	Height (m)	Ir	Ru	Rh	Pt	Pd	Au
PO+150	0.15	0.016	0.030	0.012	0.284	0.372	0.151
PO+140	0.14	0.022	0.033	0.024	0.214	0.315	0.148
PO+130	0.13	0.014	0.067	0.025	0.200	0.357	0.122
PO+120	0.12	0.014	0.120	<0.009	0.235	0.367	0.182
PO+110	0.11	0.017	0.095	0.019	0.237	0.462	0.018
PO+100	0.1	0.044	0.086	0.017	0.323	0.498	0.174
PO+90	0.09	n.d.	n.d.	n.d.	0.053	n.d.	n.d.
PO+80	0.08	0.025	0.129	0.012	0.261	0.443	1.635
PO+75	0.075	<0.010	0.059	<0.008	0.169	0.208	0.375
PO+70	0.07	0.015	0.070	0.012	1.062	0.394	n.d.
PO+65	0.065	0.024	0.048	0.015	0.221	0.379	n.d.
PO+55	0.055	0.012	0.103	0.092	0.212	0.365	0.793
PO+50A	0.05	0.010	0.042	0.024	0.161	0.217	1.073
PO+50B	0.05	0.013	0.049	0.023	0.189	0.205	12.030
PO+45	0.45	0.017	0.038	0.041	0.182	0.394	0.393
PO+40	0.04	n.d.	<0.015	0.012	0.050	0.037	0.028
PO+35	0.035	0.016	0.129	0.058	0.095	0.088	0.315
PO+30	0.03	n.d.	0.041	0.018	<0.015	0.024	0.224
PO+25	0.025	n.d.	n.d.	n.d.	n.d.	<0.011	0.199
PO+20	0.02	n.d.	n.d.	n.d.	<0.019	<0.014	0.119
PO+15	0.015	n.d.	n.d.	n.d.	n.d.	n.d.	n.d.
PO+10	0.01	<0.007	<0.029	n.d.	n.d.	n.d.	0.210
PO+5	0.005	n.d.	n.d.	n.d.	n.d.	n.d.	0.772
PO-0	0	n.d.	0.054	<0.007	n.d.	n.d.	<0.019
PO-5A	-0.005	0.010	0.033	0.015	0.101	0.071	0.074
PO-5B	-0.005	<0.009	0.031	0.013	0.089	0.074	0.069
PO-10	-0.01	0.012	0.016	<0.005	n.d.	0.036	n.d.
Detection Limit (X Series)		0.003	0.009	0.004	0.01	0.005	0.007
Quantification Limit (X series)		0.01	0.031	0.01	0.032	0.016	0.024

Table 7. PGE and Au concentrations (in ppb) in Durlston Bay samples. Height expressed in meters relative to the base of the Cinder Bed (bed DB111). "n.d." indicates below the detection limit. Values between the detection and quantification limits are reported as <x, where x is the next greatest integer.

Sample	Height (m)	Ir	Ru	Rh	Pt	Pd	Au
DB114-M	4.35	0.011	<0.023	n.d.	0.052	0.067	0.142
DB114-B	4.2	0.021	0.047	0.03	0.179	0.239	0.061
DB113-T	4.1	0.012	<0.019	n.d.	0.08	0.106	0.057
DB113-M	3.7	<0.006	n.d.	n.d.	0.051	0.047	n.d.
DB113-B	3.3	<0.006	<0.016	n.d.	0.073	0.081	n.d.
DB112-M	3.1	<0.010	<0.015	n.d.	0.174	0.182	0.619
DB112-BS	2.95	0.022	0.043	0.017	0.161	0.368	0.852
DB111-C	2.7	<0.009	n.d.	n.d.	0.06	0.094	0.268
DB111-B	1.9	<0.006	n.d.	n.d.	0.053	0.103	0.291
DB111-A	0.6	<0.008	n.d.	n.d.	0.069	0.117	0.342
DB-110-22	-0.2	0.015	<0.029	0.01	0.111	0.31	0.091
DB110-26	-0.26	0.016	0.033	0.013	0.21	0.352	0.253
DB110-30	-0.3	0.019	0.037	0.016	0.246	0.375	0.173
DB110-33	-0.33	0.015	n.d.	<0.009	0.216	0.22	0.218
DB110-36	-0.36	0.019	0.061	0.012	0.214	0.247	0.222
DB110-40	-0.4	0.023	0.037	0.017	0.196	0.328	0.598
DB-109	-0.5	0.056	0.102	0.051	0.891	1.05	1.37
DB-108	-0.74	<0.010	<0.025	n.d.	0.056	0.082	0.049
Detection Limit (X Series)		0.003	0.009	0.004	0.01	0.005	0.007
Quantification Limit (X series)		0.01	0.031	0.01	0.032	0.016	0.024

New Evidence for Impact from the Suvasvesi South Structure, Central East Finland

Fabio Donadini[1], Jüri Plado[2], Stephanie C. Werner[3], Johanna Salminen[1], Lauri J. Pesonen[1], Martti Lehtinen[4]

[1]Division of Geophysics, University of Helsinki, P.O. Box 64, FIN-00014 Helsinki, Finland (fabio.donadini@helsinki.fi)
[2]Institute of Geological Sciences, University of Tartu, EST-51014 Tartu, Estonia
[3]Freie Universität of Berlin, Institute for Geosciences, Malteserstrasse 74-100, D-12249 Berlin, Germany
[4]Geological Museum, University of Helsinki, P.O. Box 64, FIN-00014 Helsinki, Finland

Abstract. The circular Suvasvesi South structure (diameter about 3.8 km) is located in Central East Finland (62°35.8'N, 28°13'E) and correlates with the Haapaselkä open lake area, the southern of the two Suvasvesi lakes. Suvasvesi S was first noticed in satellite images and might form a crater doublet with the proven Suvasvesi N impact structure. We have previously presented evidence, such as presence of fractured target rocks and shatter-cone boulders on the eastern shore of Haapaselkä, which suggest that the Suvasvesi South is also an impact structure. During the summer 2002 we carried out a field survey in the area of the Suvasvesi lakes, which led to additional discoveries of shatter cones in boulders. We also discovered impact melt boulders in gravel pits along the roadsides, about 5 km southeast of the structure. Microscopic studies of the thin sections of impact melt and of the shatter cone boulders reveal the presence of well developed and decorated PDFs in quartz grains, maskelynite, fluidal textures between impact melt mineral clasts and kink bands in micas. Consequently, the melt rock is considered to be of impact origin. It is unlikely that the boulders with shatter cones and impact melt blocks were derived from the northern structure, because material transported from it by ice drift would not have passed this area. The outcrops on the islets of the Suvasvesi South area are heavily fractured with subvertical NNW-SSE and ENE-WSW trending; however the fracturing may be related to the Svecofennian tectonic deformations occurring in this area. Also, the outcrops show shatter cone features with a maximum of 50 cm in size. We interpret the shatter cone features to be of impact origin because of their shape and because the orientation of their apices differ from the other deformation systems. However, thin section analysis from outcrop specimens has not shown impact evidence so far. The new findings suggest the presence of an eroded impact melt layer in the southern structure. Bathymetric and airborne magnetic data point to a distinct structure of smaller dimension than the northern one. Preliminary paleomagnetic measurements on the granitic host rocks of Suvasvesi South reveal two components, of which one (steep downwards) is probably either a hard viscous remanence of present age or a Svecofennian age feature. The other one is poorly defined, but has a southwest direction similar to that isolated for the northern structure and could be related to impact.

1 Introduction

The Suvasvesi South structure is located in Central East Finland, about 50 km southeast of Kuopio, and relates to the open Haapaselkä lake, the southern part of the large Suvasvesi Lake (Fig. 1a). The Suvasvesi Lake has the form of an oval and in its open part contains the Suvasvesi N and the Suvasvesi S structures. The area of the Suvasvesi South structure consists of a maximum 29 m deep open lake, surrounded by several islands, which create a ring of about 3.8 km in diameter (Fig. 1b).

Previous studies, including airborne geophysical data, drilling and in-situ observations have confirmed that Suvasvesi North crater (Henkel and Pesonen 1992; Pesonen 1996; Werner et al. 2002) was caused by a hypervelocity impact. In addition, the aeromagnetic map (Fig. 2) shows an anomaly also corresponding to the southern lake. This is supported also by topographic and bathymetric data (Fig. 3). Moreover, shatter cones in boulders have recently been found in the Lusikkaniemi bay, on the eastern part of the suspected crater (Lehtinen et al. 2002, Öhman et al. 2003). The purpose of the present paper is to report the new discoveries obtained in the field campaign during summer 2002, which support the concept that Suvasvesi South is also an impact structure.

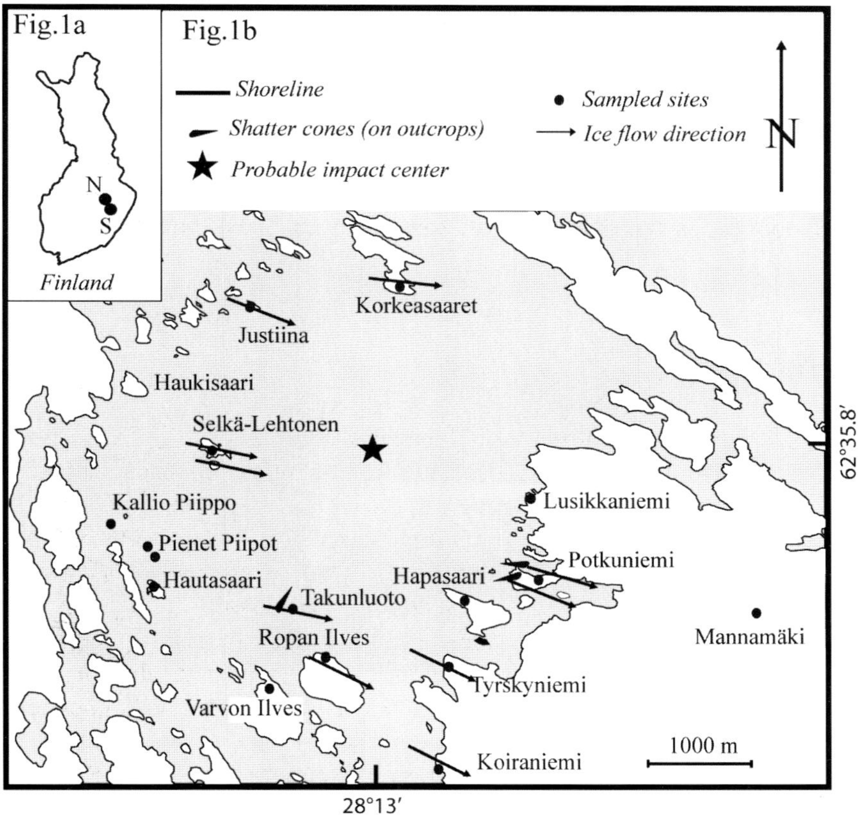

Fig. 1. (a) Location of the Suvasvesi structures. (b) Simplified topographic map of the Haapaselkä lake, indicating the sampling sites (closed circles), the directions of ten measured ice flow directions (arrow) and the orientation of the shatter cones features observed on outcrops (droplets). The probable impact center (star) corresponds to the deepest part of the Haapaselkä lake (29 m).

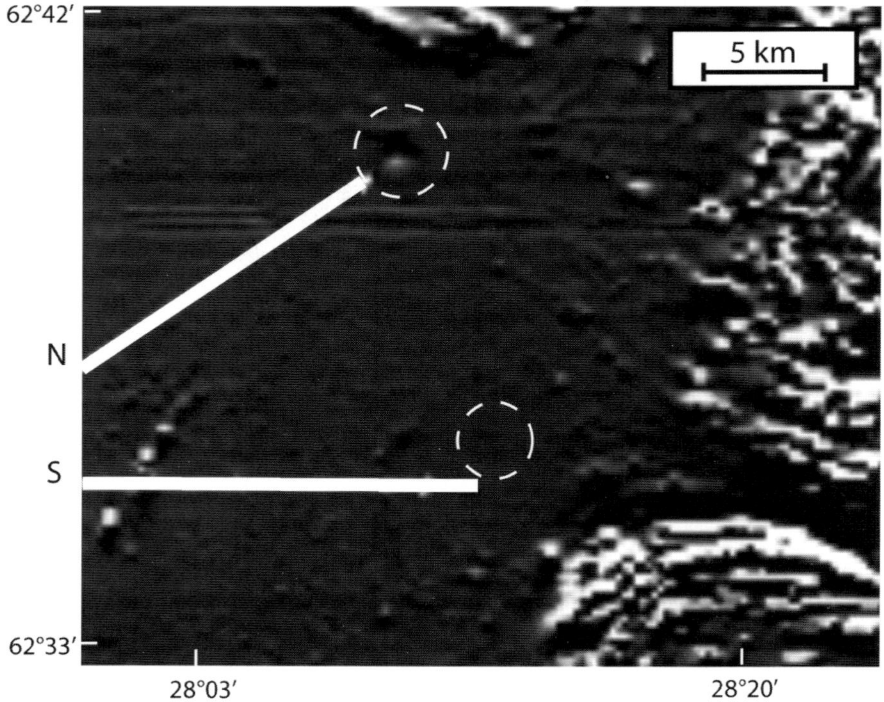

Fig. 2. Aeromagnetic map of the Suvasvesi area. The Suvasvesi North (N) and the Suvasvesi South (S) structures are indicated. The dashed circles represent the border of the structures.

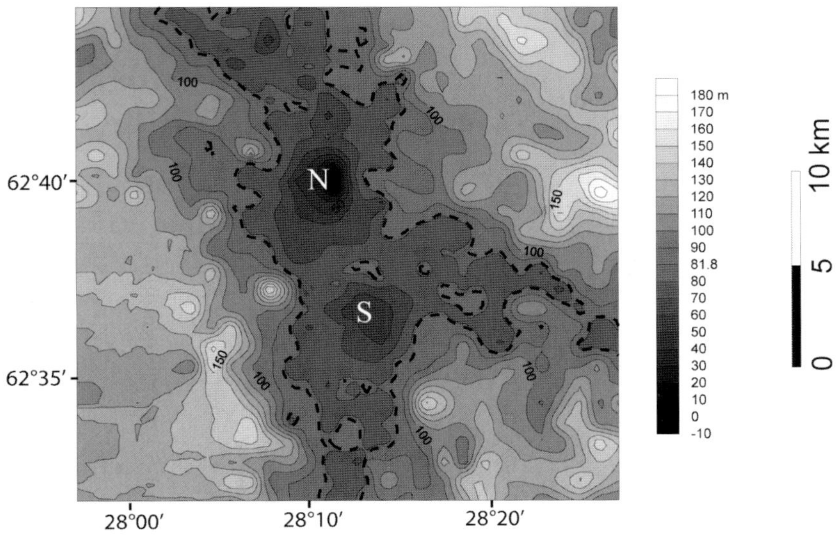

Fig. 3. Topography and bathymetry of the Suvasvesi area. The Suvasvesi North (N) and South (S) structures are indicated.

2 Geological background

The Suvasvesi area is located on the Lake Ladoga – Bothnian Bay tectonic belt, a large fault zone extending from Lake Ladoga to Raahe (Finland) with SE orientation. This fault zone formed about 1.9 Ga ago and divides the Proterozoic (to the south) from the late Archean (to the north) terranes of the Fennoscandian Shield (Fig. 4).

The rocks in the Suvasvesi South area consist mainly of Proterozoic mica schists and migmatites in its easternmost part, whereas in the western part granitoid outcrops dominate. Evidence for complex tectonic deformation can be observed in the crenulated micaschists and heavily folded migmatites. Granitoids show up more compact and massive. Pegmatites are abundant in the granitoids.

The present-day morphology of the structure is strongly modeled by Quaternary ice movements: outcrops are smoothed and rounded, and show glacial striations trending in E-SE direction.

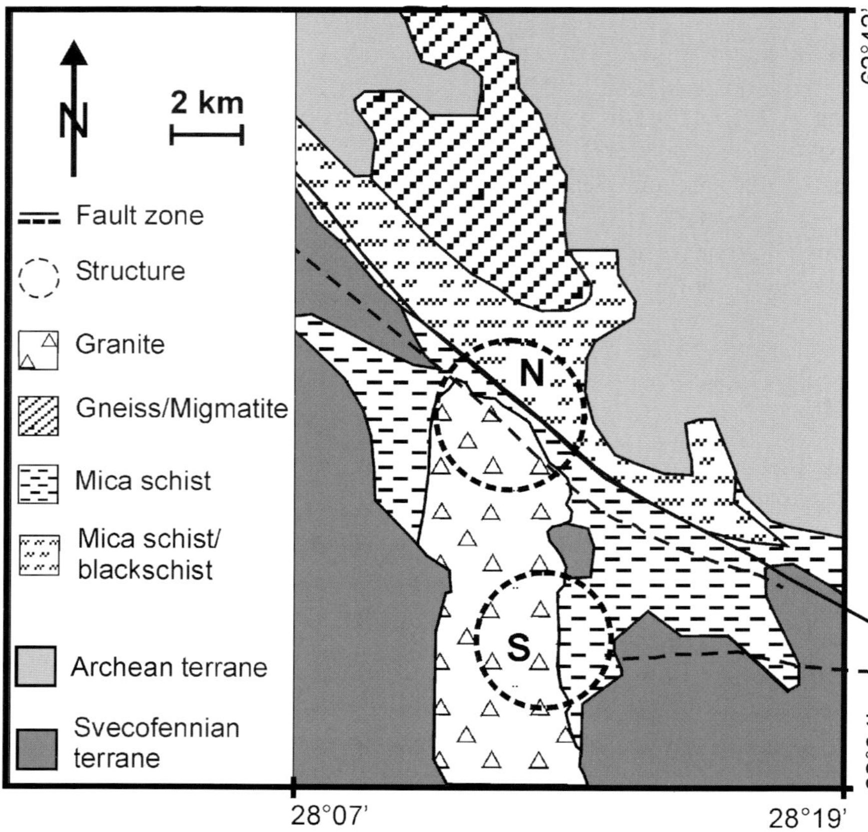

Fig. 4. Geological setting of the Suvasvesi area. The Archean (light gray) is divided from the Proterozoic (Svecofennian) terrane (dark gray) by the Lake Ladoga-Raahe fault zone. The Suvasvesi North structure (N) is located between the two terranes. The Suvasvesi South (S) structure lies on the Svecofennian terrane. Host rocks are granites and mica schists.

3 Observations and data analysis

3.1 Field work

During one week field work we sampled the western coastline of the Suvasvesi South lake and the islets closer its the center. In addition, we inspected the road cuts leading to Mannamäki hill. In total, sixteen sites were investigated (Fig. 1b). From each site, three samples for paleomagnetic investigation and thin section analysis have been taken. The orientation of ice flow lineations and fractures were also measured.

3.2 Fractures

We investigated the fractures of six different outcrops from the islands (Kallio-Piippo, Varvon Ilves, Hautasaari, Korkeasaari and Takunluoto) and from the coast (Koiraniemi) of the Haapaselkä lake area. In total we took eighty measurements, where thirty-eight are from the heavily fractured Takunluoto islet. In Fig. 5a we plot the forty-two measurements from the mentioned outcrops with exception of Takunluoto. The two dominant fracture systems are subvertical, NNW-SSE and ENE-WSW oriented. Fig. 5b shows the thirty-eight measurements from Takunluoto islet, closer to the lake center. Also in this case the same fracture systems appear and may be related either to impact event or to the strong tectonic deformation of the area. Also, three of the measurements (triangles in fig. 5b) were taken in correspondance of the shatter cone features. The readings show shallower inclination and have orientation pointing towards the lake center, and they may represent the cone apices.

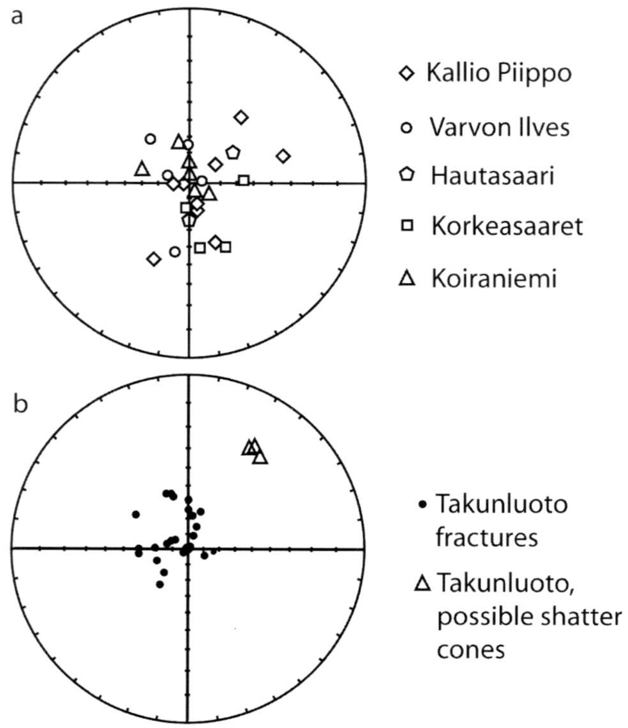

Fig. 5. (a) Stereographic projection of the directions of fractures in the Suvasvesi South area. Forty-two measurements were taken at five different sites (see legend). (b) Stereographic projection of the directions of fractures in the Takunluoto islet, close to the lake center. Three measurements (triangles) over thirty-eight show shallower direction pointing towards the lake center and may be shatter cone apices.

3.3 Lithologies

Shatter cones (10 to 20 centimetres in size) were found in mica schist (Fig. 6a) and granitoid boulders along the NW coast of Haapasaari island and at the Lusikkaniemi area (Fig. 1b). In this two sites about 5% of the boulders show shatter cones; we collected 28 shatter cone boulders. Further cutting of the rocks in the lab revealed that such boulders might consist also of breccia containing granitoid clasts of a few cm size. In such cases the cones are well-developed on the granitoid clasts.

The occurrence of shatter cone boulders is restricted to particular areas: the Lusikkaniemi and the Haapasaari region, which are 1 km away from each other. This fact is relevant if we consider that the diameter of the Haapaselkä lake is only 3.8 km. The shatter cones at both sites are similar and the cones vary between few centimeters and few tenths of centimeters in length.

In situ mica schists are heavily deformed, but no shatter cones have been observed on outcrops if these rocks. However, a migmatite outcrop in the NW edge of Potkuniemi (Fig. 1b) shows two features, which may be interpreted as shatter cones. In fact, the strike and dip of the cone apex (255°, 24°; 270°, 43°) is clearly different from the primary and secondary tectonic deformation (Fig. 6b). However, such structures could also be the result of glacial erosion, even though the linear direction of both structures points toward the center of the crater.

Also at Suolasaaret island (Fig. 1b) we found deformations of same size as the shatter cones, which differ clearly from both folding and fracturing of the migmatite. Yet, it is unclear if those are the result of glacial erosion or impact.

At Takunluoto islet, possible in situ shatter cones of 50 cm length can be observed on granitic outcrops (Fig. 6c). Also in this case, the cones are weakly developed and it can not be excluded that they could be slickenslide structures related to the heavy fracturing. However, we also note that the cone apices point towards the center of the open lake.

We found 3 boulders in the gravel pits along the road to Mannamäki hill, about 5 km eastward from the Haapaselkä lake shore line. This rock consists of a fine-grained black matrix resembling recristallized impact melt glass with fluidal texture containing cm-sized clasts of mica schist and granite (Fig. 6d). The clasts are mainly melted, vesciculated or crushed. The boulders were identified as impact melt rock, also according to the thin sections analysis (see section 3.5). Also in this case, the finding of impact melt rock was restricted to a particular place, located just after the bifurcation of the road to Mannamäki (Fig. 1b).

Fig. 6. (a) Shatter cones developed on a mica schist boulder from the Lusikkaniemi bay, eastern shoreline of the Haapaselkä open lake. (b) Possible shatter cone developed on a migmatite outcrop at Potkuniemi, east shoreline of the Haapaselkä open lake. Size of the shatter cone is about twenty centimeters.

Fig 6 cont. (c) Granite outcrop in Takunluoto islet (central part of the Haapaselkä open lake) showing a shatter cone-like feature. Size of the shatter cone is about fifty centimetres (d) Impact melt rock boulder from Mannamäki, 5 km east of the shore-line. The length of the boulder is about 20 cm. Photo: Kari A. Kinnunen.

3.4
Ice flow directions

We measured the directions of glacial striations on fourteen outcrops, in order to determine the direction of glacier movement and the consequent displacement of boulders. Fourteen measurements, taken at fourteen different places around the lake, show a concordant ESE direction (Fig. 1b). The mean value for the direction is 104 ° (Fig. 7).

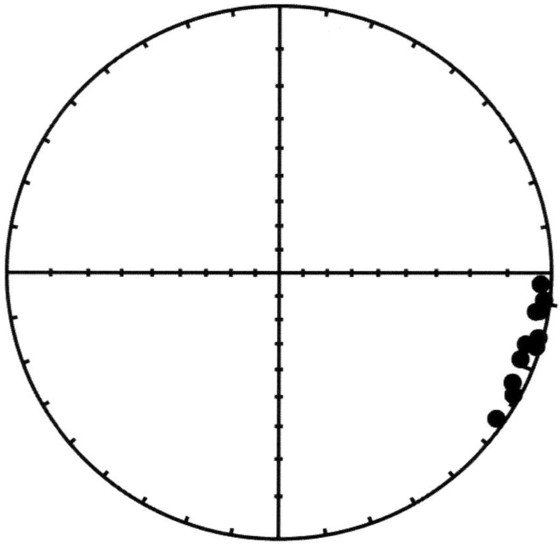

Fig. 7. Stereographic projection showing the directions of glacial striations on outcrops in the Suvasvesi South area. Fourteen measurements were taken at different sites. The mean ice flow direction is 104°.

3.5
Thin section studies

Ten thin sections were prepared from four in situ granitic samples as well as out of three shatter cone boulders and three melt rock boulders. In this study we found planar deformation features (PDFs) in quartz grains (Fig. 8a and b) and a breccia texture in the shatter cone boulders. In thin section (#1229) of melt rock, about 50% of all grains have PDFs. PDFs can be distinguished as multiple sets and are mainly decorated (Fig. 8b), probably pointing to the fact that either syngentic impact fluid inclusions are present or post impact hydrothermal phase occured. In addition, Öhman et al. (2003) measured the crystallographic orientation of the PDF lamellae using the U-stage method and found six sets of PDFs on different quartz grains. The breccia structure is shown by quartz or feldspar grains (smaller than 1 millimeter) surrounded by fine-grained matrix.

In the melt rock specimen, small patches of diaplectic plagioclase glass – maskelynite – were found. In case of twinned plagioclase, asymmetric isotropization can occur, leading to an alternation of birefringent lamellae and maskelynite lamellae (Fig. 8c). The thin sections from impact melt samples also show its fluidal character. In Fig. 8d, fluidal texture demonstrates rapid flow of melt without homogenization of the material. In the melt rock samples, all the large (diameter > 0.8 mm) grains of quartz and feldspar are shattered, and quartz grains with PDFs are very common. In thin section (#1229) of the melt rock, tens of quartz fragments and grains (about 50% of all) have PDFs. Practically all the mica flakes (biotite and muscovite) are kinked.

As a consequence, the thin sections prove the impact origin of the boulders. The thin sections of granite or mica schist samples collected from outcrops did not show any clear evidence of impact.

3.6
Petrophysics and paleomagnetism

Sixty-one unoriented specimens of melt rock, shatter cones and granites have been used for petrophysical measurements (susceptibility, remanent magnetization (NRM), density and porosity). Results are summarized in table 1. Eight oriented specimens underwent paleomagnetic analysis. The ladder derived from granitic or mica schists samples collected from Takunluoto, Tiirinluoto and Selkä Lehtonen islets. Petrophysical measurements have been carried out using the “Risto-5” apparatus described by Pesonen et al. (2001) and magnetic remanence has been measured using SQUID or spinner magnetometer.

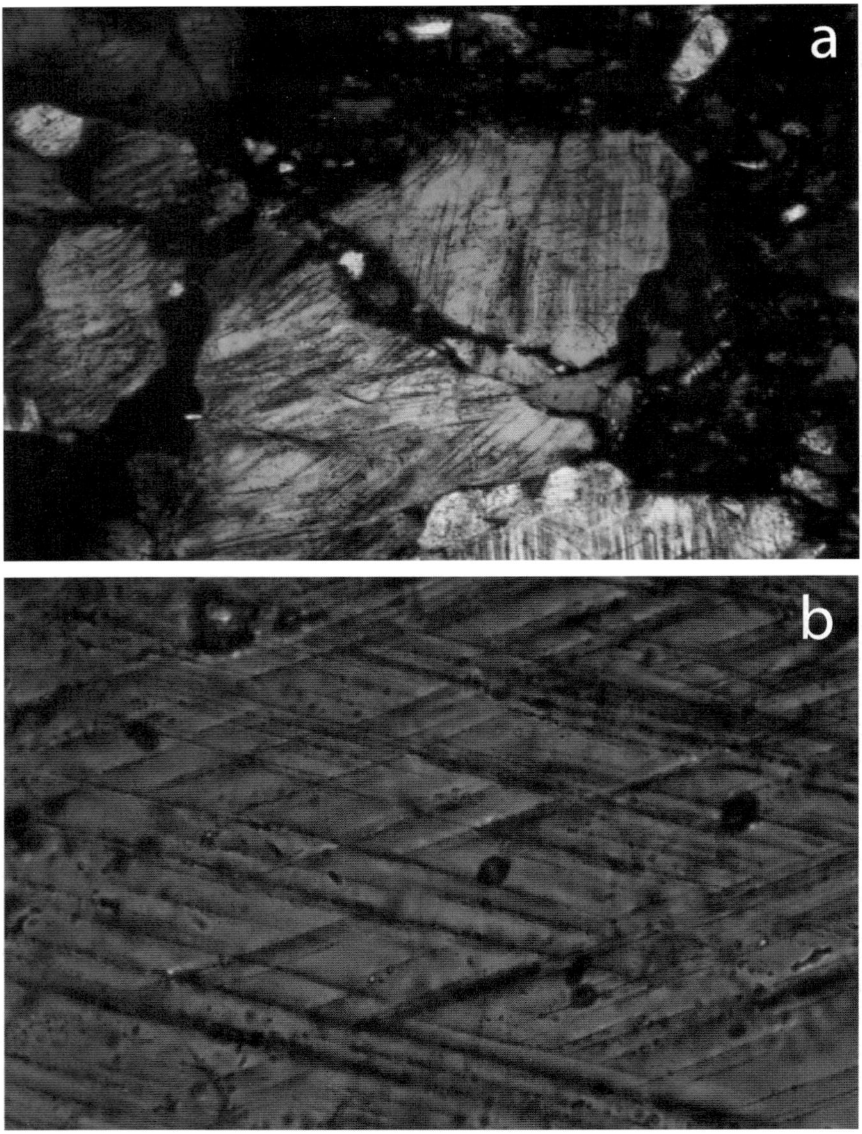

Fig. 8. (a) Multiple sets of PDFs in a shattered quartz grain surrounded by fine-grained matrix of the mica schist. The sample is from a shatter cone boulder found in Lusikkaniemi. Crossed polarizers. Width of the figure is about 1.7 mm. (b) Detail from Fig. 8a, representing two multiple sets of decorated of PDFs. Crossed polarizers. Width of the figure is 0.4 mm.

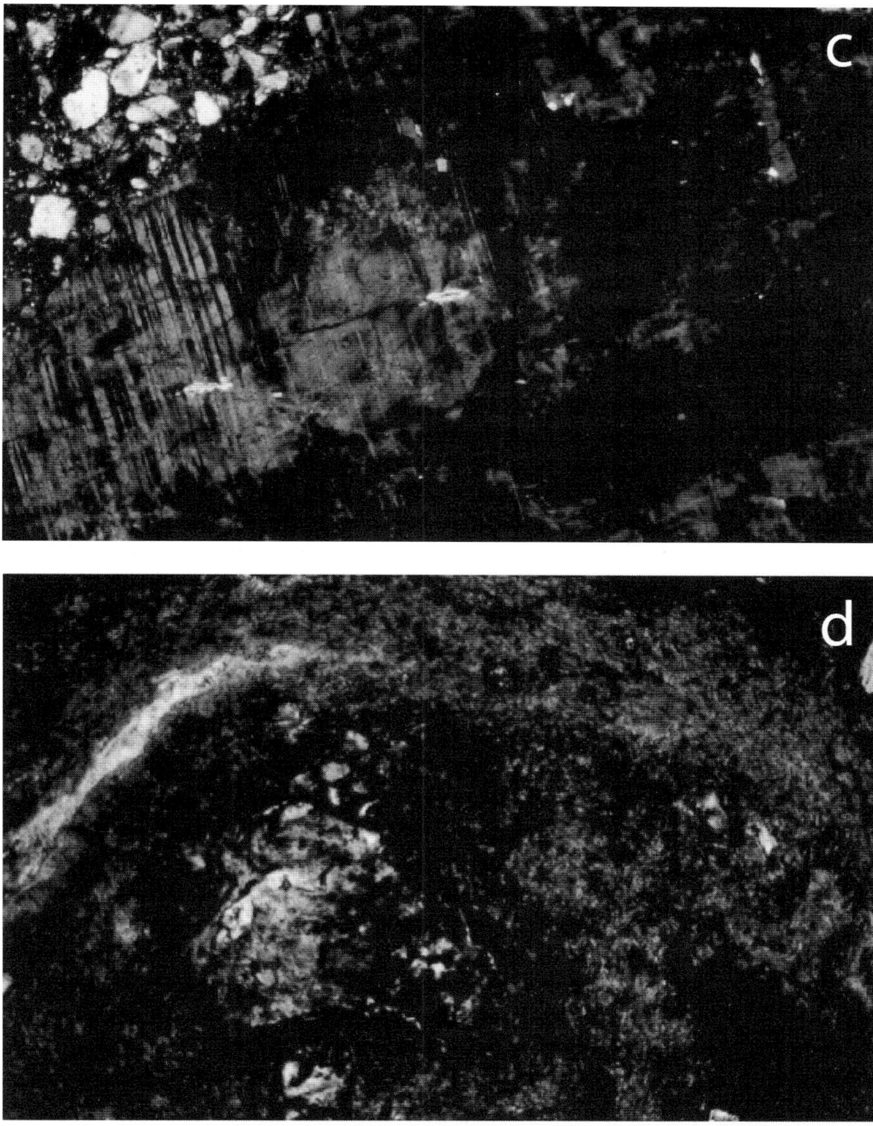

Fig. 8 (c) Maskelynite from a melt rock specimen from Mannamäki (#1229). Isotropization of twin lamellae and patchy black maskelynite after plagioclase. Crossed polarizers. Width of the figure is 1.7 mm. (d) Fluidal texture on a impact melt sample from Mannamäki. Elongated (partly melted) fragments of quartz and plagioclase follow the flow direction. Crossed polarizers. Width of the figure is 1.7 mm.

The petrophysical analysis reveals a good distinction between the density and porosity of melt rock, shatter cones and country rocks. This distinction is expected because the melt rock is fractured and porous, whereas shatter cones samples and country rocks are more massive. The density of country rocks is consistent with values for mica schist or granite (e.g., Kivekäs 1993), whereas their porosity seems to be higher than normal and reflects the fracturing. With exception of site Koiraniemi (4.4 km from lake center), all analysed specimens reveal weak magnetic properties: susceptibility less than 400 10^{-6} SI, NRM less than 5 mA/m and low Q-values (<0.3). Those values are significantly lower in comparison to the data measured by Werner et al. (2002) from similar rock specimens of Suvasvesi North. The susceptibility vs. density plot (Fig. 9a) shows a good distiction of density values between the melt rock and the shatter cones. This distinction is attributed to the high porosity of the melt rock. Similar density values are observed between the shatter cones and the country rocks. Susceptibility values of the three lithologies lie in the same interval (0-400 SI), revealing paramagnetic character. The porosity vs density plot (Fig. 9b) shows well grouping of the different lithologies. Shatter cone boulders and country rocks have different porosities but similar densities.

The remanent magnetization is often unstable, but for most of the specimen we recognize a hard component that correlates well with the present Earth magnetic field (PEF) in Finland (D=6°, I=70°) (Pesonen et al. 1995). In addition, two of the specimens seem to carry a soft magnetization of a direction pointing southwest and with negative inclination (D,I: 276°; -61.8°), which is comparable with the results obtained for Suvasvesi North by Werner et al. (2002). This magnetization may be of impact origin; however further measurements are required to confirm this possibility.

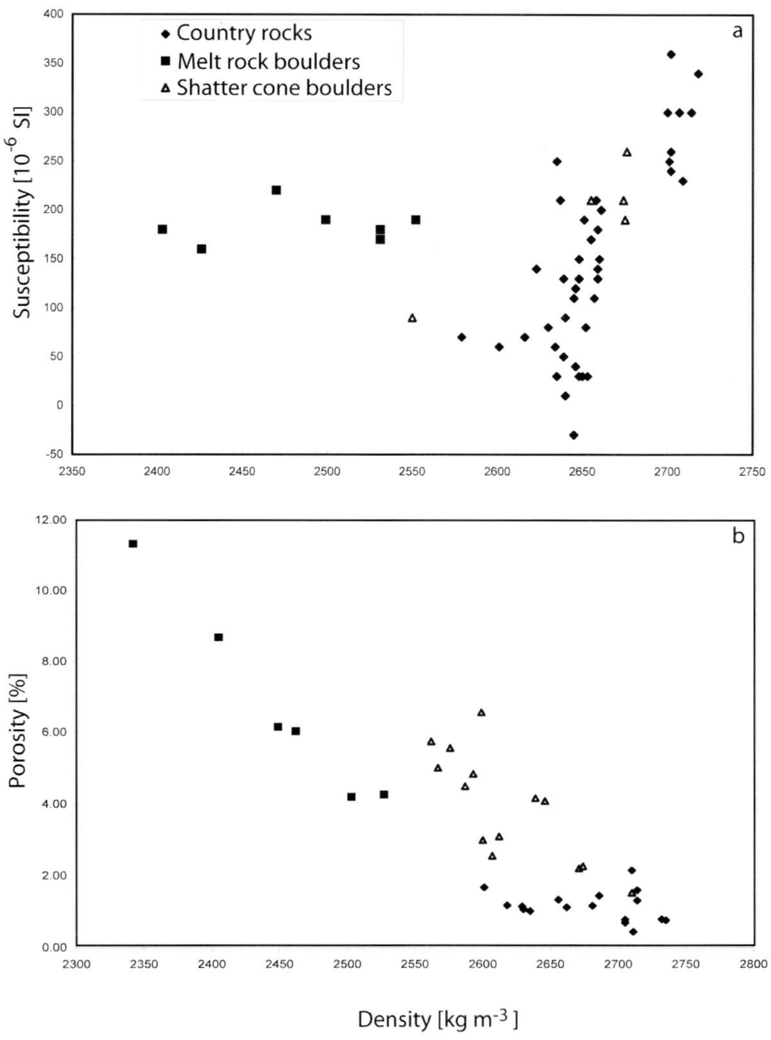

Fig. 9. (a) Susceptibility vs. density plot showing the data for impact melt rock samples (squares), shatter cones of granitoid and mica schist samples (open triangles) and granite sample (diamonds). (b) Porosity vs. density plot showing good distinction of the three different lithologies. Symbols are as same as Fig. 9a.

Table 1. Physical properties of rocks from the Suvasvesi structure

Site	n/N	δ kgm^{-3}	ϕ %	χ 10^{-6}SI	NRM mAm^{-1}	Q	Comment
			Shatter cones boulders				
Lusikkaniemi	4/11	2599	4.30	250	3.5	0.43	mica schist, PDFs in qz
Haapasaari	1/3	2664	2.33	-	-	-	granite, coarse grained
Mean	**5/14**	**2631**	**3.31**	**250**	**3.5**	**0.43**	
			Melt rock boulders				
Mannamäki	2/5	2478	6.78	193	2.6	0.35	fractured, altered, clasts of qz and plg
Mean	**2/5**	**2478**	**6.78**	**193**	**2.6**	**0.35**	
			Country rocks				
Ropan Ilves	4/30	2657	1.87*	154	3*	0.29*	granite
Justiina	2/2	2664	1.23	185	3	0.34	granodiorite
Potkuniemi	2/2	2737	0.78	290	3	0.24	granite
Pienet Piipot	1/2	2633	1.11	120	2	0.37	granite
Selkä-Lehtonen	2/2	2709	0.87	215	2	0.23	granite
Takunluoto	2/3	2624	1.29	113	2	0.34	heavily fractured granite
Tyrskyniemi	1/1	2691	1.3	220	4	0.39	mica schist
Koiraniemi	2/3	2706	0.74	287	*161*	*14.38*	mica schist, 3.1 km from lake center
Mean	**14/42**	**2677**	**1.14**	**229**	**2.7**	**0.31**	

Site	see Fig. 1b
n/N	number of samples/specimens
δ, ϕ	laboratory bulk density, porosity
χ	weak field susceptibility
NRM	intensity of original NRM; values in italic were not used for the mean
Q	Koenigsberger ratio, values in italic were not used for the mean
*	only three specimens measured

4
Biological aspects

According to French (1998) the projectile size is proportional to the diameter of the structure. In the case of Suvasvesi South (diameter 3.8 km) the projectile size would be about 190 m, releasing 1.3 x 10^{18} J energy. Assuming that the Suvasvesi craters form a doublet, the energy would increase to about 7 x 10^{18} J. From a biological point of view, such an impact would have severe consequences on the organisms living in the proximity of the impact area (e.g., Kring 1997; Cockell and Lee 2002). The bedrock of Finland nowadays consists mainly of Archean and Proterozoic basement, with fossil being rare. In particular, no fossil has ever been found from the Suvasvesi area. Hence, it is difficult to say what kind of biota may have been affected by the impact.

The reverse magnetic component, which seems to correlate with the measurements carried out for the Suvasvesi North structure might be a chemical remanent magnetization (CRM) induced by hot circulating fluids after the impact occurred (e.g., Pesonen et al. 1999). In such case, a phase of thermal biology would have been taking place. This possibility would also explain the obervation of decorated PDFs in quartz grains.

5
Conclusions

The new discoveries made during the 2002 fieldtrip to the Suvasvesi South area strengthen the hypothesis of an impact origin for this structure. In fact, the new discovery of shatter cones and impact melt rock in boulders, in comparison with the measurements of ice flow directions, show that the boulders have been transported from the southern structure (Fig. 1b). Also, thin sections analysis of the shatter cone and impact melt rock boulders found along the Suvasvesi South lake shoreline shows the presence of PDFs in quartz grains, breccia structure, fluidal structure of the melt rock as well as presence of maskelynite. The petrophysical analysis of the melt rock and of the shatter cone boulders reveal lower values of susceptibility and NRM with respect of values observed from the Suvsvesi North structure. This observation can also confirm that the boulder found in the Suvasvesi South area are not stemming from the northern structure.

The location of the boulders in restricted areas suggests either that erosion has been acting with different strength or that lithologies were not homogeneously distributed. Another interpretation may be also related to the fragility of impact rocks, which may have been distroyed during glaciations.

The shatter cones observed on outcrops are weakly developed and difficult to recognize. However, the directions of the cone apices point toward the lake center, and thus to the probable crater center.

Compared to its northern companion, the topographic and bathymetric anomaly of Suvasvesi South is smaller and shallower.

The analysis of fractures defines two families, which are identical for each outcrop investigated and may be related to the heavy tectonism occurring in the zone.

As a result, we suggest that Suvasvesi South is an eroded, shallow simple impact crater of 3.8 km diameter. Further paleomagnetic measurements may confirm that the Suvasvesi structures for a doublet. In that case the age of Suvasvesi South would be 250 Ma. From a biological point of view such impact did have effects on a local scale and possibly hydrothermal activity took place in a successive phase, creating an habitat for hydrothermal biology.

Acknowledgements

We thank Christian Koeberl and Wolf-Uwe Reimold for their detailed and constructive reviews. The field research to the Suvasvesi South structure was possible due to the financial support of the Barringer Family Fund for impact crater research.

References

Cockell CS, Lee P (2002) The biology of impact craters – a review. Biological Reviews 77: 279-310

French BM (1998) Traces of Catastrophe: A Handbook of Shock-Metamorphic Effects in Terrestrial Meteorite Impact Structures. Lunar and Planetary Institute Contribution 954, Lunar and Planetary Institute, Houston, 120 pp

Henkel H, Pesonen LJ (1992) Impact craters and craterform structures in Fennoscandia. In: Pesonen LJ, Henkel H (eds) Terrestrial impact craters and craterform structures, with special focus on Fennoscandia. Tectonophysics 216: 123-142

Kivekäs L (1993) Density and porosity measurements at the petrophysical laboratory of the Geological Survey of Finland. In: Autio S (ed) Geological Survey of Finland, Current Research 1991-1992, Geological Survey of Finland, Special Paper 18: 119-127

Kring DA (1997) Air blast produced by the meteor crater impact event and a reconstruction of the affected environment. Meteoritics and Planetary Science 32: 517-530

Lehtinen M, Pesonen LJ, Kuulusa M, Stehlik H (2002) The Suvasvesi South structure, Central Finland: New evidences for impact [abs.]. Lunar and Planetary Science XXXIII: abs. 1188 (CD-ROM)

Öhman T, Badjukov D, Raitala J, Petrova T, Stehlik H (2003) Impactites of the Päässelkä and Suvasvesi South craters, Finland [abs.]. Lunar and Planetary Science XXXIV: abs. 1571 (CD-ROM)

Pesonen LJ (1996) The impact cratering record of Fennoscandia. In: Rickman H and Valtonen MJ (eds) Worlds in Interaction: Small bodies and planets of the Solar System. Kluwer Academic Publisher, Dordrecht, pp 377-393. Also appeared in Earth, Moon, and Planets 72: 377-393

Pesonen LJ, Nevanlinna H, Leino MAH, Rynö J (1995) The Earth's Magnetic field map of 1990. Geophysica 30: 57-77

Pesonen LJ, Elo S, Lehtinen M, Jokinen T, Puranen R, Kivekäs L (1999) Lake Karikkoselkä impact structure, central Finland: New geophysical and petrophysical results. In: Dressler BO, Sharpton VL (eds), Large Meteorite Impacts and Planetary Evolution II, Boulder, Colorado, Geological Society of America Special Paper 339, pp 131-139

Pesonen LJ, Donadini F, Kuulusa M (2001) Renewal of the research and study strategy at the Division of Geophysics, University of Helsinki [abs. in Finnish]. XX Geofysiikan Päivät, abstracts and programmes, pp 167-172

Werner SC, Plado J, Pesonen LJ, Janle P, Elo S (2002) Potential fields and subsurface models of Suvasvesi North impact struncture, Finland. Physics and Chemistry of the Earth 27: 1237-1245

Kärdla Impact (Hiiumaa Island, Estonia) – Ejecta Blanket and Environmental Disturbances

Sten Suuroja[1;2] and Kalle Suuroja[1]

[1]Geological Survey of Estonia, Kadaka tee 82, Tallinn 12168, Estonia
[2]Department of Mining, Tallinn Technical University, Kopli 82, Tallinn, Estonia
(s.suuroja@egk.ee)

Abstract. The Kärdla impact occurred at ca. 455 Ma (Upper-Ordovician, Caradoc) in a shallow (ca. 100 m) epicontinental sea not far (ca. 100 km) from the erosion area on the Baltic Shield (Grahn et al. 1996). The explosion of the meteorite ca. 200 m in diameter generated a complex crater 4 km wide and more than 500 m deep on the sea bed. The crater is surrounded by elliptical ring fault, up to 15 km in diameter, within which the sedimentary target rocks are strongly deformed. The ejected matter was spread almost concentrically around the crater, within a 50-km-radius, on ca 5500 km^2. The ejected matter is found also farther away as an admixture in limestones. Most of the ejecta blanket was covered by limy mud immediately after the impact. The crater was buried somewhat later; therefore the ejecta blanket is well preserved, except the rim wall area. Rate of accumulation of deposits and its facial composition in the crater deep, rim wall area and surroundings was different during some millions of years.

The ejecta blanket lies in a succession of Upper-Ordovician carbonate rocks as a 0.01–3.5 m thick southward inclined (from 40 m b.s.l. [below sea level] in the island's northernmost point up to 190 m b.s.l. in the southernmost point) bed of silty and sandy limestones or limy silt- and sandstones. On the sea bed about 10 km northward of the island the ejecta blanket is cut by the erosion escarpment (Baltic Klint). The distal ejecta layer consists mostly of silt- to gravel-sized debris of the target rocks (mostly Cambrian siliciclastic and Paleoproterozoic metamorphic rocks). In the lower part of the bed and closer to the impact centre coarser clasts occur. Farther from the impact site, the thickness of the ejecta layer, as well as the size of the grains decreases. The size of the ejected matter decreases also from the bottom towards the top of the layer. The ejected matter contains up to 1 vol% shock metamorphosed quartz grains with PDFs. The Kärdla impact was too small to cause substantial and long-term global environmental changes and catastrophic shifts in the biosphere. Its long-term effect was restricted mostly to changes in sea bed relief and related facial changes, as well as the changes in the biotic communities of pelagic organisms caused by the latter.

1 Introduction

The Kärdla meteorite crater (Hiiumaa Island, Moonsund Archipelago, Estonia; 58°58′N, 22°46′E) (Fig. 1.) was formed in the Upper Ordovician (Salvador 1994; Remane 2000; Remane et al. 1996), earliest Caradoc time (ca. 455 Ma) in a shallow epicontinental sea not far (ca. 100 km) from the erosion area in the Baltic Shield (Puura and Suuroja 1992; Plado et al. 1996; Suuroja 2001; Suuroja et al. 2001; Suuroja 2002). The time of the impact was obtained by several authors (Suuroja et al. 1991; Lindström et al. 1992; Grahn et al. 1996) and it was proved by determination of the position of the ejecta layer in the sequence of the pre- and post-impact sedimentary rocks.

Between 1968 and 1994 more than 150 drill holes for different purposes were drilled in the area of Kärdla impact structure (geological and hydrogeological mapping (Kala et al. 1971; Suuroja et al. 1991; Suuroja et al. 1994), exploration of mineral resources (Kala et al. 1976) and other applied geological activities), and 42 of these penetrate the ejecta blanket. To study the ejecta layer, 25 cores from these drill holes were studied in more detail (Fig. 2). The drill core in the interval of the ejecta blanket was bisected using a diamond saw, photographed and described in detail macro- and microscopically (in thin sections). Special attention was paid to structural changes, changes in fauna and traces of their activity. Altogether 36 samples from 10 drill cores (F-340, F-343, F-345, F-355, F-362, F-364, F-366, F-369, F-372 and K-31) were analytically investigated (grain-size distribution and grain shape, and mineralogical analysis).

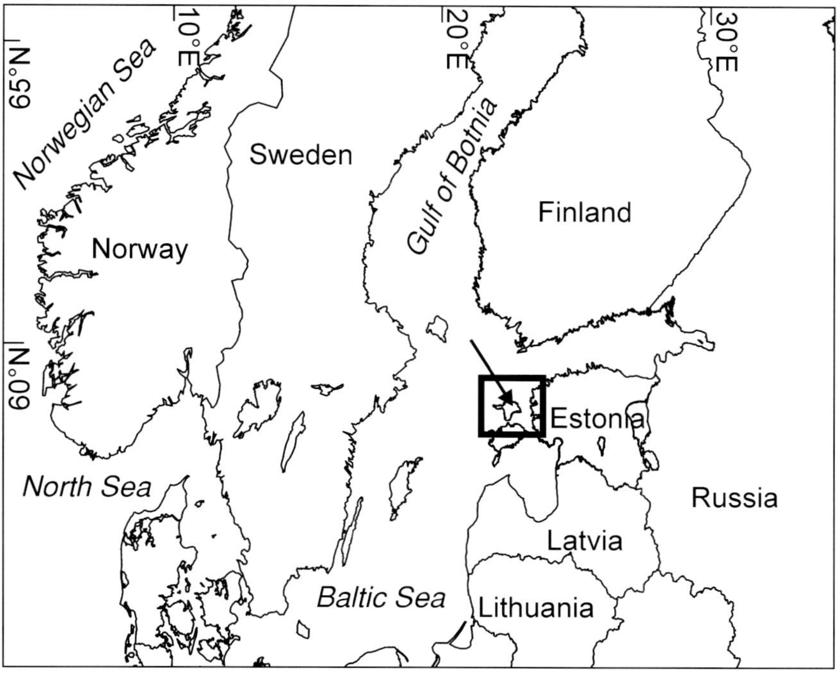

Fig. 1. Location of the Kärdla impact structure.

All of them were studied for shock-metamorphosed minerals (quartz grains with PDFs - planar deformation features) and the latter were identified in 22 samples from the above-mentioned drill cores. Insoluble residue and size, shape and mineralogical content of the grains were analyzed in the samples. The insoluble residue was produced by long-term (ca 7 days) solution of the crushed rocks in 5% hydrochloric acid (HCl) at room temperature. Grains were divided into four size groups: clay (less than 1/256 mm); silt (1/256–1/16 mm); sand (1/16–1 mm); granules (1–4 mm). By shape the grains were angular, subangular, subrounded, rounded and well rounded. The 1000 quartz grains from the insoluble residue from the size group "sand" (1/16–1 mm) were picked up under a binocular and examined for PDFs in immersion liquid under polarised light by identifying the crystallographic orientations of the planes. The most common orientations were {1013}, {1012} and {1011}. All these analyses were carried out in the Laboratory of the Geological Survey of Estonia.

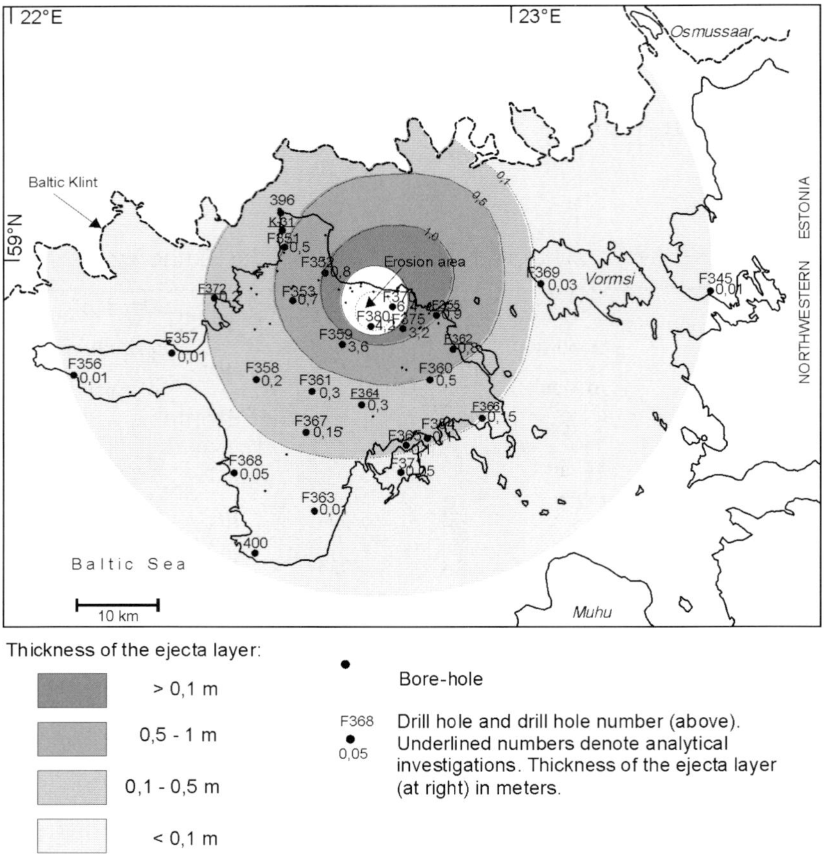

Fig. 2. Distribution of the ejecta blanket of the Kärdla crater, Estonia.

2
Geological settings of the ejecta blanket

The ejecta blanket of the Kärdla impact in the sequence of the Ordovician carbonate rocks (limestones) is marked by a sandy layer 0.01–3.5 m thick (Fig. 5, 8, 9, 10). Soon after the impact the ejecta blanket was buried under marine deposits (biodetrital limy mud) and is therefore mostly well preserved. On the top of the rim wall and outer slopes of the crater in a short time (some thousands years) after the impact the ejected matter was eroded. On the outer slope of the rim wall and in its closest surroundings, the brecciated target rocks are covered by up to 6 m thick bed of sandstones. This bed is not connected with the ejecta blanket, but formed as a result of the erosion of the ejecta on rim wall.

Table 1. Some characteristics of the ejecta blanket of the Kärdla impact.

Core	Thickness of the ejecta layer (cm)	Depth of the bottom of ejecta layer (m a.s.l.)	Distance from crater centre (km)	Core	Thickness of the ejecta layer (cm)	Depth of the bottom of ejecta layer (m a.s.l.)	Distance from crater centre (km)
F345	10	-88	42	F363	10	-162	26
F351	50	-54	13	F364	30	-141	12
F352	80	-76	7	F365	10	-149	18
F353	70	-92	9	F366	15	-145	20
F354	10	139	18	F367	15	-148	17
F355	90	-101	8	F368	50	-152	27
F356	10	-97	38	F369	30	-101	22
F357	10	-85	25	F370	640	-81	3
F358	20	-118	16	F371	50	-145	21
F359	360	-108	5	F372	20	-67	19
F360	50	-130	12	F375	320	-106	5
F361	10	-128	12	F380	420	-163	3
F362	80	-116	12				

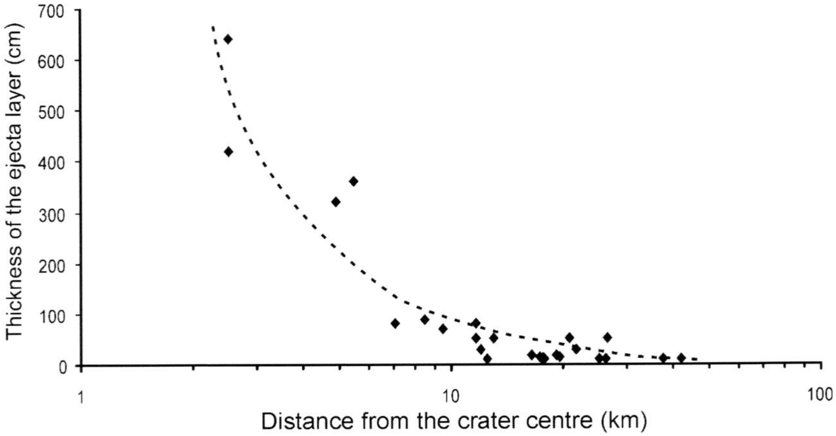

Fig. 3. Thickness of the Kärdla ejecta layer *versus* distance from the impact centre.

The ejecta connected with the Kärdla impact was distributed almost concentrically around the crater in more than 50-km radius (Fig. 2; Table 1). As an admixture in limestones the ejected silt and sand grains from the Kärdla are found farther away, too. The core section of drill hole F-340 is the farthest site (55 km to NEE from the impact centre) where the shock matter (quartz grains with PDFs) ejected from the Kärdla impact site has been found. The ejecta blanket like a bed with the distinct boundaries is spread in more than 40-km radius around the crater (42 km N-S and 38 km W-E direction; see Fig. 2). Farther the ejecta blanket may exist as rare quartz grains in limestone, but samples were not taken from the sections where the layer of ejected material was not observed with a naked eye. The distal ejecta contains clay-sized matter as well, but it cannot be recognised which part of this originates from the pre-impact target rocks and which is produced in the course of the impact. The present-day distribution of the ejecta blanket is restricted by the erosion escarpment of the Baltic Klint (Tammekann 1940) on a sea-bed 10–12 km northward of the island. Northward of this line the ejecta blanket was eroded in pre-Pleistocene time (Neogene?). Investigations of the surface upon which the ejecta blanket is deposited are of special interest (Fig. 3). In the area around the crater, in ca 4-km radius the ejecta layer as well as the part of the rim wall and crushed by the subsurface release target rocks have been removed by tsunami-like resurging waves or short-term post-impact erosion. As a result of subsurface release caused by reflection of a rarefaction wave (Melosh 1989; Dence 2002), in the crater's surroundings (up to ca. 5 km from the impact centre) up to 30 m of the subsurface layers of the target sedimentary rocks were crushed and partially or entirely removed (Fig. 5). It is difficult to establish which part of the removed rocks were displaced by the releasing of subsurface layers and which by resurging tsunami or erosion, because afterwards all of them have been eroded. Farther from the crater the share of removed and crushed target rocks decreases up to wedging out at a distance of about 6 km.

Due to long-term post-impact regional tectonic movements in this region (Puura and Floden 1997) the ejecta blanket inside the sequence of the Upper-Ordovician carbonate rocks has acquired gentle (ca 3.5 m per kilometre) southward dip. As a result of these movements the ejecta blanket presently lies in the succession of the Upper Ordovician carbonate rocks (mostly limestones) at a depth of 40 m b.s.l. in northern part of the island (drill hole 396 Tahkuna), and at a depth of 190 m b.s.l. southern part (drill hole 400 Sõru) (Fig. 2.). Farther south and deeper the run of the ejecta layer cannot be followed because in this area the drill holes are missing and content of the ejected matter in the limestones is limited, too. East- and westward of the centre of the Kärdla impact crater the ejecta blanket lies at a depth of 100–110 m b.s.l. (Fig. 4.).

In some places, especially westward of the impact site, in the areas where the ejecta blanket is 10–50 cm thick (drill holes F-351, F-352, F-358, F-366, F-367, F-368, F-372 in Fig. 2 and Fig. 3) it is impregnated with natural bitumens (Suuroja et al. 1991; Kattai et al.. 1994).

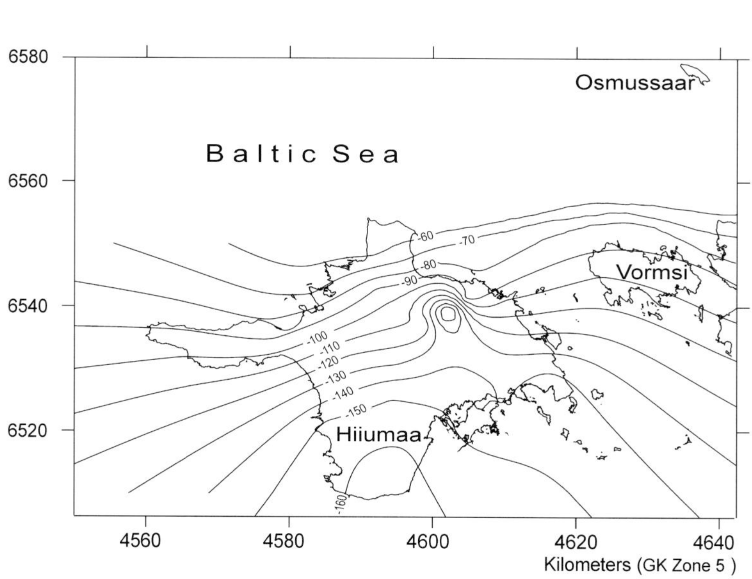

Fig. 4. The isolines of the ejecta blanket. Distance between isolines 10 m.

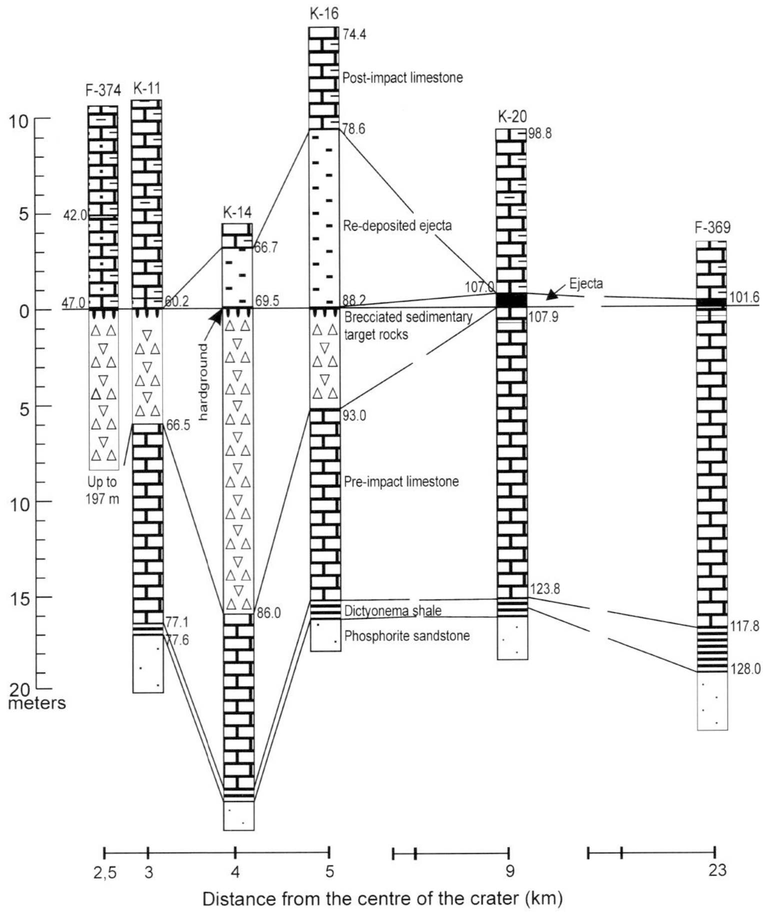

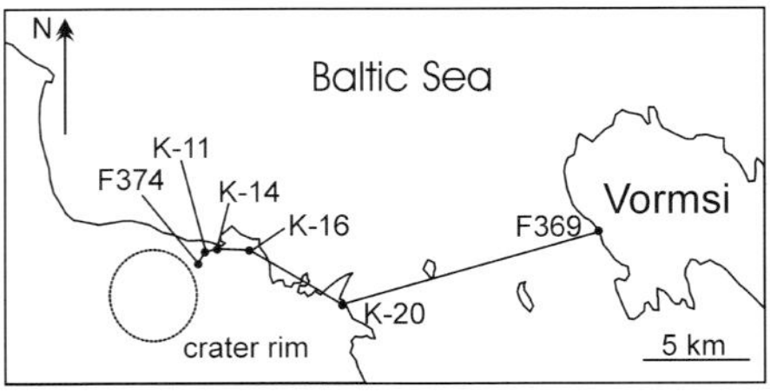

Fig. 5. W-E cross-section (drill holes F-374; K-11; K-14; K-16; K-20; F-369) of the impact-related rocks of the surroundings of the Kärdla crater.

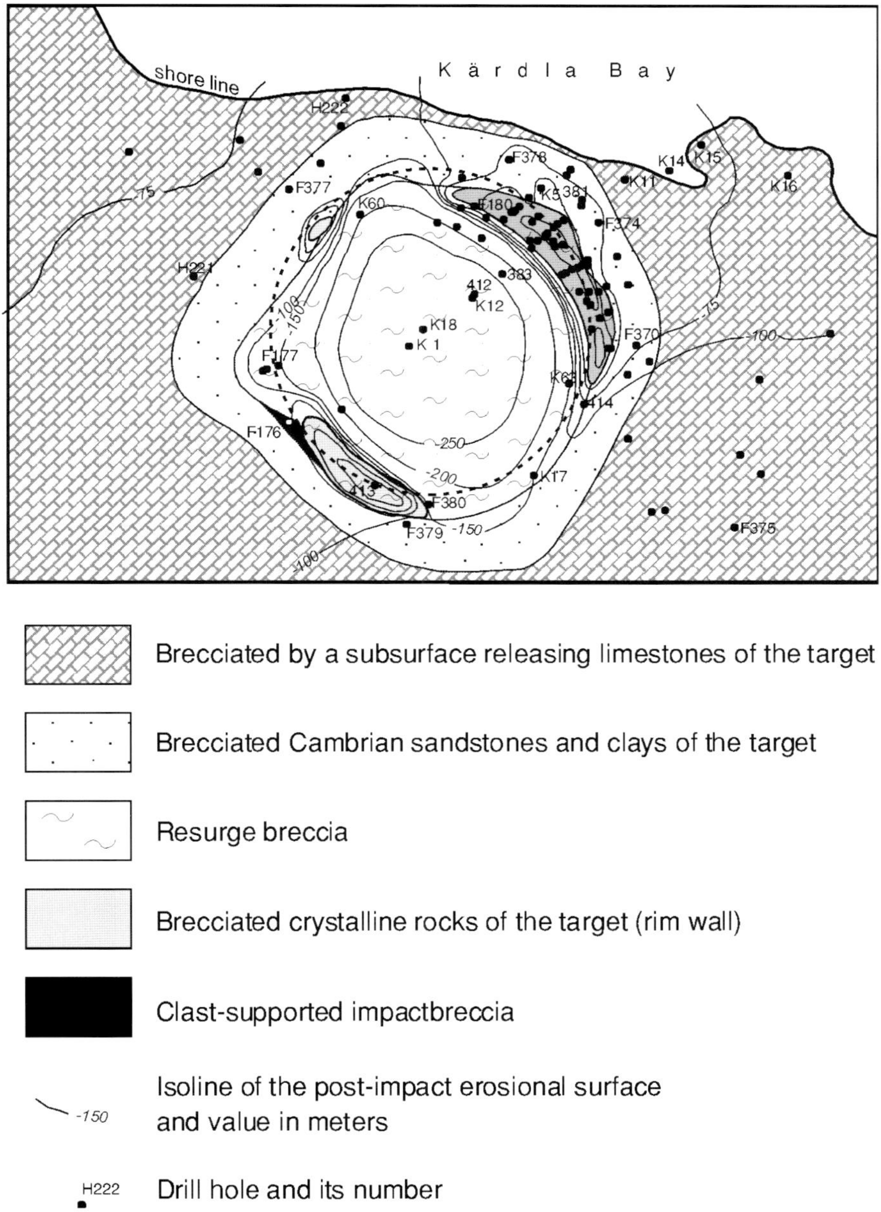

Fig. 6. The post-impact surface of erosion in the surroundings of the Kärdla crater.

3 Composition of the ejecta blanket

The ejecta blanket consists mostly of silt- to gravel-sized debris of the target rocks. In its lower part and closer to the impact centre coarser clasts (pebbles, cobbles, and blocks) occur, too. The clasts consist of the sedimentary (limestones, sandstones, siltstones, clays) and crystalline (gneisses, granitoids, amphibolites, migmatites) target rocks. In the proximal and coarse ejecta mainly clasts of the target rocks (sedimentary and metamorphic) occur, while in the distal and finer ejecta clasts of the minerals or grains of the disintegrated siliciclastic rocks (Cambrian silt- and sandstones) prevail.

Farther from the impact centre thickness of the ejecta layer (Fig. 3.) as well as the grain size of the ejected matter decrease. Within a section, the grain size of the ejecta decreases from the base to the top of the layer. In the surroundings of the impact centre (6–8 km from the centre) coarse clasts (blocks, cobbles and pebbles) are found. On the outer slope of the rim wall, at a distance of ca. 1 km from the ridge (drill hole K5) a huge (ca 40 m in diameter) block of brecciated metamorphic target rocks (granite, gneisses) was discovered (Fig. 6).

At a distance of 6–12 km from the impact centre at least two separate beds of siliciclastic rocks with sharp contacts (Fig. 9) are observed – coarser (below) and finer (above). Closer to the impact centre the ejecta layer has been partly or entirely removed, and farther away the contacts between different layers are smoother or transitional (Fig. 10).

The character of the lower contact of the ejecta layer mainly depends on the distance from the impact centre. Closer to the crater (5–8 km) the layer of crushed sedimentary target rocks are strongly eroded by the tsunami (Fig. 9), but at a distance of 8–16 km from the crater the pre-impact sea bed (unlithified carbonate mud at the time of the impact) weakly eroded by the tsunami. Farther than 20–25 km from the impact centre noticeable traces of the sea bed erosion are absent and the contact is clear and it becomes transitional at a distance of more than 30 km (Fig. 9).

At the upper boundary of the ejecta layer in this area (where it is not eroded) noticeable diversities are observed. At a distance of 5–15 km (Fig. 8, 9) the boundary is transitional rather than sharp, but at a distance of 15–30 km on the top of the layer often an impregnated (pyrite and/or phosphate) or non-impregnated wavy discontinuity surface occurs (Fig. 10). The boundary between separate parts of the ejecta layer (lower – coarse and upper – finer) is quite sharp, but less than 10 km from the centre (Fig.

7, 8) it becomes more transitional (Fig.9, 10). Two separate layers can be distinguished up to a distance of 30 km from the impact centre.

The lower, coarser bed of the ejecta layer consists of angular clasts (cobble, pebble, granules, sand) derived from the target rocks by the explosion, and they are cemented by the fine-grained (silt and clay) matrix (20–40 vol%).

In the matrix disintegrated Cambrian siliciclastic rocks (clay, silt- and sandstones) prevail but fine angular debris of crystalline basement metamorphic rocks and limestones is observed as well. From this part of the ejecta layer only two grain-size distribution and mineralogical analyses have been made, both from the drill core F-359 (5 km SW of the impact centre) from a depth of 122.6 and 122.4 m. The content of insoluble residue is 74 and 83 vol% respectively; from this, granules form 18 and 11%, sand – 46 and 48%, silt – 6 and 10%, and clay – 30 and 21%. The share of coarse material decreases with increasing depth, while total content of the insoluble residue, conversely, increases. The coarser fraction (granules) in these samples consists mainly of angular clasts of different target rocks (crystalline 75–62%; siliciclastic 16–12%; limestones 9–12%). With increasing depth the content of insoluble residue decreases and the size of clasts increases. On the grounds of the evidence of two analyses is difficult to judge on mineralogical changes of the coarse part of the ejecta layer, but it seems that with increase of the depth the content of the rocks derived from crystalline basement decreases.

Numerous shock metamorphosed quartz grains with PFs (planar fractures) and PDFs and disintegrated grains of Cambrian silt- and sandstones (up to 1% of the analysed grains of 1–1/16-mm fraction) are encountered in the ejecta layer. By shape the grains are mainly (80%) rounded or well rounded. The upper (fine-grained) part of the ejecta layer consists mainly of clay, silt and sand fractions of the disintegrated Cambrian siliciclastic rocks. Coarser clasts (pebbles, granules) of crystalline and sedimentary target rocks are very rare (Figs. 11 and 12). Farther away from the impact centre, the absolute thickness of this part of the ejecta layer decreases but its relative importance increases. The total content of insoluble residue is higher (60–80 vol%) in the middle part of the layer and decreases downwards and upwards. The upper part of the ejecta layer differs occasionally from the pre- and post impact limestones, mostly by the content of insoluble residue and by the fractional composition: in the pre- and post-impact limestones the content of insoluble residue is 5–15 vol% and it consists of more than 95vol% of clay, while in the ejecta layer the content of insoluble residue is 40–80vol%, and the content of the clay decreases to 40–60 vol%.

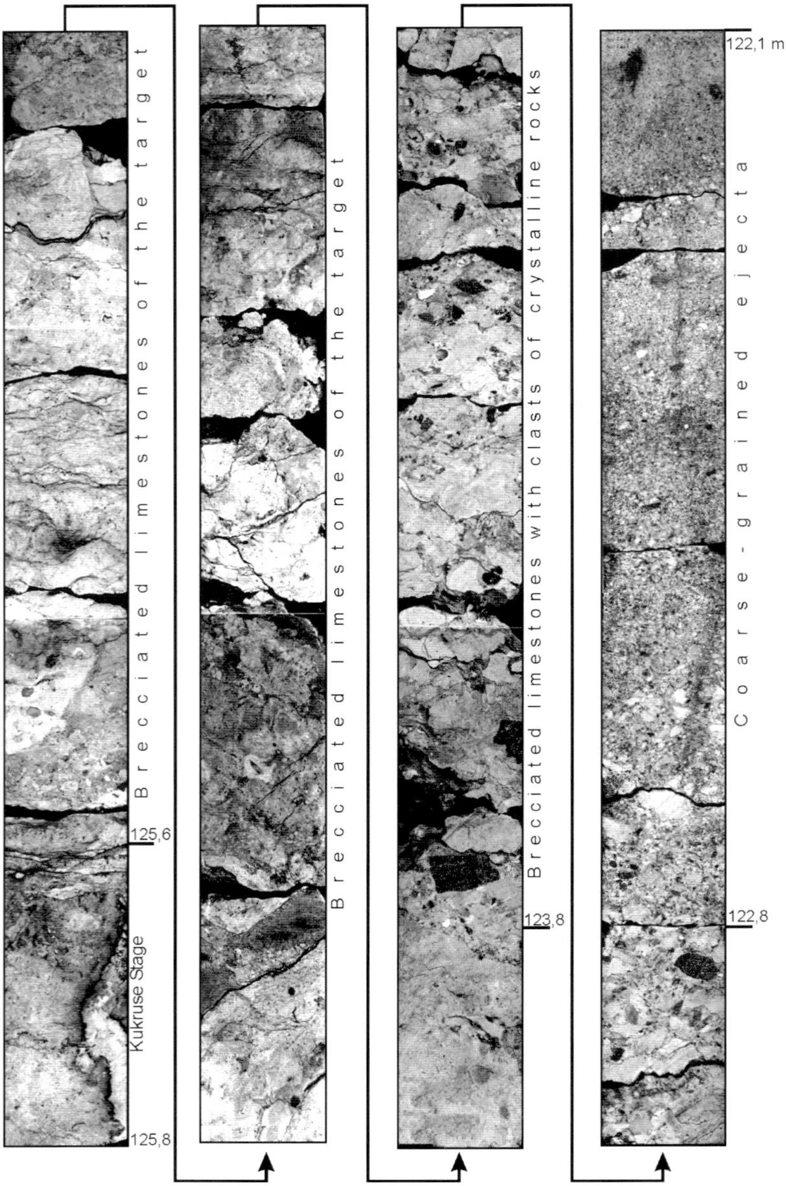

Fig. 7. Photo-log of drill core F-359 (5 km SW of the impact centre). The brecciated target limestones and the ejecta layer. 125.8-125.6 m – almost intact pre-impact Upper-Ordovician limestone of the Kukruse Stage with a hardground on the top; 125.6-123.8 m – brecciated by a subsurface release limestones of the target; 123.8-122.8 m – brecciated target limestones, containing clasts of crystalline rocks; 122.8-122.1 m – coarse-grained ejecta.

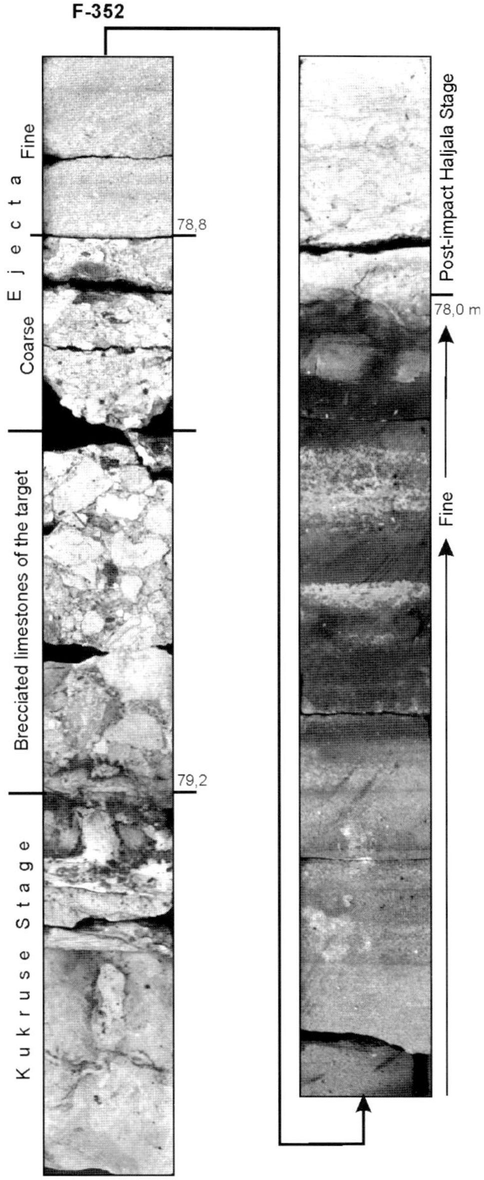

Fig. 8. Photo-log of drill core F-352 (6 km WSW from the impact centre). Dark grey – ejecta impregnated with hydrocarbons.

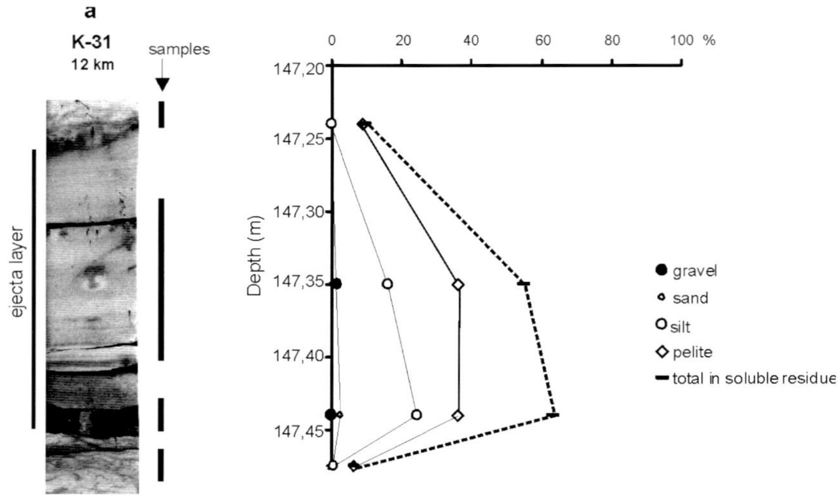

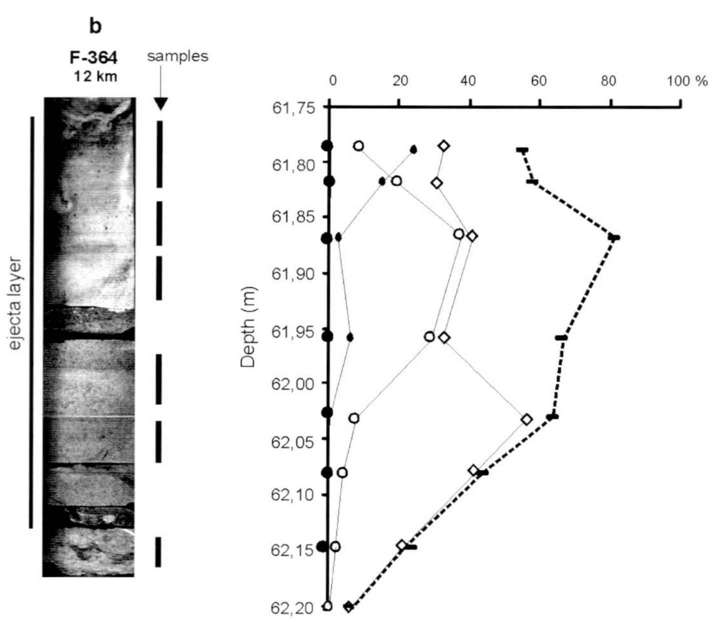

Fig. 9. Photo-logs of the ejecta layer and grain size and content of insoluble residue: a) drill core K-31; b) drill core F-364.

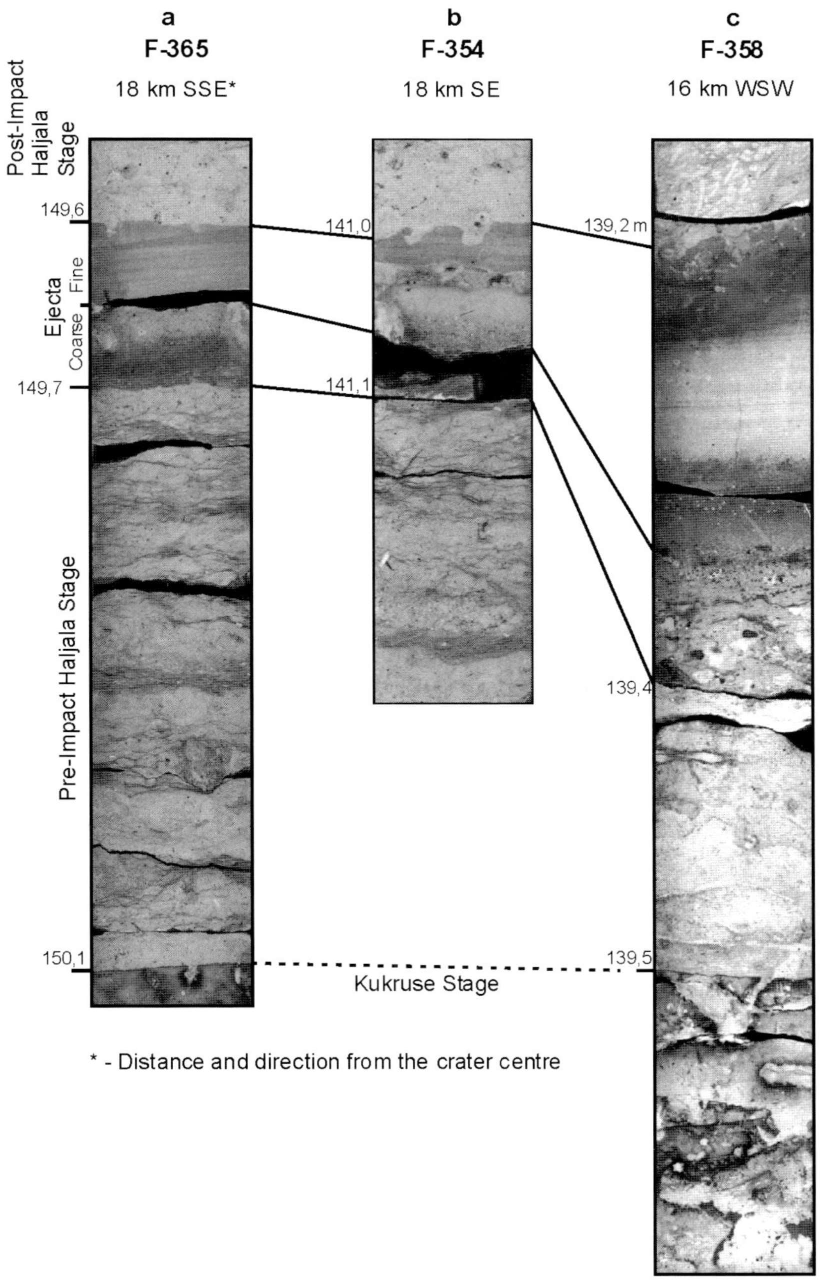

Fig. 10. Photo-logs of the ejecta layer from drill cores: a) F-365 (18 km SSE of impact centre); b) F-354 (18 km SE of impact centre); c) F-358 (16 km WSW of impact centre).

The content of carbonate (calcite) in the surrounding limestones is mostly more than 90%, while in the ejecta layer it decreases to 20–40%. Farther away from the impact centre the content of calcite in the layer increases and content and grain size of insoluble residue decrease. The mineral composition of the silt and sand fractions of insoluble residue in the layer becomes simpler farther from the impact centre, and in a vertical direction – from the top to base. In the upper part of the layer and far from the impact centre the insoluble residue consists mostly (ca. 90%) of quartz. The quartz grains with shock metamorphic features (PFs and PDFs) are abundant and the analysed fraction (1–1/8 mm) contains approximately 1% of them. In these, 6 different directions of lamella (3 per grain) are observed. According to some authors (Stöffler et al. 1975; Masitis et al. 1980; Stöffler and Langenhorst 1994; Stöffler and Grieve 1996) that may indicate a shock pressure of about 10 GPa or same as in the case of suevites from the Kärdla crater (Suuroja et al. 2002). The shocked quartz grains are mostly (95%) rounded or sub-rounded; angular and well-rounded grains are very rare.

The upper (finer) part of the ejecta layer, which precipitated from the debris-saturated water somewhat later, is separated from the lower (coarse) bed by a quite distinct boundary (Figs. 8 and 9). The substance of this sharp boundary is not clear yet, but it seems possible that deposition of the upper layer is connected with re-deposition of the primary ejecta. This presumption is supported by the observation that the upper part of the ejecta layer has sometimes fine-bedded textures. In the quartz of the ejected matter five sets (with a maximum of 3 orientations per grain) of PDFs with different crystallographic orientations are distinguished (Suuroja et al. 2001).

Ejecta deposition, except the nearest surroundings of the crater, took place on smooth sea bed at approximately constant depth (ca. 100 m). The tsunami caused by the impact did not affect the sea bed and bottom deposits farther than 10 km off the impact centre.

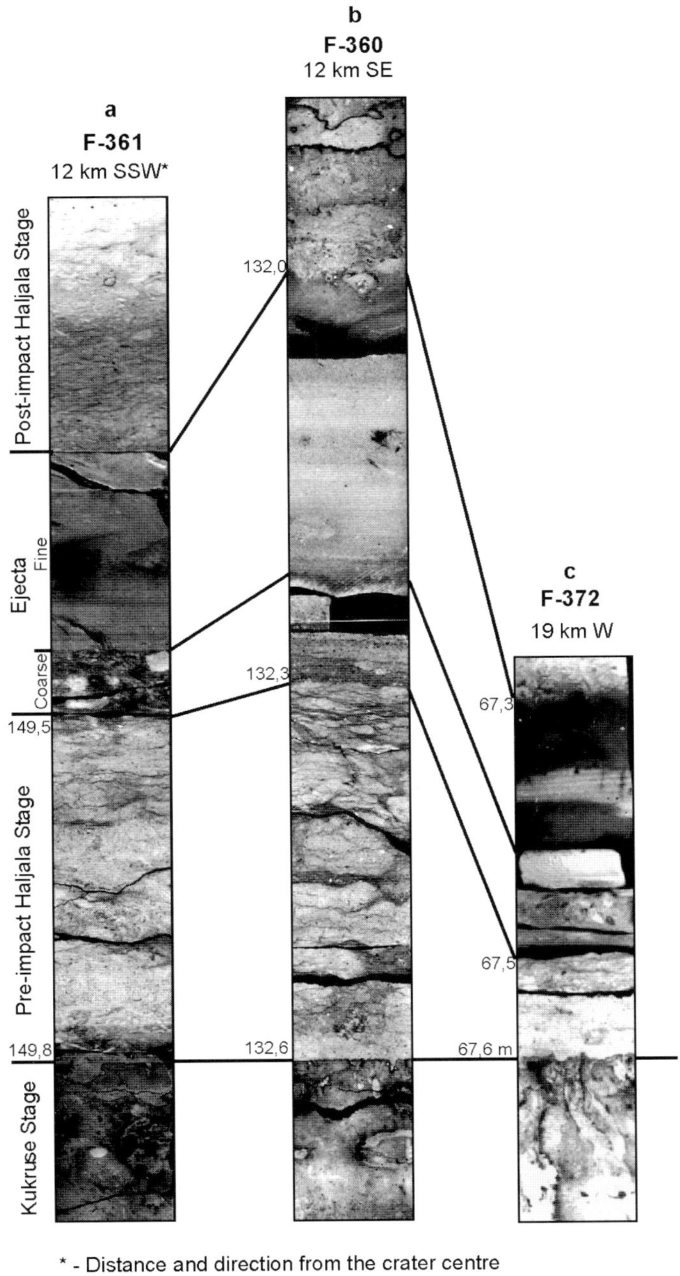

Fig. 11. Photo-logs of the ejecta layer from drill cores: a) F-361 (12 km SSW of impact centre); b) F-360 (12 km SE of impact centre); c) F-372 (18 km W of impact centre).

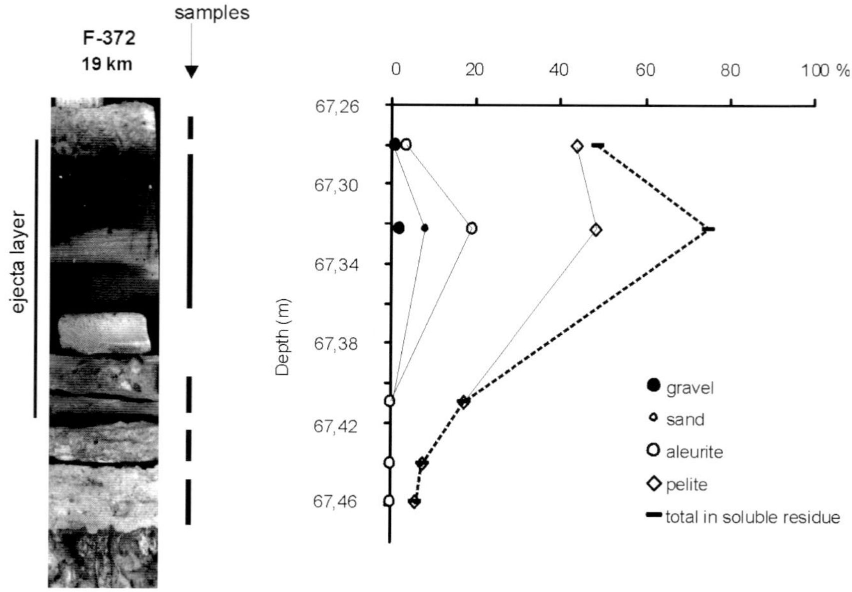

Fig. 12. Photo-log, grain size and content of insoluble residue from drill core F-372.

4 Discussion

The Kärdla impact took place 455 Ma ago (Puura and Suuroja 1992; Suuroja et al. 1994; Grahn et al. 1996) in a shallow epicontinental sea (ca. 100 m deep) where at that time, i.e., in the Upper Ordovician (Cowie and Basett 1989; Webby 1998) bioclastic debris-rich limy mud was deposited (Männil 1966; Jaanusson 1995; Nestor and Einasto 1997). In the described sea, 700–800 km from the Kärdla impact site, at approximately the same time one small-scale (Tvären) and one medium-scale (Lockne) meteorite impact took place (Lindström et al. 1992; Lindström et al. 1996; Ormö and Lindström 2000; Sturkell et al. 2000; Abels et al. 2000; Abels et al. 2003). Considering the distance and size of these meteorite impact structures it can be assumed that these were too small and the distance from the Kärdla impact site was too great to influence the deposition in the latter region.

By calculations the impactor ca. 200 m in diameter penetrated more than 100 m thick water layer and ca. 140-m-thick sedimentary cover and exploded in the uppermost part of the crystalline basement. The explosion which had a power of about 600 MT, removed more than 2×10^9 m^3 of crushed crystalline and sedimentary target rocks from the crater deep. In the result of this, a complex crater 4 km wide and more than 500 m deep, having a central uplift 130 m high and 600–700 m in diameter, was formed (Suuroja 2001). Due to the marine environment most impact-related deposits, among these the ejecta blanket, were buried and therefore are still preserved in an area of thousands of square kilometres around the crater. The ejecta blanket was eroded by short-term post-impact erosion from the rim wall and its outer slopes around the crater within ca. 4 km radius (Fig. 2).

Earlier optimistic suggestion (Põlma 1982; Hints 1997) that quartz sand in the bioclastic limestone of the Kisuvere Member (corresponding to the level of the post-impact limestones of the Haljala Stage) distributed more than 200 km east of the Kärdla crater in eastern Estonia, is in some way connected with the Kärdla impact, is not proved. Firstly, the sand from the Kisuvere Member does not contain shock metamorphosed quartz grains with PDFs, and secondly, the interval of the distribution of sand (up to 1 m) is too thick for such short-time and violent event as an impact. Also, the distance to the erosion area on the Baltic Shield, from where the siliciclastic matter was transported to the Kisuvere Member, was more than two times shorter than distance to the Kärdla impact site.

One possible site from which the shock metamorphosed quartz grains might have been transported to the Upper-Ordivician carbonate deposits of western Estonia could be the shock metamorphosed rocks of the Neugrund meteorite crater. The latter formed in the Early Cambrian (535 Ma) time and is relatively close (60 km NE) to the Kärdla impact site (Suuroja and Suuroja 2000; Suuroja et al. 2002). However, even closer to the Kärdla impact site (55 km NE of Osmussaar Island) cropped out the so-called Osmussaar breccia veins (Middle Ordovician, ca. 475 Ma) which too, contain shock metamorphosed matter (quartz grains with PDFs). The shock metamorphosed matter in the Osmussaar breccias as well as spatially and temporally closely related with them sandstones and sandy limestones of the Pakri Formation (Middle Ordovician, Kunda Regional Stage) are supposed to originate from the Neugrund impact structure area (Suuroja et al. 2003). Presence of the shock metamorphosed material of Neugrund origin in the limestones corresponding to age of the Kärdla impact is excluded, because the area of Neugrund impact structure and its surroundings, including the ejecta blanket, were at that time already buried under the cover of carbonate deposits.

It is difficult to calculate the initial volume and thickness (Fig. 3) of the ejecta blanket of the Kärdla impact because the impact took place in water, shortly after that the structure was eroded to a 6–8-km radius and finally it was buried. The calculations become even more complex considering that in the surroundings of the crater it is difficult to distinguish between the material formed in the result of subsurface release of sedimentary target rocks, in the result of resurging tsunami, in the result of deposition of the ejecta, and re-deposited matter from the eroded rim wall and the ejecta layer.

During a crater formation and excavation stage about 90% of all material excavated from a crater is deposited as a proximal ejecta (McGetchin et al. 1973; Oberbeck 1975; Melosh 1989; Koeberl and Martinez-Ruiz 2003; Dence 2002) at a distance equal to 3–4 radii of the crater. In the case of the Kärdla impact this distance may have been equal to 6–8 km and this is what we observed (Fig. 3). In Kärdla the proximal ejecta layer was strongly eroded by the resurge wave and post-impact erosion and therefore it has only partially preserved. In addition, closer to the rim wall it is difficult to differentiate between the primal proximal ejecta, re-deposited ejecta and the material carried to the deposits from the crater walls.

The origin of the up to 16 m thick bed of brecciated limestones lying at a distance of 3–6 km from the impact centre, on the top of the mostly intact (about 5 m from the 15 m thick layer of the limestones have been removed) complex of the target rocks, has been remained ambiguous (drill core K-14; Fig. 5). In the area where the upper part of this layer contains clasts of the crystalline rocks it is treated as a proximal ejecta layer. The lower part of this layer has formed in consequence of subsurface release as a result of reflection of a rarefaction wave.

The natural bitumens locally distributed in the sandy ejecta layer (Kattai et al. 1994; Suuroja et al. 1994; Suuroja 2002) are not authigenic and probably are of migratory origin. They are distributed at other stratigraphical levels and in other rock types (limestones) as well. There was a regional W-E or NW-SE direction flow of hydrocarbons which brought them not only to the island of Hiiumaa and to the surroundings of Kärdla crater, but also to some other sites on the eastern coast of the Baltic. The version of NW-SE migration is supported by the so-called "shade" of the Kärdla crater – a 15 km long oval area around the Kärdla crater where occurrences of natural bitumens (impregnation, liquid oil, asphalt) are missing. However, abundant occurrences of the natural bitumens are found on NW slope of the Kärdla crater. The exact time of migration of the hydrocarbons has not been identified but it must have been in the post-Silurian time because the natural bitumens impregnate the whole sequence of the Silurian limestones in this region (Suuroja et al. 1991).

The asymmetrical features (different height of the rim wall, elliptical shape of the ring fault) of the Kärdla structure (Suuroja et al. 2001), which according to the some authors (O´Keefe and Ahrens 1977; Deutsch and Langenhorst 1994; Artemieva 2002; Shuvalov 2003) imply an oblique impact, are not observed in the distribution of the ejecta blanket around the Kärdla crater.

The Kärdla impact was too small to cause substantial and long-term environmental changes (Ainsaar et al. 2002) and catastrophic shifts in the biosphere. However, the anomalous structure generated by the meteorite explosion (the highly uplifted rim wall and the deeply sunken crater proper) on the sea bed caused short-time anomalies in the sedimentation and changes in the biotic communities of pelagic organisms. For example, at the time (Upper Ordovician, Haljala time) when the ridge of the rim wall rose above the sea level and was eroded, graptolite-containing mud deposited in the about 300 m deep crater proper (Kala et al. 1971). Some millions of years later (Upper Ordovician, Oandu time) when around the rim wall's outer slopes one of Europe's earliest reef-like build-ups (the complex of skeletal grainstones consisting only of the fragments of shafts of the cystoids) formed, fossil-rich marls containing only the cup plates of these cystoids accumulated in the crater deep. Differences in the facies composition between the crater deep, rim wall area and surroundings of the crater lasted for about ten million years, up to middle of the Rakvere time (Caradoc, Upper Ordovician).

The ejecta blanket of the Kärdla impact event as well as many other small- and medium-scale impact events (Deutsch and Schärer 1994; Koeberl and Anderson 1996; Masaitis 1999; Gilmour and Koeberl 2000; Koeberl 2001; Koeberl and MacLeod 2002; Masaitis 2002; Gurov et al. 2002; Gurov et al. 2003; King and Petruny 2003; Valter and Plotnikova 2003) have been a good, but unfortunately of only local importance, time-marker in a biomorphic-matter-rich sequence of the marine deposits.

Acknowledgements

We are grateful to senior geologist Elmar Kala for long-time assistance. He initiated geological investigations in the area of Kärdla impact structure, and within these studies in 1969–1971 the drill holes were made, which enabled to identify the crater-like shape of the anticlinal Paluküla structure (Viiding et al. 1969). Special thanks go to Alla Popova who carried out mineralogical and grain size analyses at the Laboratory of the Geological Survey of Estonia. The Impact Programme of the European Science Foundation provided all-round support. We also thank V. Puura for helpful comments and suggestions. The investigation was partially founded by Estonian Science Foundation grants 5192 and 4417.

References

Abels A, Zumsprekel H, Bischoff L (2000) Basic remote sensing signatures of large, deeply eroded impact structures. In: Gilmour, Koeberl C (eds), Impacts and the Early Earth, Springer Verlag, Berlin Heidelberg, Lecture Notes in Earth Sciences 91: 309-326

Abels A, Plado J, Pesonen LJ (2003) The Impact Cratering Record of Fennoscandia – A close look at the database. In: Plado J, Pesonen, L (eds) Impacts in Precambrian Shields, Impact Studies vol. 2, Springer Verlag, Berlin Heidelberg pp 1-58

Ainsaar L, Suuroja K, Semidor M (2002) Long-term effect of the Kärdla crater (Hiiumaa, Estonia) on Late Ordovician carbonate sedimentation. Deep-Sea Research, Part II 49/6: 1145-1155

Artemieva N (2002) Tektite origin in oblique impacts: numerical modelling of the initial stage. In: Plado J, Pesonen LJ (eds) Impacts in Precambrian Shields. Impact Studies vol. 2, Springer Verlag, Berlin Heidelberg, pp 257-276

Cowie JW, Bassett MG (1989) Global stratigraphic chart with geochronometric and magnetostratigraphic calibration. Episodes 12 (2), Supplement.

Dence M (2002) Re-examining structural data from impact craters on the Canadian Shield in the light of theoretical models. In: Plado J, Pesonen LJ (eds) Impacts in Precambrian Shields, Impact Studies vol. 2, Springer Verlag, Berlin Heidelberg, pp 59-79

Deutsch A, Langenhorst F (1994) Geological formations in and around impact structures. In: Marfunin AS (ed), Mineral Matter in Space, Mantle, Ocean Floor, Biosphere, Environmental Management, and Jewelry, Springer Verlag, Berlin, pp 89-95

Deutsch A, Schärer U (1994) Dating terrestrial impact events. Meteoritics 29: 301–322

Gilmour I, Koeberl C (eds) (2000) Impacts and the Early Earth. Lecture Notes in Earth Sciences, Vol. 91, Springer Verlag, Berlin-Heidelberg, 455 pp

Grahn Y, Nõlvak J, Paris F (1996) Precise chitinozoan dating of Ordovician impact events in Baltoscandia. Journal of Micropalaeontology 15: 21-25

Gurov EP, Gurova EP, Sokur TM (2002) Geology and Petrography of the Zapadnaya impact crater in the Ukrainian Shield. In: Plado J, Pesonen LJ (eds) Impacts in Precambrian Shields, Impact Studies vol. 2, Springer Verlag, Berlin Heidelberg, pp 117-171

Gurov EP, Kelley SP, Koeberl C (2003) Ejecta of the Boltysh impact crater in the Ukrainian Shield. In: Koeberl C, Martinez-Ruiz F (eds) Impact Markers in the Stratigraphic Record, Impact Studies vol. 3, Springer Verlag, Berlin Heidelberg, pp 179-202

Hints L (1997) Haljala Stage. In: Raukas A, Teedumäe A (eds) Geology and mineral resources of Estonia. Estonian Academy Publishers, Tallinn, 73-74

Jaanusson V (1995) Confacies differentiation and upper Middle Ordovician correlation in the Baltoscandian Basin. Proceedings of the Estonian Academy of Sciences, Geology 44: 73-86

Kala E, Kajak K, Kajak H, Elterman G (1971) Report of the integrated geological mapping of Island Hiiumaa at a scale of 1:200 000 (in Russian). Geological Survey of Estonia, Keila

Kala E, Suuroja K, Tassa V (1976) Report of the prospecting granitic rock of Paluküla (in Russian). Geological Survey of Estonia, Tallinn, 94 pp

Kattai V, Lokk U, Suuroja K (1994) The distribution of natural bitumens in Estonia. Bulletin of the Geological Survey of Estonia 4/1: 12-16

King DT, Petruny LW (2003) Application of stratigraphic nomenclature to terrestrial impact-derived and impact related materials. In: Koeberl C, Martinez-Ruiz F (eds) Impact Markers in the Stratigraphic Record, Impact Studies v. 3, Springer Verlag, Berlin Heidelberg, pp 41-64

Koeberl C (2001) The sedimentary record of impact events. In: Peucker-Ehrenbrinck B, Schmitz B (eds) Accretion of Extraterrestrial Matter throughout Earth's History. Kluwer Academic/Plenum Publishers, pp 333-378

Koeberl C, Anderson RR (1996) Manson and company: Impact structures in the United States. In: Koeberl C, Anderson RR (eds) The Manson Impact Structure, Iowa, Anatomy of an Impact Crater, Geological Society of America, Special Paper 302: 1-29

Koeberl C, MacLeod K (eds) (2002) Catastrophic Events and Mass Extinction: Impacts and Beyond. Geological Society of America, Special Paper 356, 746 pp

Koeberl C, Martinez-Ruiz F (2003) The Stratigraphic Record of Impact Events: A Short Overview. In: Koeberl C, Martinez-Ruiz F (eds) Impact Markers in the Stratigraphic Record, Impact Studies v. 3, Springer, Berlin-Heidelberg, pp 1-40

Lindström M, Floden T, Puura V, Suuroja K (1992) The Kärdla, Tvären and Lockne craters – possible evidences of an Ordovician asteroid swarm. Proceedings of the Estonian Academy of Sciences, Geology 41: 45-53

Lindström M, Sturkell EFF, Törnberg R, Ormö J (1996) The marine impact crater at Lockne, central Sweden. GFF 118: 193-206

Männil R (1966) Evolution of the Baltic Basin during the Ordovician (in Estonian with English summary). Valgus, Tallinn, 248 pp

Masaitis, VL (1999) Impact structures of northern Eurasia: The territories of Russia and adjacent counties. Meteoritics and Planetary Science 34: 691-711

Masaitis VL (2002) The middle Devonian Kaluga impact crater (Russia): new interpretations of the marine setting. Deep-Sea Research. Part II 49: 1157-1169

Masaitis VL, Danilin AN, Maschak MS, Raykhlin AI, Selivanovskaya TV, Shadenkov E (1980) Geology of the astroblemes (in Russian), Nedra Press, Leningrad, 231 pp

McGetchin TR, Settle M, Head JW (1973) Radial thickness variation in impact crater ejecta: Implications for lunar basin deposits. Earth and Planetary Science Letters 20: 226-236

Melosh H J (1989) Impact Cratering. A Geologic Process, Oxford University Press, 245 pp

Nestor H, Einasto R (1997) Ordovician and Silurian sedimentation basin. In: Raukas A, Teedumäe A (eds) Geology and mineral resources of Estonia. Estonian Academy Publishers, Tallinn, 192-204

Oberbeck VR (1975) The role of ballistic erosion and sedimentation in lunar stratigraphy. Reviews of Geophysics and Space Physics 13: 337-362

O'Keefe JD, Ahrens TJ (1977) Meteorite impact ejecta: Dependence of mass and energy loss on planetary escape velocity. Science 198: 1249-1251

Ormö J, Lindström M (2000) When cosmic impact strikes the sea bed. Geological Magazine 137: 67-80

Plado J, Pesonen LJ, Elo S, Puura, V, Suuroja K (1996) Geophysical research on the Kärdla impact structure, Hiiumaa Island, Estonia. Meteoritics and Planetary Science 31: 289-298

Puura V, Floden T (1997) The Baltic Sea drainage basin – a model of a Cenozoic morphostructure reflecting the early Precambrian crustal pattern. Sveriges geologiska undersökning 86: 131- 137

Puura V, Suuroja K (1992) Ordovician impact crater at Kärdla, Hiiumaa Island, Estonia. Tectonophysics 216: 143-156

Põlma L (1982) Comparable lithology of the Ordovician carbonatic rocks of the North and Central Baltic (In Russian). Valgus, Tallinn 152 pp

Remane J. (2000) International Stratigraphic Chart. International Union of Geological Sciences.

Remane J, Bassett MG, Cowie JW, Gohrbandt KH, Lane HR, Michelson O, Wang N (1996) Revised guidelines for the establishment of global chronostratigraphic standards by the International Commission on Stratigraphy (ICS). Episodes 19: 77–81

Salvador A (ed) (1994) International Stratigraphic Guide: A guide to stratigraphic classification, terminology, and procedure. Second edition. International Union of Geological Sciences (IUGS) and Geological Society of America and (GSA), 214p.

Shuvalov (2003) Displacement of Target Material During Impact Cratering. In: Koeberl C, Martinez-Ruiz F (eds) Impact Markers in the Stratigraphic Record, Impact Studies vol. 3, Springer, Berlin-Heidelberg, pp 121–135

Stöffler D, Grieve RAF (1996) Classification and nomenclature of impact metamorphic rocks: a proposal to the IUGS Subcommission on the Systematic of Metamorphic Rocks. Lunar and Planetary Science [abs.] 25: 1347-1348

Stöffler D, Langenhorst, F (1994) Shock metamorphism of quartz in nature and experiment: I. Basic observation and theory. Meteoritics 29: 155-181

Stöffler D, Gault DE, Wedekin J, Polkovski G (1975) Experimental hypervelocity impact into quartz sand: Distribution and shock metamorphism of ejecta. Journal of Geophysical Research 80: 4062-4077

Sturkell E, Ormö J, Nõlvak J, Wallin Å (2000) Distant ejecta from the Lockne marine-target impact crater, Sweden. Meteoritics and Planetary Science 35: 929-936.

Suuroja K (2001) Kärdla Meteorite Crater. Geological Survey of Estonia, Tallinn, 38 pp

Suuroja K (2002) Natural resources of the Kärdla impact structure, Hiiumaa Island, Estonia. In: Plado J, Pesonen, L (eds) Impacts in Precambrian Shields, Impact Studies v.2, Springer Verlag, Berlin-Heidelberg, pp 295-306

Suuroja, K., Suuroja, S. (2000). Neugrund Structure – the newly discovered submarine early Cambrian impact crater. In: Gilmour I, Koeberl C (eds) Impacts and the Early Earth. Springer Verlag, Berlin-Heidelberg. Lecture Notes in Earth Sciences, Vol. 91: 389-416

Suuroja K, Kadastik E, Mardim T (1994) The geological mapping of Island of Hiiumaa at a scale of 1:50 000. Six maps with explanatory note (in Estonian with English summary). Geological Survey of Estonia, Keila, 206 pp

Suuroja K, Koppelmaa H, Niin M, Kivisilla J (1991) Geological mapping of the crystalline basement on Island of Hiiumaa at a scale of 1:200 000. Two maps with explanatory note (in Russian). Geological Survey of Estonia, Keila, 291 pp

Suuroja K, Suuroja S, All T, Floden T (2001) Kärdla (Hiiumaa Island, Estonia) – the buried and well-preserved Ordovician marine impact structure. Deep-Sea Research Part II 46/2: 1121-1144

Suuroja S, All T, Plado J, Suuroja K (2002) Geology and magnetic signatures of the Neugrund impact structure, Estonia. In: Plado J, Pesonen, L (eds) Impacts in Precambrian Shields, Impact Studies vol. 2, Springer Verlag, Berlin Heidelberg, pp 277-294

Suuroja K, Kirsimäe K, Ainsaar L, Kohv M, Mahaney W, Suuroja S (2003) The Osmussaar breccia in northwestern Estonia – evidence of ca 475 Ma Earthquake or an Impact? In: Koeberl C, Martinez-Ruiz F (eds) Impact Markers in the Stratigraphic Record, Impact Studies vol. 3, Springer, Berlin-Heidelberg, pp 333–347

Tammekann A (1940) The Baltic Glint. I. Morphography of the Glint. Publicationes Instituti Universitatis Tartuensis Geographici 24: 104 pp

Valter A, Plotnikova L (2003) Biostratigraphic indications of the age of the Boltysh impact crater, Ukraine. In: Koeberl C, Martinez-Ruiz F (eds) Impact Markers in the Stratigraphic Record, Impact Studies vol. 3, Springer Verlag, Berlin-Heidelberg, pp 163-178

Viiding H, Kala E, Pobul E (1969) The mystery of Paluküla find solution (in Estonian with English summary). Eesti Loodus 8: 464-474

Webby BD (1998) Steps toward a global standard for Ordovician stratigraphy. Newsletter of Stratigraphy 36: 1-33

Sediments and Impact Rocks Filling the Boltysh Impact Crater

Eugene P. Gurov[1], Simon P. Kelley[2], Christian Koeberl[3], Natalia I. Dykan[1]

[1]Institute of Geological Sciences National Academy of Sciences of Ukraine, 55–b Oles Gonchar Street, Kiev 01054, Ukraine
[2]Department of Earth Sciences, Open University, Milton Keynes MK 6AA, England (s.p.kelley@open.ac.uk)
[3]Department of Geological Sciences, Althanstrasse14, A–1090 Vienna, Austria

Abstract. Ar-Ar dating of impact melts extracted from boreholes into the crater floor indicate that the Boltysh impact crater formed on the Ukrainian Shield at 65.17 ± 0.64 Ma, an age that is indistinguishable from that of the Cretaceous-Tertiary (K/T) boundary and formation time of the giant Chicxulub impact crater. Unfortunately almost all information relating to the drilling along with much of the actual core has been lost. We have studied the remaining core to in an attempt to illuminate the post-impact evolution of this critical crater.

The Boltysh crater formed in sub-aerial conditions of the central part of the Ukrainian Shield. Immediately after the impact, unconsolidated clastic sediments derived from the crater walls were deposited, and succeeded by slow sediment accumulation at the bottom of an anoxic crater lake into the Paleogene and even the Neogene. The crater wall was breached during marine transgression in the mid-Eocene and there followed a period of marine sedimentation. Marine regression during the Late Oligocene/Early Neogene led to the return of sub-aerial conditions and later a shallow fresh water lake again filled the crater. A final covering of loess in the Quaternary erased all surface traces of the crater which remains completely buried apart from occasional exposures of the ejecta in river valleys.

The crater sediments are rich in the remains of flora and fauna that teemed in the lake, and the excellent sediment preservation of sediments in the now buried crater provides an exceptional opportunity for the investigation of the burial history of a terrestrial impact crater and the evolution of the Ukrainian Shield area surrounding the crater area over a period of 65 million years.

1
Geology and age of the Boltysh impact crater

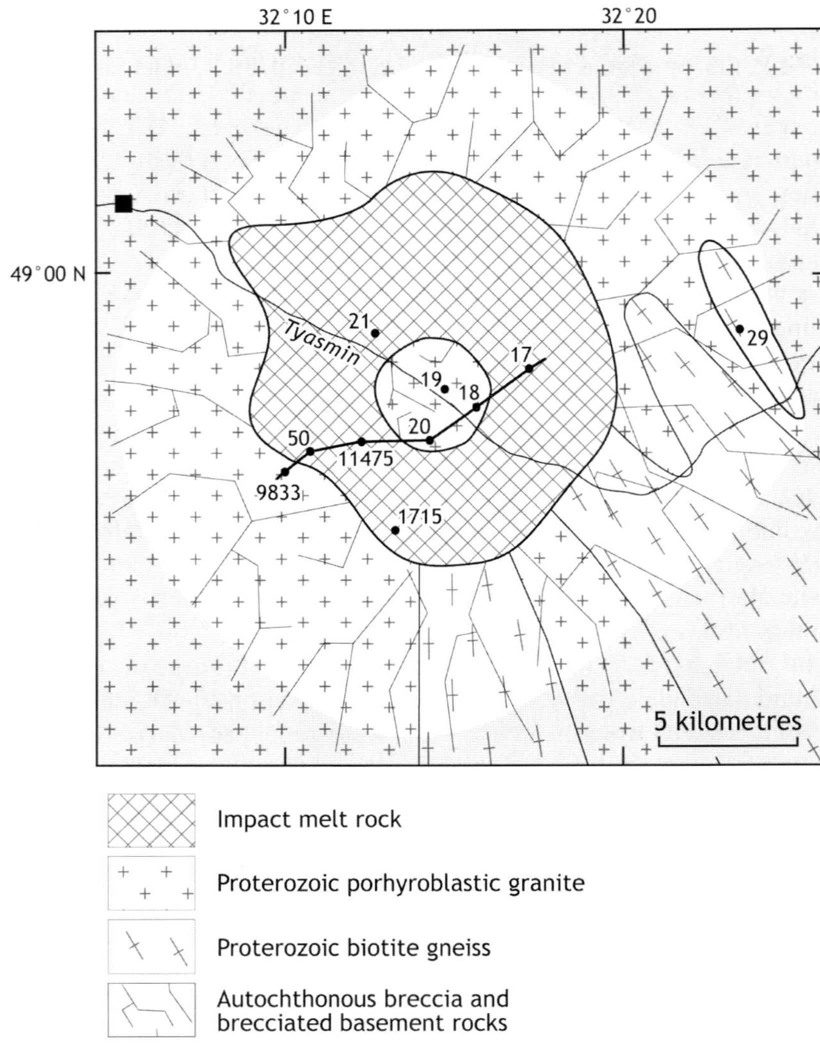

Figure. 1. A schematic map of the Boltysh impact structure. The annular impact-melt sheet occupies the inner crater around the central uplift. The post-impact sediments are omitted. The locations of some drill cores and position of cross-section I–I are indicated. (Fig. 2). The estimated outcrop of the eroded original rim is shown by a darker shading.

The Boltysh crater was formed in the crystalline basement of the Ukrainian Shield, which at the time of impact probably partly covered by a thin veneer of the late Cretaceous fine-grained siliciclastic and carbonate rocks which have been preserved only as fragments in breccias associated with the impact (Valter and Plotnikova 2003). The basement rocks of the region are Proterozoic porphyroblastic granites known locally as Kirovograd type, with ages of ca. 1550 Ma, and older biotite gneisses (ca. 1850 – 2220 Ma, Shcherbak et al. 1978). In previous work, we described the age of Boltysh (Kelley and Gurov 2002) and its ejecta which covered an area of at least 25,000 km^2 (Gurov et al. 2003), here we describe the crater fill of impact melt rocks and post impact sediments.

In geologically recent times the crater and its post-impact deposits became covered with Quaternary sediments up to 30 m thick and now have a very little surface expression. However, the deeper structure of the crater and the composition of crater filling sediments is known through numerous drill cores and geophysical investigations, which were undertaken in the 1960's and 1980's in the search for hydrocarbons (Fig. 1). We have reconstructed the evolution of post-impact sedimentation within the crater using the remaining core and descriptions of core from other boreholes in the structure.

The Boltysh impact structure is a complex crater, 24 km in diameter, with a central high, and a maximum depth to the crater floor of about 1 km around the central high. The crater floor is defined by transition from fractured granites to allochthonous fine grained breccia and lies at a depth of 1065 m in borehole No. 11475 (Figs. 1, 2), not all of the other boreholes penetrated to this depth. The height of the central high (seen in boreholes No. 18, 19, and 20) is elevated 60-80m relative to the surface of impact melt in borehole No. 50 and is covered by a thin veneer of suevite. The inner crater (about 12 km in diameter, Figs. 1, 2) is filled with impact melt rocks, suevites, and lithic breccias (Gurov and Gurova 1991; Gurov at al. 2003). The impact melt rocks form an annular sheet, 12 km in diameter and up to 220 m thick, surrounding the central uplift. The surface of the melt sheet seems to be sub-horizontal in boreholes No. 21 (581.5 m), No. 50 (593.5 m), No. 11475 (576 m) and No. 9814 (575 m), a strong evidence that impact-induced melt formed a "lake" in the deepest part of the crater immediately after the impact. The thicknesses of suevite covering the impact melt ranges from just 12 m in hole No. 11475, to 22 m in hole No. 50 west of the central uplift and as much as 97 m in hole No. 17 in the east (Fig. 2).

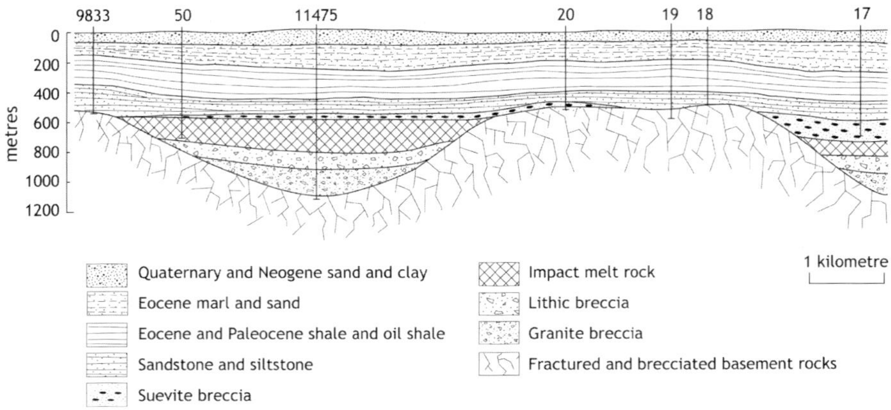

Figure. 2. Cross–section through the central part of the Boltysh structure (I–I), showing locations of boreholes and depths of sedimentary lithologies. The location of the cross–section is indicated in Fig.1.

Two main types of impact melt rocks make up the melt sheet (Fig. 3) (Grieve et al. 1987; Gurov and Gurova 1991). The lower part of the sheet is formed of melt rocks with a glassy matrix that occur in the intervals 653 – 736 m in the core No. 50 and 657 – 791 m in the core No. 11475. The upper part of the sheet is composed of microcrystalline impact melt rocks and a variably thick layer of suevite (Fig. 3).

Impact melt rocks of the lowermost horizon are microporphyritic, with phenocrysts of plagioclase and hypersthene in a glassy matrix (Fig. 4a). The glass varies from fresh yellowish isotropic glass to intensively devitrified brown glass with spots of a weak birefringence and microlites of pyroxene. Plagioclase (labradorite composition) forms prismatic, often hollow crystals that are up to 1 mm long. Two generations of hypersthene form in the melt. Prismatic phenocrysts up to 1.5 mm long form the first generation, and later 0.1mm microlites are found in the glass. The content of hypersthene of the first generation corresponds to enstatite content from 30 to 50 %, while its content in the latest hypersthene is from 50 to 73 % (Grieve et al. 1987; Gurov and Gurova 1991). Xenoliths and xenocrysts in the melt are generally quartz with rare ballen structure and partially melted granite clasts. Shocked quartz with PDFs is preserved in the marginal zone of the melt sheet up to around 25 m from its lower contact with breccias.

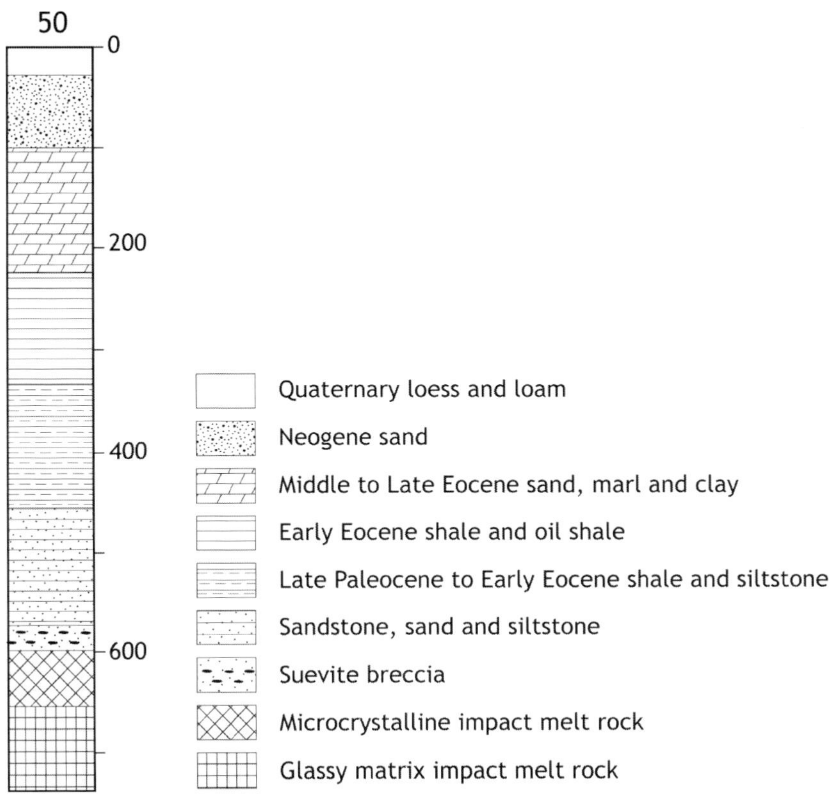

Figure. 3. A stratigraphic column of drill core No. 50 (drilled to 736 m). The location of the drill hole is shown in Figs. 1 and 2.

Impact melt rocks of the upper horizon are microporphyritic rocks with fine grained matrix containing microlites of feldspars and biotite, the latter forming pseudomorphs of pyroxene. Feldspars are prismatic zoned crystals with cores of andesine and, where present, rims of anorthoclase (Grieve et al. 1987). The matrix is composed of fine-grained to cryptocrystalline aggregates of feldspars and quartz forming spherulitic and microprismatic structures (Fig. 4b). The impact melt rocks contain the numerous xenoliths of shocked quartz. Xenoliths of highly shocked and selectively melted granites, up to 2 m in diameter, occur in the interval from 645 to 620 m in core No. 50.

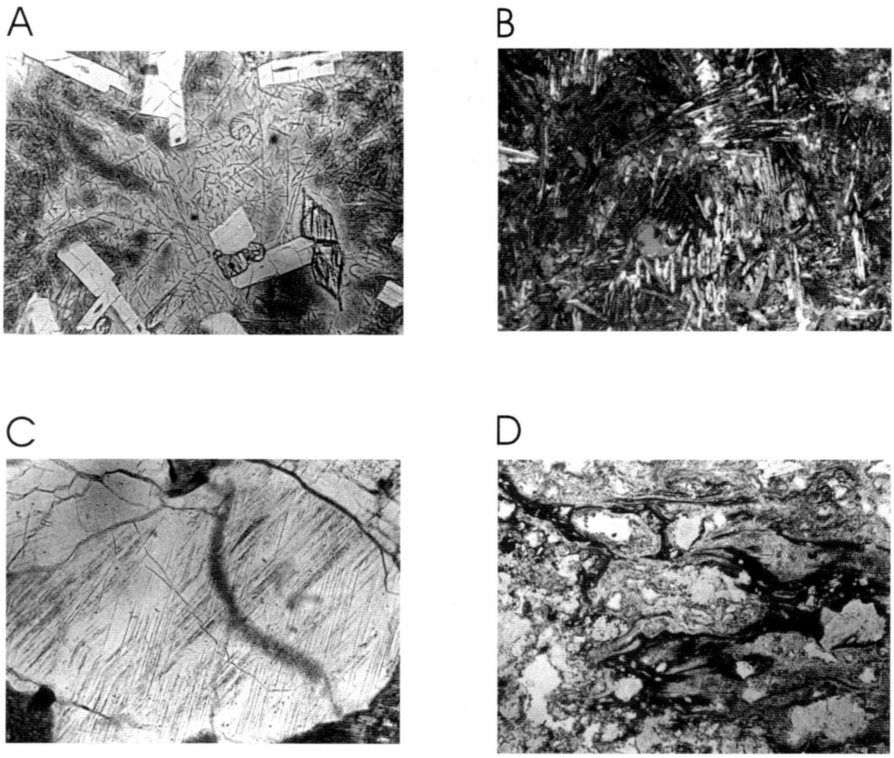

Figure. 4. Structures within impact–melt rocks, breccias, and suevites of the Boltysh structure (drill core No. 50).

a. A Glass rich matrix impact–melt rock with microlites of plagioclase and pyroxene. The matrix is partly devitrified (dark spots) and contains microlites of pyroxene (sample 698 m) Field of view is 1.7 mm wide, viewed with parallel polars.

b. A Microcrystalline impact–melt rock. The main components of the matrix are prismatic microlites of feldspars (sample 635 m). The field of view is 3.4 mm, viewed with crossed polars.

c. Several systems of PDFs seen in quartz from a lithic breccia sample underlying impact–melt rocks (drill core 11475, sample 859 m) Field of view is 1.1 mm wide, viewed with parallel polars.

d. Fluidal red colored glass in suevite with numerous clasts (drill core No. 50, sample 577 m). Field of view is 2.0 mm wide, viewed through parallel polars.

Suevites occur in core No. 50 in the interval from 593.5 to 573 m, and they are composed of glass and granite clasts cemented by fine-grained detrital material of the same composition. No sedimentary xenoliths occur in suevites, of in impact melt rocks. The aerodynamic shapes of some glassy fragments seem to indicate that the suevites are fallback rocks deposited on the surface of impact melt sheet. The upper horizon of the suevites is intensely altered.

The composition of the main types of impact melt rocks and suevites in the crater are shown in Table 1. They exhibit relatively constant compositions that are similar to the composition of the crystalline rocks of the crater basement, roughly in the proportion 5 parts of the Kirovograd granites to 1 part of biotite gneisses of the target rock (Gurov and Gurova, 1991; Masaitis et al., 1980). Note however, that the Fe_2O_3/FeO ratio varies from 0.50 in granites of the crater basement, to 0.32 in the melt rocks with glassy matrix. The degree of iron oxidation is even higher at 1.10 in the microcrystalline melt rocks, and 1.37 in the suevites. The red color of the suevites in core No. 50 in the interval from 573 m to 583 m results from hematite formation during alteration and devitrification of glass.

Another significant difference between the target rocks and impact rocks is that the potassium content appears to be lower in the impact melt rocks, especially in melt rocks with a glassy matrix, whereas in suevites have higher potassium contents. It is possible that the differences in the potassium content are due to its partial redistribution during the melt sheet formation (Gurov and Gurova 1991).

Concentrations of some trace elements in impact melt rocks of the Boltysh crater have been analyzed in an attempt to detect trends which might reflect the composition of the projectile (Grieve et al. 1987; Gurov et al. 1986). Preliminary analyses show an enrichment of the melt with Cr and Ni relative to the basement granites (Table 1), but further investigations will be necessary to confirm these trends before any clear conclusions can be drawn.

The various styles of impact rock in Boltysh all exhibit prominent signs of shock metamorphism in their constituent minerals, including ballen structures and planar deformation features (PDFs) in quartz (Fig. 4c), PDFs in feldspars (Gurov et al.1979; Gurov and Gurova 1991; Masaitis 1974; Masaitis et al. 1980; Ryabenko et al. 1982; Valter and Ryabenko 1977), diaplectic glasses of feldspars, and finally kink bands and PDFs in biotite (Gurov 1977; Gurov and Gurova 1991). In addition, coesite has been detected in allochthonous breccia from core No. 11475 at a depth of 908 m (Gurov et al., 1980).

The peripheral zone of the crater is a shallow annular depression around the inner crater. The border between the peripheral zone and the deep inner crater is defined by a sharp change of slope in the crater floor (Fig. 2). The depth of the crater floor in the annular depression ranges from around 500 m near the limits of the inner crater, and decreases slowly to the partially eroded outer rim. The floor of the peripheral depression consists of brecciated basement rocks. The original crater rim is intensively eroded (and material was probably deposited within the inner crater), and only the remnants of the rim are exposed on the surface at the northwestern limit of the structure in the Tyasmin river valley (Fig. 1). In this area the river cuts a steep valley through the shattered porphyroblastic granites of the crater rim (Fig. 5).

Figure. 5. Exposures of cataclased porphyroblastic granites of the intensively eroded original rim of the Boltysh structure. The Tyasmin river cuts a valley through the rim. The height of rock exposure is around 5 m.

The Boltysh crater is filled with post–impact sediments described in detail below, mainly based upon cored samples from drill hole No. 50 in the west of the crater. The initial post–impact filling reflects deposition in a closed fresh–water basin, and the main constituent sediments are argillites, siltstones, sandstones, sands, and oil shales.

Until recently, radiometric dating of Boltysh impact-melt rocks was limited to determinations by the K-Ar and fission track methods. The earliest investigations by the fission track method of two samples of glassy impact-melt yielded an age of 65.04 ± 1.10 Ma (1σ errors) (Kashkarov et al. 1998). Recent $^{40}Ar/^{39}Ar$ dating analyses of seven Boltysh impact-melt rocks from borehole No. 50, using a laser step-heating technique, help to constrain the crater age to 65.17 ± 0.64 Ma (the error is the 95% confidence level, using the square root of the MSWD to enhance errors; cf.

Ludwig 1999) (Kelley and Gurov 2002). Moreover, recent work on the biostratigraphic position of the Boltysh ejecta is in good accordance with radiometric dating of the impact-melt rocks (Valter and Plotnikova 2003). Valter and Plotnikova (2003) showed that the position of the ejecta outside the crater constrains the age of the impact to between Upper Maastrichtian (based upon flora and fauna in sedimentary dykes in the basement underlying the impact ejecta) and Earliest Paleocene (based upon nannoplankton and plankton foraminifera of zones NP1-NP4 in clay and carbonate rich sands overlying the ejecta).

2 Post–crater sedimentation in the Boltysh impact crater

The complete preservation of the Boltysh impact structure and its post–crater sediments offers an excellent opportunity to study a continuous section revealing the history of crater filling from the K/T boundary to the Quaternary. The present data are based mainly on study of post–crater sediments of the drill core No. 50, located about 5 km to the southwest from the crater center (Fig. 1,3). The total thickness of post–impact sediments in this core is 573.5 m. Sediments overlie 22 m of fall back suevite which in turn overlie the impact melt sheet. Similar thicknesses of the post–crater filling have been measured in the deepest central part of the structure around the central uplift in other drill cores such as No. 11475 (549 m), No. 17 (580.4 m), and No. 21 (534 m). The thickness of post–crater sediments is reduced above the top of the 4-km-diameter central high, where 466 m of sediments were found in drill core No. 18, 497 m in core No. 19, and 467 m in core No. 20. The shallow inclination of the post–impact sediments in the peripheral zone of the crater has been determined from studies of borehole samples (Bass et al. 1967; Valter and Ryabenko 1977; Gurov and Gurova 1991; Gurov et al. 2003). For example, the thickness of post–impact sediments is just 70 m in bore hole No. 29, situated 1.5 km from the E–N–E limit of the eroded crater (Fig. 1). The predominant type of post–crater sediments here are coarse-grained sandstones with no evidence of macrofossils. Those rocks contain weakly rounded gravel grains indicating that the earliest sediments were water lain in the peripheral zone of the crater. The crater floor in this peripheral zone comprises autochthonous monomict breccia and brecciated allochthonous basement rocks.

Post-impact sedimentation appears to have initiated in the deepest parts of the inner crater around the central uplift. The detritus that accumulated there was derived from poorly consolidated impact rocks from the crater rim and the inner slopes of the crater, consisting of suevites, breccias, and brecciated basement rocks. Fall–back suevites are not well represented in the peripheral part of the crater, but they probably covered the whole area in the immediate aftermath of the impact since they occur in the deepest part of bore holes in the central part of the crater overlying the surface of impact–melt sheet. The rocks of the central high may have been another source of the clastic material during the early stage of sedimentation. The uppermost part of the central high, exposed to the atmosphere after impact, is composed of altered and weathered lithic breccias (drill cores No. 18, 19) and suevites (drill core No. 20). Alteration effects seen in rocks of the central uplift consist of partial replacement of feldspars by clay minerals and chlorite.

Immediately after its formation, the Boltysh impact structure probably formed a dry circular basin surrounded by a high rim. Post-impact sedimentation began in the deepest area of about 90 km^2 within the inner crater around the central high onto the hot surface of impact melt and suevites (Table 2).

The first post–crater deposits occur in the drill core No. 50 on the surface of glass–rich suevites (Fig. 3). Strongly weathered and altered suevites form the uppermost horizon of impactites at depths of 573–576.5 m in core No. 50. These uppermost impact-related rocks consist of fragments of greenish–gray weathered glass, shocked crystalline rocks and minerals. Minerals in suevites are moderately to highly shocked quartz and feldspars. The greenish color of the rocks is related to the thin grained chlorite in glass and cement. Glass particles from 0.5 – 1 mm up to 20 – 30 mm are the main component of suevites. Altered suevites pass down to more dense suevites with xenoliths of rose and reddish flattened devitrified glass clasts up to 5–7 cm in diameter in the interval 576.5–583 m (Fig. 4d.), and some glass clasts are vesicular. The red coloration of the devitrified sueviteswas probably the result of interaction with surface water, and no fresh glass was preserved in this upper layer. However, it is unclear whether the upper levels were subject to surface water percolation, suffered some redistribution by water, or merely lay beneath the surface of post impact crater lake. The ratio of Fe_2O_3 to FeO in these altered suevites ranges from 1.37 to 1.6 compared to 0.57 in the less altered impact–melt rocks below (Grieve et al. 1987; Gurov et al. 1986; Gurov and Gurova 1991). The suevites exhibiting red glass grade down into less altered rocks with gray and dark gray glasses in the interval of 583–585 m.

The lowermost sediments of obvious post–crater origin (573–547 m in drill hole No. 50) are sands and sandstones with thin interlayers of sedimentary breccia.

A 1 m thick basal layer of polymict sandstone overlies the surface of suevite. This rock is an immature poorly sorted sandstone with angular grains, cemented by poikilitic calcite crystals up to 1 cm in diameter. Grain size is from around 0.1 mm up to 0.7 mm in diameter, while rare grains are up to 1 mm in diameter, and is composed potassium feldspar, plagioclase, quartz, rarely biotite and chlorite. Rare quartz grains contain planar deformation features (PDFs) (Fig. 6a, b). The majority of water-lain sediments in the deepest levels of the No. 50 borehole are siltstones, sands, and sandstones (Fig. 6c).

Interlayers of sedimentary breccia and gravel also occur in the lowest deposits in the interval 570 – 569 m. Granitic clasts in breccia are up to 1 cm in size. Granites often are cataclased, but shock metamorphic effects rarely occur in the clasts. Although there appears to be no evidence preserved for life in the earliest crater lake, the first particles of visible carbonaceous material, probably washed in from vegetation on the periphery of the crater, appear at a depth of 530 m (Table 2).

Continued strong interaction between the waters of the lake and underlying impact deposits seem to be indicated by thin layers of chemogenic carbonate rocks with high phosphorus contents (Gurov et al. 1985), which were deposited on the floor of this early crater filling lake (Table 3). The highest content of P_2O_5 – 5.92 wt% - was determined in fine grained carbonate rock from the interval 507 m. The composition of a 10 cm thick fine–grained carbonate rock which occurs at 462.3 m, core No. 50, is 57.4 mole% siderite, 15.4 mole% calcite, 11.6 mole% silica, and 10.7 mole% fluorapatite (normative composition was calculated by CIPW method; cf. Cross et al. 1903). The presence of siderite and fluorapatite was confirmed by X–ray diffraction patterns (reflections of fluorapatite: 0.2814, 0.2766 and 0.2614 nm) (Gurov et al. 1985). The fine–grained carbonate masses with small clasts of quartz and rare feldspar grains are visible in thin section of the rock.

Siltstones, clays and argillites with interlayers of sand, sandstones, marls and breccia occur in the interval 460–330 m and particles of preserved plant material are abundant in all rock types in this interval. A notable breccia layer 0.5 m thick occurs at 453 m, consisting of angular, partly rounded clasts of granite and predominantly quartz and feldspars, in carbonate cement. Clasts are from 1 mm up to 10–12 mm in size.

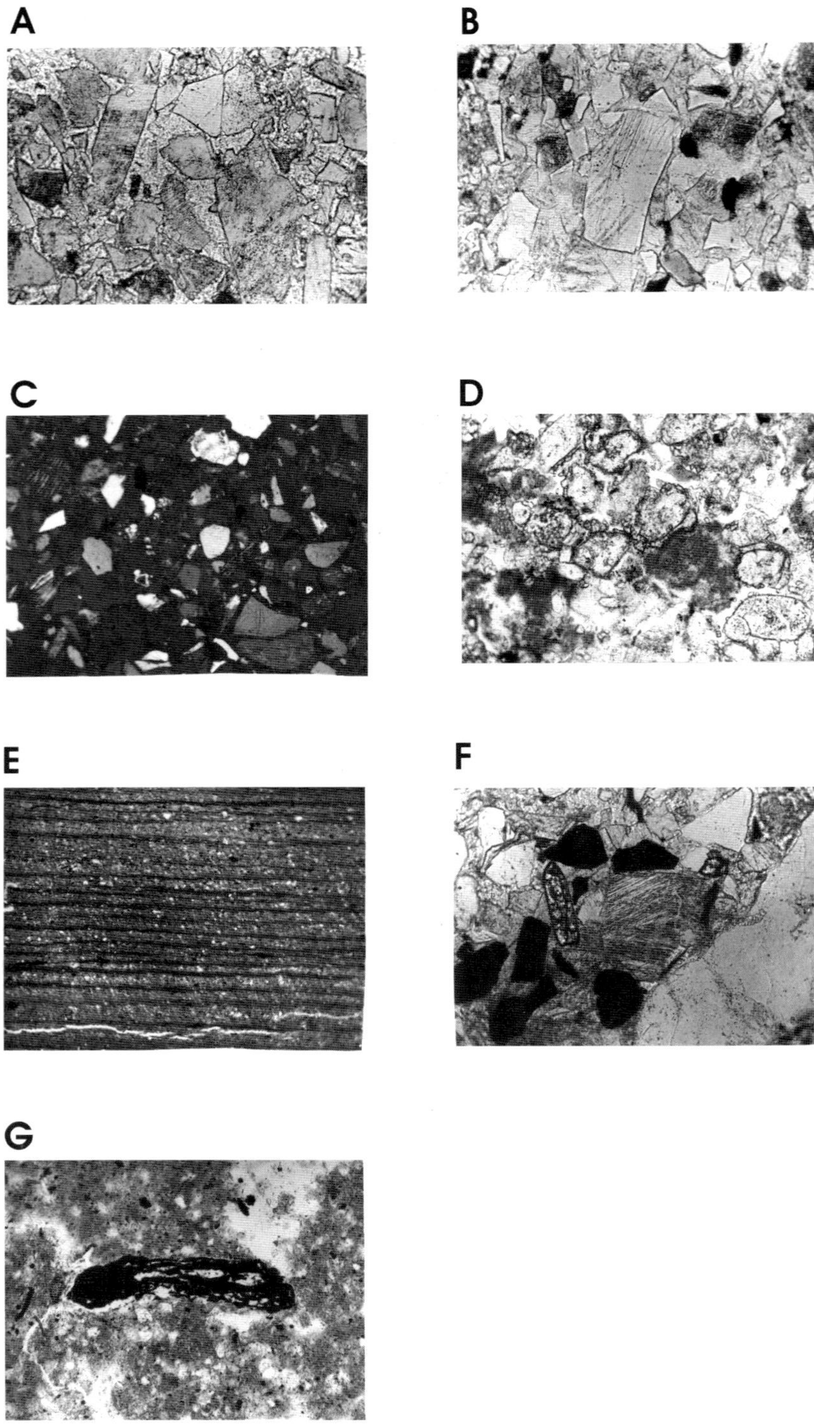
A
B
C
D
E
F
G

Figure. 6. Microphotographs of post–crater sediments of the Boltysh impact crater (drill core No. 50).

a. A coarse–grained sandstone from the basal horizon of post–impact sediments. Note that all mineral grains have angular form and are poorly sorted (sample 573 m, field of view 4.8 mm, parallel polars).

b. Quartz grains with PDFs in inequigranular sandstone (sample 573 m, field of view 1.8 mm, parallel polars).

c. A medium–grained sandstone. Detrital grains have predominantly angular shapes, though some grains exhibit partial rounding (sample 472 m, field of view 4.5 mm, parallel polars).

d. Ostracode shells are the main component of this sediment, one of several similar horizons in the crater fill (sample 385.5, field of view 2.6 mm, parallel polars).

e. An example of a varved clay with periodically repeated thin beds (sample 373 m, field of view 4.1 mm, parallel polars).

f. Detrital quartz grain with the PDFs in breccia layer (sample 362352 m, field of view 1.8 mm, parallel polars).

g. Fragments of vegetation preserved by burial in argillite (sample 322 m, field of view 2.6 mm, parallel polars).

The first recognizable imprints of macro fauna living in the lake also occur at the depth of 462 m, including ostracodes, molluscs and fish. The most abundant fauna are ostracods, which sometimes even form complete layers several millimetres thick (Fig. 6d). In one horizon at 415m depth several species of ostracods have been identified: *Bairdirdophillata simplicatilis* Mandelstam et Lübimova, *Xestoleberis triangularis* Mandelstam, *Aequacytheridea atroxa* Mandelstam, *Cytherells ovata* Roemer, *Cyteridea sp.* and *Cypridea sp.* Interestingly the ostracods found in the Boltysh crater appear 1.5 to 2 times smaller than ostracoda of equivalent age in other regions (Scheremeta 1981). These forms have been described in the Maastrichtian and Campanian – Maastrichtian marine deposits of Moldova, Turkmenia, Tadjikistan, Russia, Ukraine and North America (Lübimova et al. 1960; Mandelstam 1959; Scheremeta 1981).

The dominant rock type in the interval 409–370 m, core No. 50, is varved clay (Table 2), consisting of alternating bands of dark and light gray clay each about 0.12–0.18 mm thick (Fig. 6e). However, two interlayers of fine–grained sandstone occur at the depth 397.7 – 397.0 m and at 395.8 – 395.5 m. The predominant component of these sandstones are ostracods, while feldspars and quartz occur only rarely. Grain size of sandstones is from 0.08 to 0.15 mm and up to 0.20 mm. If the varved clay has a seasonal origin, 39 m of these sediments must have been deposited over a period of around 280 thousand years. Thin grained carbonate rocks and marls form occasional interlayers in the intervals 407 to 399 m though the proportion sandy material in those rocks only reaches 5 – 10 vol%.

The first layers of oil shale appear at a depth of 400 m, in core No. 50 (Table 2). Stanislavsky (1968) determined the Paleocene age for these sediments based upon flora in "greenish–gray sapropelites" and siltstones.

A layer of fine–fragmental breccia occurs in the interval 360.0 – 351.6 m which differs from sedimentary breccias of the lowermost section of post–crater sediments. The breccia is composed of quartz, feldspar and mica grains, granitic rock fragments up to 8 mm diameter, and rose colored particles of altered glass up to 2 mm diameter, cemented by carbonate and poorly consolidated fine–grained matrix. Shocked quartz has been observed at 352 m (Fig. 6f). This breccia deposit seems most likely to have resulted from erosion cutting into impact deposits surrounding the crater, though there remains the possibility that it may have been the result of a later impact (such as Zeleny Gai 30 km to the east).

Thin interlayers of zeolite crystals (5 to 35 micrometers in diameter) occur in siltstones at around 347 m, in core No. 50. Alternation of thin interlayers of zeolite (0.05 to 0.15 mm) and carbonate (0.15 to 0.30 mm) is visible in thin sections. An earlier investigation of similar sedimentary zeolites in post–crater sediments of the Boltysh crater (Ryabenko et al. 1982) XRF and differential thermal analysis on the zeolite indicated that is clinoptilolite, and is the only known occurrence of zeolite in impact fill sediments. Their chemical composition is presented in Table 3.

Argillites, siltstones and oil shales occur in the interval 350 – 220 m (Table 3) with thin interlayers of carbonate rocks. The predominant rocks in the series are bituminous argillites and clays with high organic matter contents in the form of abundant fragments of preserved flora and fauna (Fig. 6g). Content of $C_{org.}$ in bituminous argillites and siltstones is from 10 to 35 % and more: intervals 312 m – 22 %, 269 m – 35 %, 255 m – 23 %, 227 m – 17 %. The silica content in those rocks ranges from 32.37 % to 41.33 % (wet chemistry data). Thickness of oil shale layers reaches 7–8 m, and they are regarded as a significant resource, representing up to 3–4 billion tons of oil (Bass et al. 1967; Grieve and Masaitis 1994).

The oil shales sometimes part in thin plates, and contain the remains of algae and sub–aerial flora. Imprints of ferns predominate, but rare occurrences of the leaves of dioecious and monoecious trees are also observed in the sediments (Stanislavsky 1968). The sub–aerial flora was probably transported into the basin by streams and rivers or blown in from the surrounding area. An Early Eocene age of this strata of "gray sapropelites" was proposed based upon a correlation of local flora with Ypresian flora of the Parisian basin (Stanislavsky 1968).

The abundant faunal remains in the argillite and oil shale layers in the intervals at 319 m, 314 – 300 m, 296 – 295 m (core No. 50) are mainly fish, ostracods, and molluss. P.G. Danilchenko (Vasilyev and Selin 1970) recognized several fish including: *Baltischia brevicanda gen. et sp. nov., Parachanopsis longulus gen. sp. nov., Beryx sp., Lirolepis sp*. The fish fauna is dominated by new, previously unknown species, some of which are similar to Cretaceous fishes (Bass et al.1967; Vasiliev and Selin 1970).

In summary, the lowermost 350 m of the crater-fill sediments were deposited in a closed fresh–water basin over a period of about 15 Ma from formation of the crater at the end of the Cretaceous until the Middle Eocene (Table 2). The early history of the crater was brought to an end when the Mid- to Late Eocene transgression breached the crater rim. The ocean transgressed from the north and the boreal sea basin located SW of the East–European craton (Bondarchuk et al. 1960; Syabryay, 1963). During this marine transgression the crater became a coastal bay. Quartz sands, marls, and clays were deposited at this time and now occur in the interval 220 to 150 m depth in the core No. 50. The Middle Eocene age of these deposits was determined by comparison with the sediments of surrounding area (Syabryay, 1963). The remains of seeds of *Potamogeton* and *Limnocarpus* are present in these deposits, but do not indicate the age any more precisely (Stanislavsky 1968).

The ejecta layer around Boltysh crater was intensively eroded by the Middle Eocene transgression and both the thickness and area covered by the ejecta were significantly reduced (Gurov et al., 2003). In some boreholes, Middle Eocene sands directly overlie the remains of Boltysh ejecta (Bryansky et al. 1978).

Transgression continued during the Late Eocene (Syabryay, 1963) and the majority of the area of the Ukrainian Shield was flooded at this time as evidenced by layers of marl and sand 40 – 60 m thick within the Boltysh crater. The estimated age of these sediments is determined by correlation with the Late Eocene deposits on the northern slopes of the Ukrainian Shield (Stanislavsky 1968).

Marine conditions persisted in the region during the Late Eocene and greenish glauconite–bearing fine–grained sands (Kiev series) up to 94 m thick are present in drill core No. 50 in the interval 133–100 m. Marine conditions continued to dominate the Ukrainian Shield outside the crater into the Late Oligocene, and Oligocene sediments (Kharkov series in the local Ukrainian stratigraphy) in the Boltysh crater are dominated by sand bodies several meters thick.

Regression occurred as local sea levels fell across the Ukrainian Shield during the Late Oligocene and the Early Miocene. At this time, the majority of the Ukrainian Shield was transformed into a slightly elevated plain (Bondarchuk et al. 1960; Didkovsky 1975). Sub-aerial conditions prevailed in the region surrounding of the Boltysh crater area during Miocene and Pliocene times. The crater was flooded by the lake, and the base of the depression was probably still about 100 m below the surrounding areas. Sedimentation in the crater continued under continental conditions in the Neogene during which sands and clays 60–100 m thick were deposited (around 60m in borehole No. 50). The predominant type of Neogene sediment is fine-grained quartzose sand containing some glauconite.

The formation of the recent drainage system began on the Ukrainian Shield in the Late Pliocene and continued to the Early Pleistocene. The valleys of the Tyasmin and Ingulets rivers were formed at that time (Bondarchuk et al. 1960; Didkovsky 1975). The Tyasmin river and its tributaries dissected the Ukraine Plain and formed a complex relief of the area. The maximum elevation of the divides the present day river valleys is up to 60–70 m.

The latest and uppermost deposits in the region are loess and loam, deposited during the Quaternary glaciation. It seems likely that the Boltysh crater deposits were exposed beneath a glacier near its southern margin (Bondarchuk et al. 1960). Although no traces of the glaciation have been observed in the crater, erosional remnants of the original rim of the crater are present in Quaternary loess near the SW edge of the crater, as a thin layers of pebbles and fragments of Kirovograd granites.

Loess and loam cover all interstream areas, as they cover much of the Ukrainian Shield, but those deposits are absent in the river valleys. The thickness of loess reaches to 20–30 m between valleys and 20 m in drill core No. 50. Rare natural exposures of loam and loess occur in the crater.

Trace element compositions of a set of samples of sedimentary infill of the Boltysh crater, from corehole No. 50, are given in Table 4. The analyses were obtained by instrumental neutron activation analysis at the University of Vienna (for a current description of the analytical method, see Son and Koeberl 2005). The analytical data indicate fairly typical sedimentary compositions, and even at the bottom of the core there do not seem to be any significant anomalies in the siderophile element contents, which would indicate KT ejecta deposition.

Today the Boltysh crater morphology is only poorly expressed in the local topography and neither the crater structure nor its outline are visible on aerial or space images.

3 Discussion and conclusions

The Boltysh crater formed close to the KT boundary, and was subsequently filled by a crater lake and then slowly filled by post impact sediments. The crater now contains up to 580 m of sediments overlying suevites and impact melts in the 12 km diameter inner crater. Post-impact sediments thin towards the eroded crater rim where they overlie brecciated Proterozoic basement rocks. Sedimentation in the crater began by creep and collapse of unstable impact ejecta deposits from the rim and inner slopes of the crater.

Boreholes in the western part of the 12 km diameter inner crater indicate a relatively level impact melt surface around 580 m below the present day surface, with a patchy covering of fallback suevite from 1m to around 20 m thick. However, in the east, the melt surface appears to be deeper and covered in around 100 m of suevite. The only remaining core was derived from hole No. 50 in the western part of the inner crater (Fig. 1) and it was upon this vertical profile that we based most of our analysis of sedimentation in the Boltysh crater.

In borehole No. 50, the impact melt surface at 593.5m depth is overlain by 22 m of altered suevite. The earliest sediments are sands, sandstones, and sedimentary breccias with a thickness of around 25 m. Rare macrofossils appear in the core from 530 m upward. Though those earliest remnants appear to be flora derived from the surrounding areas.

The sediments become less clast rich higher up the sequence, and a thick series of siltstones, clays, and oil shales, about 240 m thick was deposited in a closed freshwater basin formed by the crater. An ostracod rich fauna including some molluscs and fish first colonized the lake. Later, the lake became anaerobic and algae thrived in the water during the Early Eocene, subsequently forming the predominant component of oil shales (Bass et al. 1967). Subaerial flora were also deposited in the lake and were represented by ferns, dioecious, monoecious trees and conifers. The remnants of subaerial flora seem to have been transported into the lake basin from the surrounding area by rivers or in the wind.

The Middle Eocene transgression on the northern slopes of the Ukrainian Shield resulted in erosion of the Boltysh crater rim and flooding of this basin by sea water. Sand, marl and clay were deposited over earlier sediments in the crater during the Mid- to Late Eocene.

The final period of water-lain sediment deposition within the Boltysh crater occurred during the Neogene when strata of sand and clay up to 100 m thick were deposited in the crater lake, at a time when the sea had regressed and Boltysh was a fresh water basin once again. Loess and loam were deposited on the territory of the Ukrainian Shield during the Quaternary glaciation finally filling the crater and obscuring any remnant of the circular structure in the present day landscape.

Acknowledgments

EG was supported by the Austrian Academy of Science, the University of Vienna, Austria, and the European Science Foundation IMPACT program. CK is supported by the Austrian Science Foundation (grant P17194-N10). We are grateful to D. Mader (University of Vienna) for help with the sediment analyses.

References

Bass YuB, Ghalaka AI, Grabovsky VI (1967) The Boltysh oil shales. Razvedka i ochrana nedr 9: 11-15 (in Russian)

Bondarchuk VG (ed) (1960) Atlas of the Paleogeographic Maps of the Ukrainian and Moldavian SSR. Academia Nauk Ukrainian SSR Press. Kiev, 78 maps (in Russian)

Bryansky VP, Zlobenko VG, Ryabchun VK (1978) Breccia rocks of the Paleogene from the Boltysh basin area. Geologichesky Zhurnal 38 (2): 135-138 (in Russian)

Cross W, Iddings JP, Pirsson LV, Washington HS (1903) Quantitative Classification of Igneous Rocks. Chicago, IL: University of Chicago Press, 286 pp

Didkovsky VYa, Kulinichenko VG (eds) (1975) Stratigraphy of the Ukrainian SSR 10. Neogene. Academia Nauk Ukrainian SSR Press, Kiev, Ukraine, 271 pp (in Ukrainian)

Grieve RAF, Masaitis VL (1994) The economic potential of terrestrial impact craters. International Geological Review 36: 105-151

Grieve RAF, Reni G, Gurov EP, Ryabenko VA (1987) The melt rocks of the Boltysh impact crater. Contributions to Mineralogy and Petrology 96: 56–62

Gurov EP (1977) Planar deformation features in biotite from shock–metamorphosed rocks of impact meteorite craters of the Ukrainian Shield. Zapisky Vsesoyusnogo mineralogisheskogo obshchestva 106 (2): 715–719 (in Russian)

Gurov EP, Gurova EP (1991) Geological Structure and Composition of Rocks in Impact Structures. Naukova Dumka, Kiev, Ukraine, 160 pp (in Russian)

Gurov EP, Gurova EP, Rakitskaya RB (1979) About orientation of PDFs in quartz from rocks of impact meteorite craters. Zapisky Vsesoyusnogo mineralogicheskogo obshchestva 108 (5): 578–584 (in Russian)

Gurov EP, Gurova EP, Metalidi SV (1985) Phosphorus in the sediments of the Boltysh depression. Geologichesky Zhurnal No 1: 125–127 (in Russian)

Gurov EP, Kelley SP, Koeberl C (2003) Ejecta of the Boltysh impact crater in the Ukrainian Shield. In: Koeberl C, Martinez–Ruiz F (eds) Impact Markers in the Stratigraphic Record. Impact Studies, vol. 3, Springer, Heidelberg, pp 179–202

Gurov EP, Kolesov GM, Gurova EP (1986) Composition of impactites of the Boltysh astrobleme. Meteoritica 45: 150–155 (in Russian)

Gurov EP, Val'ter AA, Rakitskaya RB (1980) Coesite in rocks of meteorite explosion craters on the Ukrainian Shield. International Geological Review 22: 329-332

Kashkarov LL, Nazarov MA, Kalinina GV, Lorenz KA, Kononkova NN (1998) Fission track dating of the Boltysh impact crater, Ukraine [abs]. Lunar and Planetary Science 29: abs. #1257 (CD–ROM)

Kelley SP, Gurov EP (2002) Boltysh, another end–Cretaceous impact. Meteoritics and Planetary Sciences 37: 1031–1043

Lübimova PS, Kazmina TA, Reshetnikova MA (1960) Ostracoda from the Mesozoic and Cenozoic sediments of the Western–Siberian Lowland. Leningrad: Gostoptechizdat Press, 374 pp (in Russian)

Ludwig KR (1999) Isoplot/Ex 2.01: a geochronological toolkit for Microsoft Excel, Berkeley Geochronology Center, California, U.S.A, 50 pp

Mandelstam MI (1959) Ostracoda from the Paleogene of Middle Asia. In: Microfauna of the USSR, X, N 136, Leningrad: Gostoptechizdat, 442–544 (in Russian)

Masaitis VL (1974) Some ancient meteorite craters in the territory of USSR. Meteoritica 33: 64-68 (in Russian)

Masaitis VL, Danilin AN, Mashchak MS, Raykhlin AI, Selivanovskaya TV, Shadenkov YM (1980) The geology of astroblemes. Nedra, Leningrad, 231 pp (in Russian)

Ryabenko VA, Valter AA, Gurov EP, Gurova EP, Lasarenko EE, Rakitskaya RB, Serebrennikov AI and Efimenko VV (1982) Geology and petrology of explosive meteorite craters. Naukova Dumka, Kiev, Ukraine, 226 pp (in Russian)

Scheremeta VI (1981) Ostracods fron the Paleogene of Ukraine. Lvov: Lvov University Press, 258 pp (in Russian)

Shcherbak NP, Zlobenko VG, Zhukov GV, Kotljvskaya FI, Polevaya NI, Komlev LV, Kovalenko NK, Nosok GM, Pochtarenko VI (1978) Catalogue of isotopic data of the Ukrainian Shield (Shcherbak N.P. ed.). Naukova Dumka, Kiev, Ukraine, 224 pp (in Russian)

Son TH, Koeberl C (2005) Chemical variation within fragments of Australasian tektites. Meteoritics and Planetary Science 40: 805-815

Stanislavsky FA (1968) Age and stratigraphy of the sapropelites of the Boltysh depression. Geologichny Zhurnal 28 (2): 105-110 (in Russian)

Syabryay VT (ed) (1963) Stratigraphy of the Ukrainian SSR 9. Paleogene. Academia Nauk Ukrainian SSR Press, Kiev, Ukraine, 319 pp (in Ukrainian)

Valter AA, Plotnikova L. (2003) Biostratigraphic indications of the age of the Boltysh impact crater in Ukraine. In: Koeberl C, Martinez–Ruiz F (eds)

Impact Markers in the Stratigraphic Record. Impact Studies, vol. 3, Springer, Heidelberg, pp 163–178

Valter AA, Ryabenko VA (1977) Explosion craters of the Ukrainian Shield. Naukova Dumka, Kiev, Ukraine, 154 pp (in Russian)

Vasilyev IV, Selin YuI (1970) New data about the paleontologic characteristic of the productive series of the Boltysh deposit of oil shales. Doclady Academii Nauk UkrSSR N 12, 1059–1060 (in Russian)

Table 1. Composition of impact melt rocks, suevites and target rocks of the Boltysh impact crater (major elements – in wt. %, wet chemistry data; Cr, Ni and Co – in wt.%, INAA data)

Component	1	2	3	4
SiO_2	66.57	68.51	67.78	70.27
TiO_2	0.50	0.39	0.32	0.40
Al_2O_3	14.67	13.91	14.22	13.72
Fe_2O_3	2.20	1.88	0.85	1.38
FeO	1.60	1.70	2.68	2.76
MnO	0.07	0.08	0.06	0.04
MgO	1.37	1.49	1.29	1.08
CaO	2.37	1.99	2.05	1.40
Na_2O	3.07	2.99	3.36	2.87
K_2O	5.18	4.22	3.73	4.88
Rb_2O_3 10^{-4}ppm	188	179	186	250
P_2O_5	0.20	0.18	0.15	0.15
H_2O^-	0.79	0.79	0.32	0.20
H_2O^+	1.35	1.65	2.85	1.10
total	98.94	99.78	99.66	100.25
Number of analyses	11	18	30	40
Fe_2O_3/FeO	1.37	1.10	0.32	0.50
Cr, 10^{-4}	22	20	28	2
Ni, 10^{-4}	51	77	50	4
Co, 10^{-4}	5	4	4	2
Number of analyses	1	4	5	3

1 – average of suevites, core 50 – the interval 577–590 m, core 11475 – the interval 563–570 m, and core 17 – the interval 653–677m; 2 – average of microcrystalline impact melt rocks, core 50 – the interval 595–645.8 m, core 11475 – the intervals 558–561 m and 582 – 640 m; 3 – average of glassy matrix impact melt rocks, core 50 – the interval 652–736 m and core 11475 – the interval 657 m – 791 m; 4 – average of 40 basement rocks (Gurov et al. 1986; Gurov and Gurova 1991)

Table 2. Changing environments during infilling of the Boltysh Crater

Depth (m)	Approx. Age	Sediments/Environment
20	Quaternary	Loess deposited over the surface obscuring last indications
100-20	Miocene - Pliocene	Sub-aerial conditions, shallow lake deposits – sands
133 - 100	Late Oligocene/Early Neogene	Regression – sub-aerial and fresh water conditions return in the crater
150 - 133	Late Eocene	Layers of marl and sand deposited in marine conditions
220-150	Mid-Late Eocene	Local transgression on Ukrainian shield, marine incursion into the crater as crater rim is eroded
330-220	Early Eocene	Deep anaerobic lake. Numerous oil shale layers – lake fauna included fish, molluscs and ostracods
347		Zeolite crystal layer
365 - 360		Impact breccia layer
400		First oil shale
409-370		Dominantly varved clay layering
460 - 330	Late Paleocene to Early Eocene	Siltstones and shales -first remains preserved of fauna from a fresh water lake
530		First preserved macrofossils – flora derived from surrounding crater rim
573-460	K/T - Paleocene	Sands and siltstones with thin interlayers of breccia – no macrofossils observed in lower layers
593.5-573	K/T	Suevite
593.5	K/T	Impact melt

Table 3. Composition of carbonate– and zeolite– bearing rocks from the sedimentary fill of the Boltysh impact crater, in wt%, by wet chemistry.

Component	1	2	3
SiO_2	2.70	11.60	60.70
TiO_2	0.21	–	0.53
Al_20_3	4.06	12.82	11.53
Fe_2O_3	3.04	1.91	1.19
FeO	35.60	1.52	1.83
MnO	1.26	0.42	tr.
MgO	3.51	0.83	2.02
CaO	14.64	35.55	2.84
Na_2O	0.22	0.40	2.07
K_2O	0.28	0.68	2.58
P_2O_5	4.57	5.92	–
CO_2	29.28	26.74	–
H_2O	0.26	1.08	14.59
Total	99.63	99.52	100.0

1 – siderite–bearing rock (hole No. 50, 462.3 m);
2 – apatite–bearing carbonate rock (hole No. 50, 507 m);
3 – clinoptilolite rock (hole No. 18, 416 m)
(Ryabenko et al. 1982)

Table 4. Trace element composition of samples of the sedimentary infilling of the Boltysh crater, core No. 50 (sample numbers include depth in meters – see Table 2 for sample type), by instrumental neutron activation analysis.

Sample	50 /220	50/ 249.2	50 /275	50 /296	50 /332	50 /335	50 /368	50 /382	50 /403	50 /438	50 /453	50/ 478.5	50 /498	50 /571
Na (wt%)	0.54	0.62	0.39	0.41	0.83	0.58	0.58	0.78	0.95	0.75	0.93	1.61	1.33	1.10
K (wt%)	3.06	2.38	1.14	1.59	2.32	1.66	2.11	1.65	1.83	2.11	2.76	3.16	2.97	3.66
Sc	11.6	9.96	6.98	8.11	9.21	9.51	10.0	9.57	10.4	92.6	11.7	6.71	11.4	11.1
Cr	56.9	40.6	28.4	36.0	45.1	39.5	49.2	48.0	38.9	42.5	41.7	30.6	45.1	4.02
Fe (wt%)	2.14	2.34	3.53	3.21	3.77	3.16	3.57	2.76	3.52	6.23	3.38	5.38	3.78	2.76
Co	7.42	7.59	6.62	8.06	8.12	11.5	10.8	8.82	9.39	10.8	9.71	5.39	9.14	3.40
Ni	41	29	29	25	29	42	24	39	30	27	23	41	28	23
Zn	142	121	134	146	94	118	163	92	113	104	122	87	187	210
As	1.42	1.46	2.06	3.08	1.32	4.09	2.65	2.21	3.83	6.49	4.66	1.92	3.48	2.91
Se	11.2	10.7	13.2	16.6	6.19	10.0	8.29	5.26	6.82	6.94	8.80	4.55	7.91	0.46
Br	2.3	3.9	10	7.3	2.2	7.8	2.7	0.3	0.5	2.2	1.1	0.7	1.3	0.7
Rb	188	151	114	146	214	130	170	145	144	170	172	182	204	218
Sr	93	67	360	290	158	166	318	361	474	338	79	437	137	76
Zr	392	342	133	116	148	210	177	178	222	252	403	541	366	360
Sb	0.08	< 0.11	0.04	< 0.10	< 0.11	< 0.11	0.09	0.25	0.05	0.07	< 0.10	< 0.11	< 0.14	0.20
Cs	3.60	2.90	2.45	3.17	4.12	2.73	4.05	3.09	3.27	4.15	3.59	2.50	4.40	5.24
Ba	758	727	571	436	580	571	507	552	610	873	609	1250	636	1145
La	113	99.5	64.0	64.4	55.6	84.2	62.3	77.9	88.3	78.2	114	82.9	132	186
Ce	209	176	117	115	111	173	124	153	154	155	212	154	250	343
Nd	91.4	79.9	55.0	53.2	49.5	69.2	51.0	65.0	69.7	65.0	94.1	60.5	106	155
Sm	14.5	13.3	9.49	8.51	7.82	12.2	8.56	11.2	11.4	12.8	15.4	10.9	17.2	25.9
Eu	2.02	1.72	1.14	1.08	0.94	1.36	0.98	1.12	1.18	1.02	1.58	1.38	1.59	1.70
Gd	15.0	10.5	6.7	8.8	7.4	9.8	7.6	8.8	8.3	9.2	14.0	10.4	14.7	17.5
Tb	1.92	1.51	1.12	1.06	0.94	1.60	1.12	1.37	1.32	1.62	1.93	1.51	2.05	3.15
Tm	0.63	0.58	0.55	0.41	0.38	0.69	0.43	0.50	0.64	0.83	0.96	0.77	0.94	1.35
Yb	4.28	4.10	3.08	2.86	2.22	4.36	2.93	3.33	3.93	5.52	6.24	4.73	5.57	8.60
Lu	0.67	0.58	0.41	0.41	0.37	0.67	0.46	0.52	0.62	0.89	0.91	0.69	0.81	1.20
Hf	9.94	7.91	2.98	2.70	3.84	3.26	3.55	2.76	4.93	3.82	12.0	15.3	9.62	8.40
Ta	1.62	1.43	0.71	0.91	1.40	1.09	1.29	1.28	1.19	1.37	1.53	1.45	1.62	2.15
Ir (ppb)	< 0.4	<0.4	< 0.2	< 0.6	< 0.4	< 0.6	< 0.3	< 0.6	<0.2	< 0.7	< 0.5	< 0.5	<0.5	< 0.5
Au (ppb)	< 0.6	< 0.6	0.5	< 0.6	0.3	0.5	0.2	< 0.6	0.3	0.5	0.2	< 0.6	0.3	< 0.7
Th	51.3	43.8	21.0	26.5	31.9	33.0	27.9	31.2	27.0	33.9	42.1	37.9	49.2	61.3
U	7.28	7.44	4.75	6.16	6.07	10.4	6.59	7.82	8.66	21.0	8.15	5.21	8.73	14.9
K/U	4203	3199	2400	2581	3822	1596	3202	2110	2113	1005	3387	6065	3402	2456
Th/U	7.05	5.89	4.42	4.30	5.26	3.17	4.23	3.99	3.12	1.61	5.17	7.27	5.64	4.11
La/Th	2.20	2.27	3.05	2.43	1.74	2.55	2.23	2.50	3.27	2.31	2.71	2.19	2.68	3.03
La(N)/Yb(N)	17.8	16.4	14.0	15.2	16.9	13.1	14.4	15.8	15.2	9.6	12.3	11.8	16.0	14.6
Eu/Eu*	0.42	0.44	0.44	0.38	0.38	0.38	0.37	0.34	0.37	0.29	0.33	0.40	0.31	0.24

Stones in the Sky: From the Main Belt to Earth-Crossing Orbits

Daniel Benest

Observatoire de la Côte d'Azur, Observatoire de Nice, B.P. 4229, F-06304 Nice Cedex 4, France (benest@obs-nice.fr)

Abstract. It is now well known that impacts play a great role in the formation of planetary bodies, and have some influence during their evolution -- particularly on the biosphere. The story begins officially in the early 19th century, with the discovery of the first asteroids and the recognition that stones can fall from the skies. Through the findings of the Kirkwood gaps, of Hirayama families and finally the importance of the 3:1 and other resonances, the processes that send Main Belt asteroids on trajectories crossing Earth's orbit have been gradually understood. A little paradox remained: computations show that the resonances become completely empty in less than a billion years, but observations reveal that celestial bodies orbiting in such Earth-crossing trajectories still exist; this has been explained by collisions in the Main Belt, the products of which are sometimes injected into nearby resonances and then continue to supply the population of possible Earth-crossers; non-gravitational forces may also inject bodies into resonances. Besides, hazards due to collisions between such Earth-crossers and our planet have been recognized. The example of asteroid 4179 Toutatis, discovered in 1989 (at the Observatoire de la Côte d'Azur), very near the 3:1 resonance, is presented.

1 Introduction to asteroids

"Stones do not fall from the sky, because there are not any stones in the sky!"

So said XVIIIth century's academics; it is said that they were exasperated to receive so many darkened stones called 'fallen stones': "These are thunderstones!", they repeated, "Once and for all, these are only thunderstones!"; we can understand their lost of patience - amateur geologists being then more numerous. Nevertheless, this academic attitude is only a parenthesis in history and, during other times, it was admitted that stones could fall from the sky. Many citations from ancient and modern literature testify to this. Without going back to the Holy Bible ("Josua", ch. 10, v. 11), I would like to cite:

"Je viens vous annoncer une grande nouvelle.
"Nous l'avons en dormant, Madame, échappé belle;
"Un monde, prés de nous, a passé tout du long,
"Est chu tout au travers de notre tourbillon;
"Et s'il eût en chemin rencontré notre Terre,
"Elle eét été brisée en morceaux comme verre."
(Molière, "Les femmes savantes" -- 1672 --, Acte 4, sc. 3, v. 1265-1270)

together with, more recently, Jules Verne and his "La chasse au météore" (1908) where he tells humoristically of the 'hijacking' of a golden asteroid; nowadays, many stories are told or written, cartooned or filmed (like, e.g., *Armageddon*, a Michael Bay movie -- 1998) about the hazards from fallen stones.

It is now well-known that heavy stones can actually fall on our planet (and, more generally, on any celestial body), leave scars which are impact craters, and have even important consequences on the biosphere. But it was only in the early XIXth century that actual falls of stones from the sky were scientifically established. However, the determination of the origin of these stones remained to be done; for, as the shooting star showers were soon recognised to be linked with comets, a question was left open: where do the largest impactors come from?

It is also in the very early XIXth century that the first asteroids were discovered; and, subsequently, these bodies will give the answer. As soon as the beginning of the XVIIth century, Kepler himself was astonished by the large interval between the orbits of Mars (at roughly 1.6 AU from the Sun)[1] and Jupiter (at 5.2 AU); but Kepler had also shown (in fact he believed he had shown) that there exist only 6 planets orbiting the Sun, using the 5 regular polyedra -- the only 5 to exist. Later, in the XVIIIth century, Titius, then Bode, proposed an empirical relation that approximated the series of intervals between two successive planets: for example, the interval between the orbits of Earth and Mars (1.6 - 1 = 0.6 AU) is twice the one between Venus and Earth (1 - 0.7 = 0.3 AU); here too, there was a large `hole' between the red planet and the king of gods, as if a planet was missing. However, people were not ready to accept this idea.

Fortunately, the discovery of Uranus (by Sir William Herschel in 1781) -- which 'obeyed' to the Titius-Bode relation -- liberated the situation: so planets could exist that were not yet discovered ... therefore let us search for the ‘missing planet’, which would theoretically orbit at roughly 2.8 AU from the Sun (for the interval between this planet and Jupiter -- 5.2 - 2.8 = 2.4 AU -- would be twice the one between Mars and itself -- 2.8 - 1.6 = 1.2 AU). Very rapidly after, on the 1st of January 1801, Piazzi ‘discovered’ the planet, which had been named Ceres.

Alas! The next year saw the discovery of another little planet (or "minor planet", or "asteroid" - this latter name having been proposed by Herschel himself), then another one in 1804, then another one (1807) and, from 1845, numerous ones were discovered. The count nowadays is up to 150,000, among which 58,092 (in 2003, May the 6th) have an orbit sufficiently determined for them to have received a number -- and, for tens of thousands, a name. The majority of these tiny heavenly bodies orbit the Sun between Mars' and Jupiter's orbit, more precisely between 2 and 3.5 AU: this is what we call the "Asteroid Main Belt" (fig. 1); others orbit beyond Neptune's orbit, in the so-called "Kuiper Belt" -- or Edgeworth-Kuiper --, we are not interested here in these very far objects. Their orbits are generally a little inclined with respect to the ecliptic plane (the Earth orbital plane), and have a low eccentricity; we will see that this low eccentricity protects them from any catastrophic fate but that, when this eccentricity increases, catastrophe may occur.

[1] AU = Astronomical Unit = the semi-major axis of the heliocentric Earth orbit = about 150 millions km.

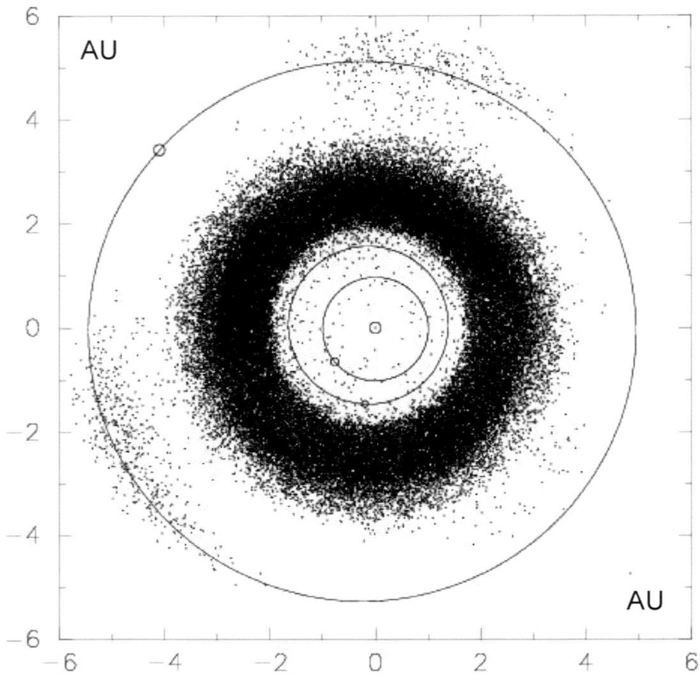

Fig. 1. Heliocentric position -- projection onto the ecliptic plane -- on the 6th May 2003 of the 58,092 numbered asteroids (at least the ones nearer then 6 AU from the Sun), together with the orbits of Earth, Mars and Jupiter -- and the position of the planets at that date. The Main Belt is clearly visible, but there are also some minor planets inside Mars' orbit, even inside Earth's orbit.

2 Kirkwood gaps and mean motion resonances

Around 1860-65, the American astronomer Daniel Kirkwood built a diagram of the distribution of asteroids according to their orbital semi-major axis. This distribution could be expected to be regular, with a maximum peak or plateau (for example around 2.8 AU) and two wings decreasing gently and regularly down to very low values (for example below 2 AU and beyond 4 AU). But what he observed is shown in figure 2: an irregular distribution, with important ups and downs and, particularly, values of the semi-major axis for which the number of corresponding asteroids is very low: this is what we call "Kirkwood gaps".

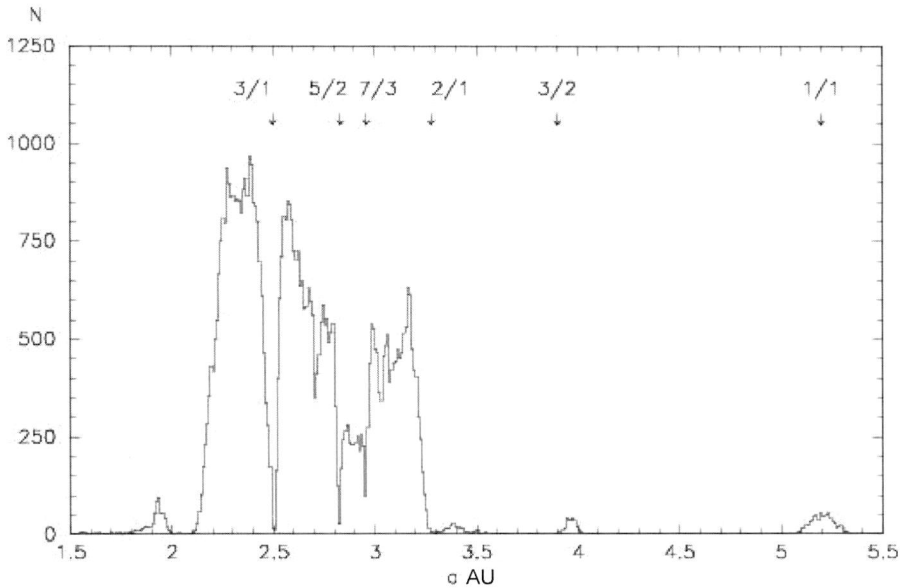

Fig. 2. Distribution, according to their orbital semi-major axis *a*, of the 58,092 numbered asteroids (at least the ones whose semi-major axis is between 1.5 and 5.5 AU); the sample step is 0.01 AU, which means that the ordinate is the number N of minor planets between, for example, 3.46 and 3.47 AU.

Kirkwood remarked that the gaps were related to phenomena of RESONANCES: there is resonance between two periodic motions when the ratio between the two periods is a fraction, and preferably a simple one (for example 2/3, 3/4 or 1/2). Then, when the ratio between the orbital periods of an asteroid and a planet is a simple fraction, there is a so-called MEAN MOTION RESONANCE between the 2 celestial bodies. Let us recall that the 3rd law of Kepler allows us to compute the orbital period T (the time needed by the planet to complete 1 orbit around the Sun) from the semi-major axis a: a^3 is proportional to T^2 (moreover, the quantity called "mean motion" is proportional to the inverse of the period, so that finally these resonances do not depend on any other orbital element than the semi-major axis); besides, the orbital period of Jupiter is well known to be almost a dozen years. Therefore, an asteroid whose orbital period is nearly 4 years is at or very near the 3/1 resonance with Jupiter; this means that the minor planet orbits 3 times around the Sun during one, and only one, revolution of the giant planet. The most interesting resonances are indicated on Figure 2.

But what could be the effect of a resonance?

QUALITATIVELY, let us take an analogy with a child's swing. It is well known that it is more efficient to push when the apparatus is at the maximum elongation on its trajectory and begins to go down, so that the oscillations are amplified (or at least maintained): energy has been transferred to the swing; if the push occurs at each oscillation, this is a 1/1 resonance; if the pusher is a little tired, and pushes only one time over two [or over three], then the resonance is a 2/1 [or 3/1] one; but if the pushes occur anywhere on the trajectory, we are not at a resonance, and the energy transfer is globally null. In the same way, an asteroid in resonance with Jupiter gains more and more energy (this energy is lost by Jupiter, but the mass of the giant planet is so great that this loss does not have any consequence -- at least during the lifetime of the Solar System); on the other hand, a minor planet far from any resonance does not gain any energy.

QUANTITATIVELY, how does this increase in energy act on the asteroid? Astronomers, and particularly celestial mechanicians, have searched for the solution for a long time. Let me pass over all the adventures occurred during the quest, and give you the final result: this leads to an increase of the orbital energy, whose consequence is an increase of the orbital eccentricity (without a change in the semi-major axis, at least in this first stage). And when the eccentricity increases, the orbit becomes more and more flattened, so that the perihelion becomes closer and closer to the Sun (down to graze the surface of our star) and the aphelion becomes farther and farther away, up to cross the orbit of Jupiter [we are here in an ultra-simplified model of the Solar System, which comprises only the Sun, Jupiter and the minor planet], as shown in Figure 3. Therefore, the final fate of the asteroid may be a collision with either the Sun or Jupiter, except if a strong perturbation (during a close approach with one of the two massive bodies) throws the minor planet onto an escape trajectory -- escape from the Solar System -- through an increase of the semi-major axis.

But our actual Solar System comprises much more than three bodies, and a more realistic model must include at least the other giant planets (Saturn, Uranus and Neptune) together with the main telluric planets: Mars, Earth and Venus (Mercury is often neglected); therefore, a less important increase in eccentricity is sufficient for the asteroid's trajectory to cross the orbit of Mars (Figure 4), and, later, Earth's one ... Here again, this configuration increases hazards of collision, or strong perturbation caused by close approach, with the telluric planets, especially our own one. An example is given in Figure 5.

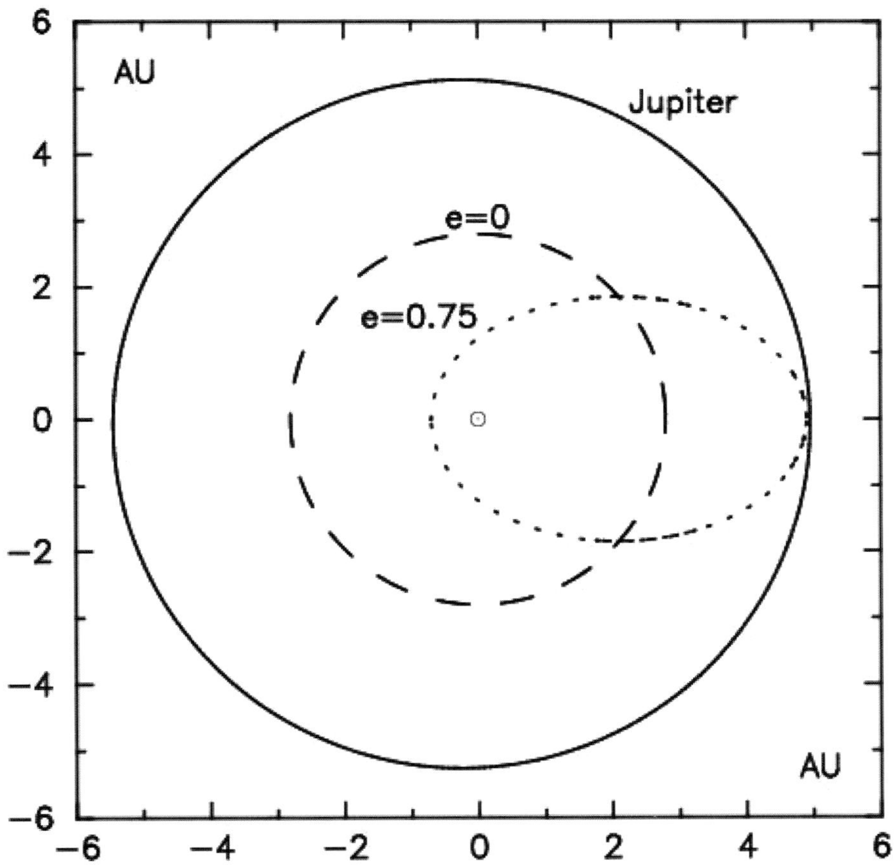

Fig. 3. Increase of the orbital eccentricity of an asteroid in resonance with Jupiter (three-body model Sun-Jupiter-asteroid; the Sun lies at the center); when the value of the eccentricity becomes near to 1, there are hazards for a collision, or strongly perturbative close approach, with one of the two massive bodies.

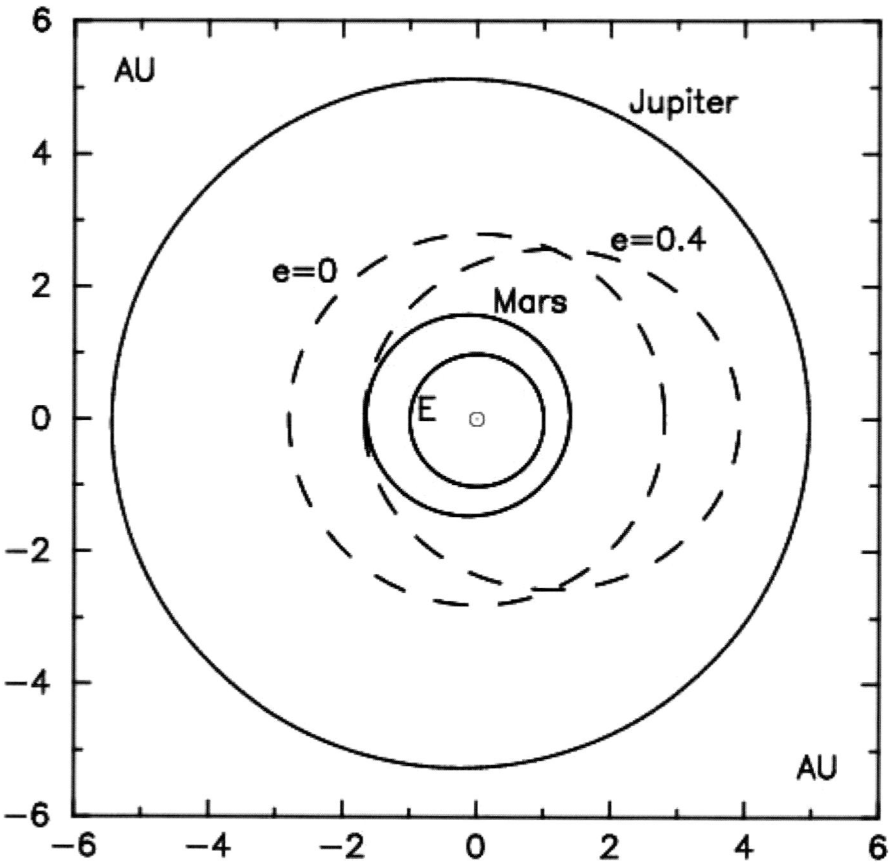

Fig. 4. Same as Figure 3, but for a more complete model: Venus, Earth, Mars, asteroid, Jupiter, Saturn, Uranus and Neptune; an increase of eccentricity up to 0.4 is sufficient for hazards of collision, or a strongly perturbative close approach, with Mars in front line and subsequently the Earth.

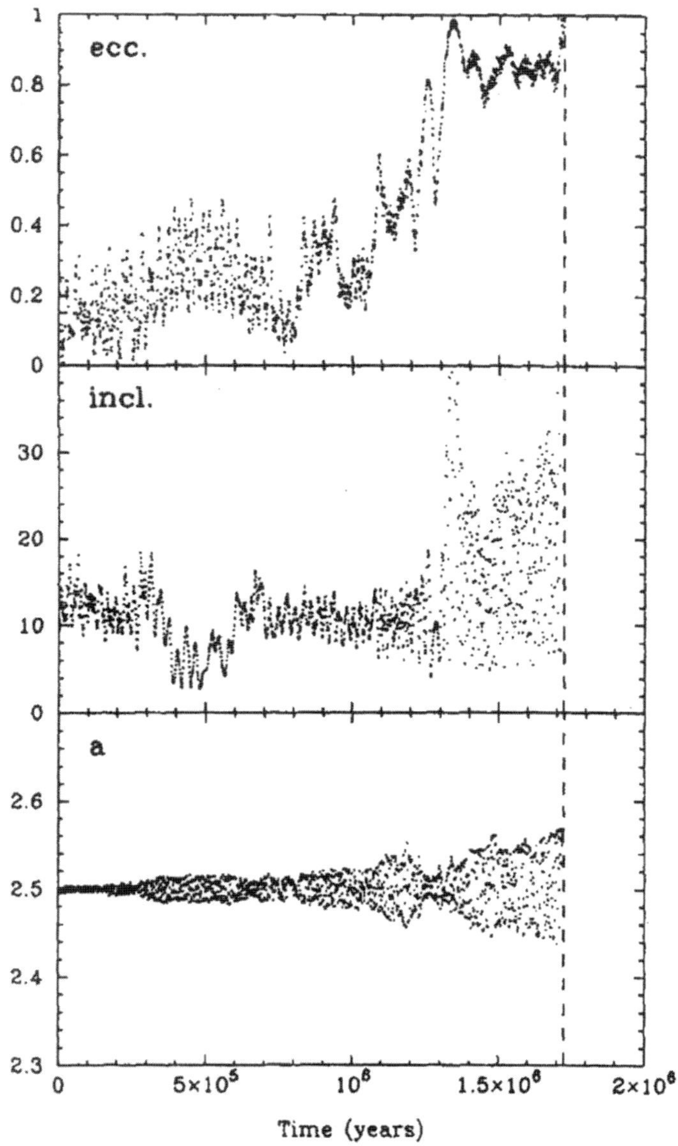

Fig. 5. Evolution with time of the eccentricity, inclination and semi-major axis of a fictitious asteroid in the 3/1 resonance with Jupiter; its final fate is to plunge into the Sun (from Morbidelli and Benest 1999; Morbidelli and Froeschlé 1998).

3
Secular resonances and Kozai resonance

Mean motion resonances are not the only ones to act on the dynamics of the asteroids. We know two other kind of resonances, and they depend on other orbital elements than the semi-major axis: the direction of the perihelion (its "argument") and the axis of the nodes, i.e., the line of crossing of the ecliptic and the asteroid's orbital plane (its "longitude of the ascending node"); or, more precisely, the precession of these two elements, i.e., their slow rotation under the influence of gravitational perturbations from the other planets [the precession of Mercury's perihelion is famous for its role in the history of the generalisation of Newton's gravitation by Einstein].

It may occur that the period of the precession of the perihelion of the minor planet -- i.e., rotation of the orbit in its plane -- is equal to the one of another planet: the strongest resonances of this kind are with Jupiter [1] and Saturn [2]. It may occur too that the period of the precession of the nodes -- rotation of the orbit in space -- is equal to the one of another planet: the most common resonances of this kind are with Saturn [3]. These resonances are called SECULAR RESONANCES, due to the slowness of the precession motions (up to several millions years), and they are noted traditionally ν_5 (case [1]), ν_6 (case [2]) and ν_{16} (case [3]).

It may occur that the periods of the precessions of the perihelion and of the nodes of the tiny body are equal: this is the KOZAI RESONANCE, which relies on variations of the eccentricity and the inclination of the minor body's orbit; this kind of resonance is much more active for comets than for asteroids.

Moreover, all these resonances may combine, and the global effects are:
1- increase of the eccentricity, then, after close encounter with a major planet,
2- variation of the semi-major axis of the asteroid; an example is given in Figure 6. The final result is that any minor planet in resonance will leave it in less than 100 millions years, which is so much less than the age of the Solar System that there must not have existed any asteroids in these resonances for a very long time.
But ...

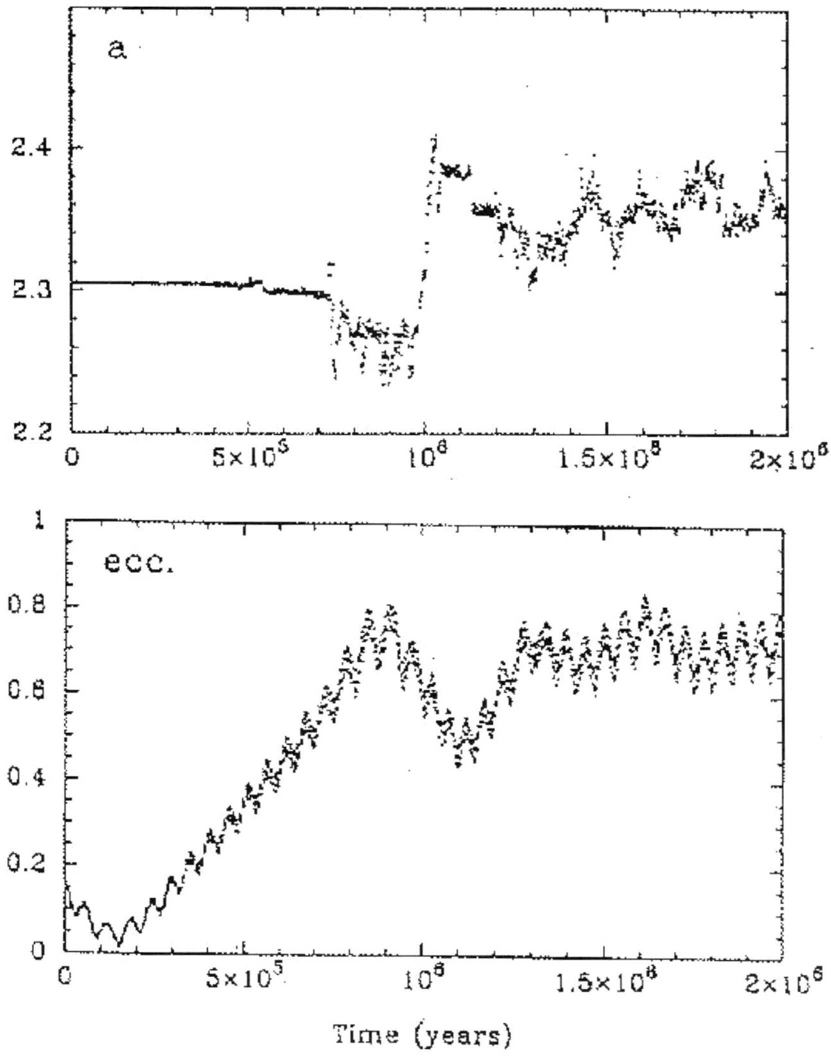

Fig. 6. Evolution with time of the semi-major axis and the eccentricity of a fictitious asteroid in the resonance ν_6 (the model includes all planets from Venus to Neptune); the increase of the eccentricity is irreversible, even after having left the resonance (from Froeschlé 1999 and Morbidelli and Froeschlé 1998).

But, nowadays, several hundreds asteroids approach the orbit of the Earth (see Figure 1): they are called NEAs -- for Near-Earth Asteroids -- (in french *géocroiseurs*), which change the old labels of "classes" Amor,

Apollo and Aten[2] (Figure 7) and others, like the IEAs (Inner-Earth Asteroids), whose aphelion is less than 0.983 AU -- but only a very few such bodies are known.

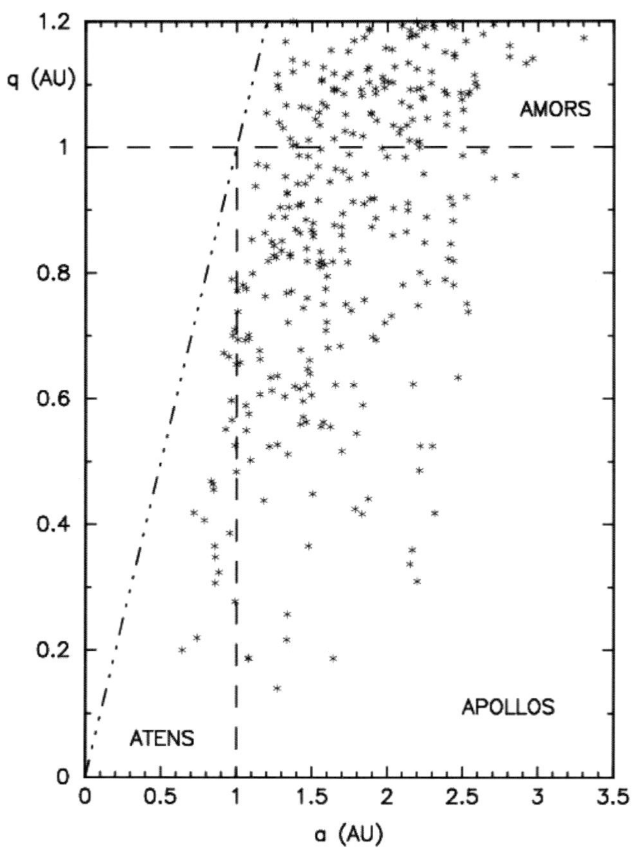

Fig. 7. Repartition in the (a,q) plane [semi-major axis, perihelion distance] of 248 NEAs known among the 58,092 numbered asteroids: for the Amors $a > a_T$ and $q > Q_T$, for the Apollos $a > a_T$ and $q < Q_T$, for the Atens $a < a_T$ and $Q > q_T$ (where Q is the aphelion distance of the asteroid, and a_T is the Earth's semi-major axis, q_T = 0.983 AU and Q_T= 1.017 AU its perihelion and aphelion distances).

[2] named after the minor planets 1221 Amor and 1862 Apollo (discovered in 1932) as well as 2062 Aten (discovered in 1976).

Let us take an example: the NEA (of Apollo class) 4179 Toutatis, discovered in 1989, has been known to be very near the 3/1 mean motion resonance with Jupiter (Figure 8) and to cross the orbit of the Earth every around 4 years -- with the collision hazard you can imagine, for its size is relatively important (about 4 and 2.5 km, for it is probably a double). But it did not tell us what it is doing in this resonance, which it must have left more than 4.4 billions years ago, as we assume that it was born there during Solar System formation - and knowing that resonances become completely empty in less than 100 millions years.

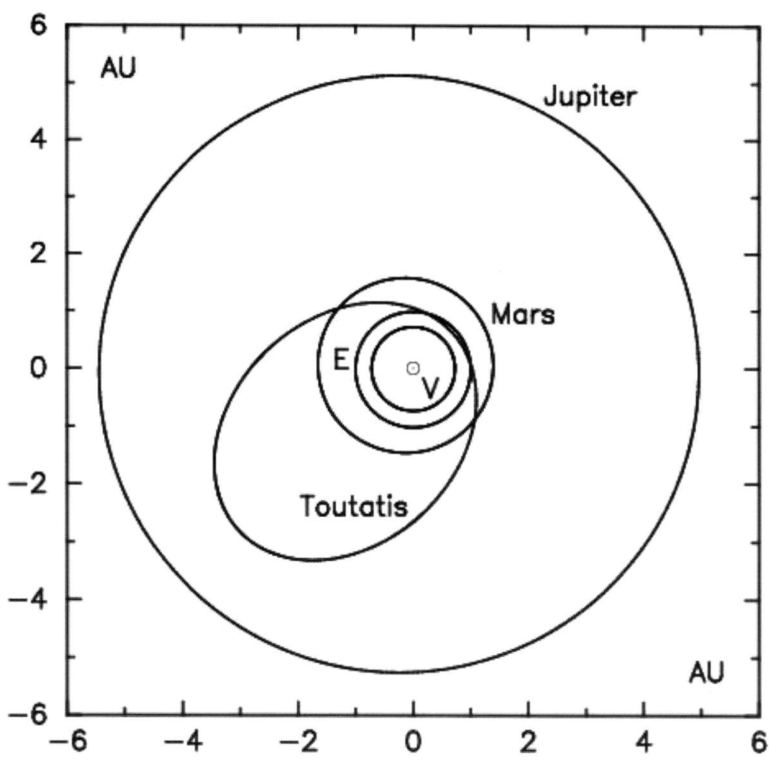

Fig. 8. Orbit of the NEA 4179 Toutatis, together with the orbits of Venus, the Earth, Mars and Jupiter.

4
Asteroid families

For solving this problem, we have to go back to the beginning of the XXth century, when the Japanese astronomer, Kiyotsugu Hirayama, discovered (in 1918) that some asteroids of the Main Belt had orbital elements rather close to each other; moreover, by eliminating the effects of the planetary perturbations through filtering and averaging techniques, he obtained very close orbital elements called PROPER ELEMENTS. Then he named FAMILIES those groups that have the same proper elements, and he suggested that the members of a group could be fragments of a larger body which had been destroyed during some catastrophe. Doing so, Hirayama gave rebirth, but at a much less scale, to an idea proposed by Olbers in the beginning of the XIXth century about the origin of the asteroids: these bodies could come from the destruction of a telluric planet lying at 2.8 AU from the Sun. Olbers' hypothesis has been abandoned since then (we think rather that the asteroids are planetesimals which could not accrete into a planet, essentially due to the perturbations from the young Jupiter), but Hirayama's proposition turned out to be very fruitful.

Nowadays, we have detected about twenty such families; note that the two asteroids visited by the *Galileo* spacecraft en route to Jupiter, 951 Gaspra (in 1991) and 243 Ida (in 1993), each belong to a family. Ida's size approaches 50 km, and it is a fragment! – thus, we can assume that its parent body was much larger: at least several hundred kilometers in diameter.

Which kind of natural phenomenon can destroy such a large stone? Well, simply a collision, a strong enough collision with either another large stone, or a very speedy little stone. Well, it is now thought that perturbations from Jupiter -- and, at a lesser degree, from other planets -- increase the relative velocities of tiny bodies and therefore encounters between such speedy bodies become dangerous and destructive; and there are still nowadays numerous collisions in the asteroid Main Belt: what becomes of the fragments? If they are "far" from a resonance, nothing happens and they continue to orbit wisely in the Main Belt. But if they are too "near" to a resonance, there is a risk for them to "fall" into the resonance and then have the fate described above.

This is why there are still nowadays objects in resonances: COLLISIONS SUPPLY, MORE OR LESS REGULARLY, THE RESONANCES; and this is why Toutatis is always in the 3/1 resonance: it arrived there only recently!

Figure 9 presents this scenario of origin, evolution and final destiny of NEAs: the NEA population has remained constant for about 3 billion years, but destruction processes (and fragmentation of the fragments through collisions between the NEAs themselves) limit an NEA lifetime to less than 100 millions years. A continuous supply is therefore needed: collisions in the Main Belt, and transport processes through resonance, produce NEAs continuously.

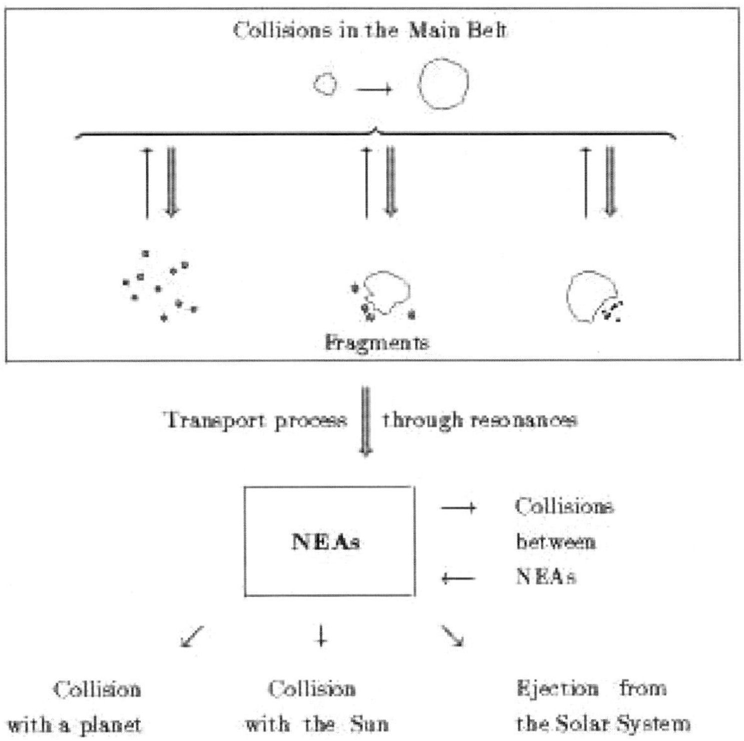

Fig. 9. A scenario for the life of NEAs.

5 The Yarkovsky effect

Recently, astronomers have needed to complete the scenario to explain several facts. New, faster and more powerful computational methods, appearing in the last years of the XXth century, make it appear that the dynamical lifetime of the NEAs is ten times shorter than previously estimated; this implies that the resonances have to be resupplied more efficiently than previously thought. Now, besides, it has been found that the number of candidates for collisions in the Main Belt seems too low to provide this resupply. Moreover, the magnitude distribution of the NEAs does not fit very well with distributions observed inside the families or predicted by theoretical considerations. Therefore, collisions seemed not to be the unique explanation for the existence of a constant NEA flow.

Finally, it is well known that part of the meteorites were not fresh collisional fragments when they were injected into a resonance: their surface age, given through the duration of their exposure to cosmic rays, is often several billion years. Even a large body like Eros has such a surface age (2 Ga), which has been estimated with crater counts thanks to the NEAR spacecraft, very much higher than its dynamical NEA's lifetime; so the collision which created Eros occurred long before its injection into any resonance.

Then we need a process which changes the orbital elements over millennia, such that a little body (but not the large ones) slowly drifts and hazardously "falls" into a resonance. Such a process exists. It is a non-gravitational effect due to the thermal emission of the asteroid's surface in a direction other than sunwards, because of the rotation of the asteroid. The region heated by the Sun radiates infrared and, when the asteroid rotates, this radiation results in a weak thrust which slowly changes the semi-major axis of the asteroid's orbit. Depending on whether its rotation is direct or retrograde (with respect to its heliocentric revolution), the semi-major axis increases or decreases and the orbital period, which varies inversely (third Kepler law), crosses sooner or later with the value corresponding to a resonance (e.g., around 4 years for the 3/1). This is called the Yarkovsky effect.

Nevertheless, we still need collisions because we need fragments: the drift in semi-major axis is estimated to be around 10^{-4} AU per million years for a 1-km body, and decreases roughly proportionally to the size, so that the larger asteroids are not affected. Therefore, it is necessary that the fragment population is continuously replenished to resupply the resonances at a sufficient rate. This can be done, thanks to the Yarkovsky effect which allows us to cast our net over a very much wider space.

6 Conclusion

There has been much work over the centuries to understand the origin of these stones which fall from the heaven - not for the regular falling star showers, which have been recognized to originate in comets by Chladni at the end of the XVIIIth century, but for the sporadic ones, particularly the large precursors of cratering events. The role of the resonances in the Main Belt of Asteroids were definitively established in the 1980's and, as we have seen, the scenario has been completed by the Yarkovsky effect (and by some more subtle other effects) in the last decade.

Of course, we have to keep in mind that not all the impactors come from asteroids. A small fraction of the risk comes from comets (remember the Tunguska event in 1908), mainly from short-period ones, but in a lesser amount, from long-period ones. But this, as Kipling said, is another story.

References

Benest D, Froeschlé C, Gonczi R (1994) Stochasticity of the Apollo asteroid 4179 Toutatis. In: Kozai Y, Binzel R, Hirayama T (eds) Seventy-five years of Hirayama asteroid families: The role of collisions in the Solar System history. Astronomical Society of Pacific Conference Series vol. 63, pp 7-14

Benest D, Froeschlé C (eds) (1992) Interrelations between physics and dynamics for minor bodies in the Solar System. Editions Frontières vol. C49, 651 pp

Benest D, Froeschlé C (eds) (1998) Impacts on Earth. Springer, Heidelberg, Lecture Notes in Physics vol. 505, 223 pp

Froeschlé Chr (1999) Les "géocroiseurs" (astéroïdes qui frôlent la Terre) et l'origine des météorites. In: Benest D, Froeschlé C (eds) Astéroïdes, météorites et poussières interplanétaires. Editions ESKA, pp 81-116

Gronchi GF (2002) Generalized averaging principle and Proper Elements for NEAs. In: Benest D, Froeschlé C (eds) Singularities in gravitational systems - Applications to chaotic transport in the Solar System. Springer, Heidelberg, Lecture Notes in Physics vol. 590, pp 179-211

Kozai Y, Binzel R, Hirayama T (eds) (1994) Seventy-five years of Hirayama asteroid families: The role of collisions in the Solar System history. Astronomical Society of Pacific Conference Series vol. 63, 303 pp

Morbidelli A (2002) Modern Celestial Mechanics - Aspects of Solar System dynamics. Taylor and Francis - Advances in Astronomy and Astrophysics vol. 5, 368 pp

Morbidelli A, Benest D (1999) Ordre et chaos dans le Système Solaire. In: Benest D, Froeschlé C (eds) Invitation aux planètes. Editions ESKA, pp 65-87

Morbidelli A, Froeschlé C (1998) Origin and dynamical transport of near-earth asteroids and meteorites. In: Benest D, Froeschlé C (eds) Impacts on Earth. Lecture Notes in Physics, vol. 505, Springer, Heidelberg, pp 31-53

Morbidelli A, Vokrouhlicky D (2003) The Yarkovsky-driven origin of near-Earth asteroids. Icarus 163: 120-134